MAMMALIAN TRP CHANNELS AS MOLECULAR TARGETS

The Novartis Foundation is an international scientific and educational charity (UK Registered Charity No. 313574). Known until September 1997 as the Ciba Foundation, it was established in 1947 by the CIBA company of Basle, which merged with Sandoz in 1996, to form Novartis. The Foundation operates independently in London under English trust law. It was formally opened on 22 June 1949.

The Foundation promotes the study and general knowledge of science and in particular encourages international co-operation in scientific research. To this end, it organizes internationally acclaimed meetings (typically eight symposia and allied open meetings and 15–20 discussion meetings each year) and publishes eight books per year featuring the presented papers and discussions from the symposia. Although primarily an operational rather than a grant-making foundation, it awards bursaries to young scientists to attend the symposia and afterwards work with one of the other participants.

The Foundation's headquarters at 41 Portland Place, London W1B 1BN, provide library facilities, open to graduates in science and allied disciplines. Media relations are fostered by regular press conferences and by articles prepared by the Foundation's Science Writer in Residence. The Foundation offers accommodation and meeting facilities to visiting scientists and their societies.

Information on all Foundation activities can be found at http://www.novartisfound.org.uk

Novartis Foundation Symposium 258

MAMMALIAN TRP CHANNELS AS MOLECULAR TARGETS

2004

John Wiley & Sons, Ltd

Published in 2004 by John Wiley & Sons Ltd,
The Atrium, Southern Gate,
Chichester PO19 8SQ, UK

National 01243 779777
International (+44) 1243 779777
e-mail (for orders and customer service enquiries): cs-books@wiley.co.uk
Visit our Home Page on http://www.wileyeurope.com
or http://www.wiley.com

Other Wiley Editorial Offices

John Wiley & Sons Inc., 111 River Street, Hoboken, NJ 07030, USA

Jossey-Bass, 989 Market Street, San Francisco, CA 94103-1741, USA

Wiley-VCH Verlag GmbH, Boschstr. 12, D-69469 Weinheim, Germany

John Wiley & Sons Australia Ltd, 33 Park Road, Milton, Queensland 4064, Australia

John Wiley & Sons (Asia) Pte Ltd, 2 Clementi Loop #02-01, Jin Xing Distripark, Singapore 129809

John Wiley & Sons Canada Ltd, 22 Worcester Road, Etobicoke, Ontario, Canada M9W 1L1

Wiley also publishes its books in a variety of electronic formats. Some content that appears in print may not be available in electronic books.

Novartis Foundation Symposium 258
ix+275 pages, 33 figures, 18 tables

British Library Cataloguing in Publication Data

A catalogue record for this book is available from the British Library

ISBN 0 470 86254 8

Typeset in 10½ on 12½ pt Garamond by Dobbie Typesetting Limited, Tavistock, Devon.
Printed and bound in Great Britain by T. J. International Ltd, Padstow, Cornwall.
This book is printed on acid-free paper responsibly manufactured from sustainable forestry, in which at least two trees are planted for each one used for paper production.

Contents

Symposium on Mammalian TRP channels as molecular targets, held at the Novartis Foundation, London, 25–27 March 2003

Editors: Derek J. Chadwick (Organizer) and Jamie A. Goode

This symposium was based on a proposal made by William P. Schilling

Participants

Indu S. Ambudkar Secretory Physiology Section, Gene Therapy and Therapeutics Branch, National Institutes of Dental and Craniofacial Research, 20892 Bethesda, MD 20892-1190, USA

K. S. Authi Centre for Cardiovascular Biology and Medicine, King's College London, New Hunt's House, Guy's Campus, London SE1 1UL, UK

Greg Barritt Department of Molecular Biochemistry, School of Medicine, Faculty of Health Sciences, Flinders University, GPO Box 2100, Adelaide, SA 5001, Australia

David J. Beech School of Biomedical Sciences, Worsley Building, University of Leeds, Leeds LS2 9JT, UK

Christopher D. Benham Neurology Centre of Excellence for Drug Discovery, GlaxoSmithKline, New Frontiers Science Park, Third Avenue, Harlow CM19 5AW, UK

Brian Cox Novartis Horsham Research Centre, Wimblehurst Road, Horsham, West Sussex RH12 5AB, UK

Patrick Delmas Intégrations des Informations Sensorielles (UMR 6150), CNRS, Faculté de Médecine, IFR Jean Roche, Bd Pierre Dramard, 13916 Marseille cedex 20, France

Andrea Fleig Center for Biomedical Research, The Queen's Medical Center & University of Hawaii, 1301 Punchbowl Street — UHT 8, Honolulu, HI 96813, USA

Marc Freichel Institut für Pharmakologie und Toxikologie, Universität des Saarlandes, Unikliniken Homburg, Gebäude 46, 66421 Homburg, Germany

Donald L. Gill Department of Biochemistry and Molecular Biology, University of Maryland School of Medicine, 108 North Greene Street, Baltimore, MD 21201, USA

Klaus Groschner Department of Pharmacology and Toxicology, Karl-Franzen-University Graz, Universitätsplatz 2, A 8010 Graz, Austria

Thomas Gudermann Philipps-Universität Marburg, Fachbereich Humanmedizin, Institut für Pharmakologie und Toxikologie, Karl-von-Frisch-Str 1 (Lahnberge), 35033 Marburg, Germany

Roger Hardie Department of Anatomy, University of Cambridge, Downing Street, Cambridge CB2 3DY, UK

Diana L. Kunze Rammelkamp Center for Education and Research, Metrohealth Medical Center, Department of Neuroscience, Case Western University, Cleveland, OH 44109-1998, USA

Su Li Novartis Horsham Research Centre, Wimblehurst Road, Horsham, West Sussex RH12 5AB, UK

Craig Montell Department of Biological Chemistry, The Johns Hopkins University School of Medicine, 725 N. Wolfe Street, Baltimore, MD 21205, USA

Shmuel Muallem Department of Physiology, University of Texas Southwestern Medical Center, Dallas, TX 75235-9040, USA

Bernd Nilius Department of Physiology, Campus Gasthuisberg, KU Leuven, Leuven, Belgium

Anant B. Parekh Laboratory of Molecular and Cellular Signalling, University Laboratory of Physiology, Parks Road, Oxford OX1 3PT, UK

Reinhold Penner Center for Biomedical Research, The Queen's Medical Center & University of Hawaii, 1301 Punchbowl Street — UHT 8, Honolulu, HI 96813, USA

Chris Poll Novartis Horsham Research Centre, Wimblehurst Road, Horsham, West Sussex RH12 5AB, UK

James W. Putney (*Chair*) Calcium Regulation Section, National Institute of Environmental Health Sciences, P O Box 12233, Research Triangle Park, North Carolina, NC 27709, USA

Andrew M. Scharenberg University of Washington, Pediatrics, Division of Immunology, 1959 NE Pacific Street, PMB 356320, Seattle, WA 98195, USA

William P. Schilling Rammelkamp Center for Education and Research, Metrohealth Medical Center, Department of Physiology and Biophysics, Case Western Reserve University, Cleveland, Ohio, OH 44109-1998, USA

Colin W. Taylor Professor of Cellular Pharmacology, Department of Pharmacology, University of Cambridge, Tennis Court Road, Cambridge CB2 1PD, UK

Rudi Vennekens (*Novartis Foundation Bursar*) Institut für Pharmakologie und Toxikologie, Universitat des Saarlandes, Unikliniken Homburg, Gebäude 46, D-66421 Homburg, Germany

John Westwick Novartis Horsham Research Centre, Wimblehurst Road, Horsham, West Sussex RH12 5AB, UK

Mike X. Zhu Neurotechnology Center and Department of Neuroscience, Ohio State University, 169 Rightmire Hall, 1060 Carmack Road, Columbus, OH 43210, USA

[illegible]

James W. Putney [illegible] Health Sciences, [illegible]
[illegible]

Andrew M. Scharenberg University of Washington, [illegible]
Immunology, 1959 NE Pacific Street, [illegible]

William P. Schilling [illegible]
[illegible] Medical Center, Department of [illegible]
Western Reserve University, [illegible]

[illegible]

[illegible]

[illegible]

[illegible]

Chair's introduction

James W. Putney Jr

Calcium Regulation Section, National Institute of Environmental Health Sciences, 111 TW Alexander Drive, PO Box 12233, Research Triangle Park, North Carolina, NC 27709, USA

The purpose of this meeting is to discuss, and hopefully advance our knowledge of the TRP ion channel superfamily. In the past few years, there has been an explosion in interest in these relatively new arrivals to the ion channel field. To appreciate the problems we face in studying TRP channels, it is helpful to think about how our current knowledge of the function of these ion channels has developed. Our current understanding is based on two somewhat unrelated strategies. The first, more classical approach involves examination of a physiological function and identification and understanding of the ion channels that subtend that function. Examples would include the ligand-gated vanilloid receptor (TRPV1) or the calcium-transporting ion channels (TRPV5 and 6); these channel genes were discovered as a result of searches for specific channels with very specific functions. However, this approach applies to only a minority of the TRP genes. Rather, the majority of these genes were discovered by homology cloning, initially without a clear understanding of their physiological functions. The most obvious case in point in this category would be the TRPCs.

And so clearly one very important goal is to determine the true physiological functions of TRP channels. From the structure of the TRP channels, and given clues from their invertebrate counterparts, efforts have focused on functions of TRPs as channels regulated by either hormone or neurotransmitter agonists, or by environmental signals. In most agonist-regulated cells there are at least three general categories of regulated ion channels. Firstly, there are store-operated or capacitative Ca^{2+} entry channels (SOCs). Secondly, there are channels which are activated through direct interaction with receptors or more commonly are receptors themselves for ligands. Included in this category would be channels directly activated by environmental changes such as temperature. Thirdly, there is increasingly evidence for a large number of channels which can be activated through the generation of unknown second messengers, the so-called second messenger-operated channels. In addition to these broadly defined distinct modes of activation, each of these categories can be sub-divided further into various sub-categories. We know, for example that SOCs come in different flavours. The original store-operated current, I_{CRAC}, is carried by highly Ca^{2+}-selective channels with specific and well characterized electrophysiological properties. But on the basis of ion selectivity and single-channel behaviour, it is clear that in other cell types, endothelial cells for example, other molecular entities with different properties likely comprise the SOCs. There is increasing evidence, especially in smooth muscle, for store-operated channels which are Ca^{2+} permeable but which are in fact non-selective cation channels.

For the TRP channels, there are examples which fall into all of these categories. TRPV1 is an excellent example of a ligand-gated channel. There are second-messenger gated channels, especially among the TRPCs, although it is not yet entirely clear in all cases which second messengers activate them. There is also considerable evidence in the literature that various TRPs *can* be activated by store depletion, although whether this actually happens physiologically is controversial. And there is no doubt that additional interesting modes of regulation will be discovered, such as nutritional regulation of TRPMs.

In the formal presentations as well as in our discussions, we would especially like to highlight some of the more controversial issues in the TRP field. A few examples: are any of the TRPs really store operated? Some of the leading candidates are the TRPCs and TRPV6. What other signals underlie activation of TRPCs, especially TRPC3/6/7? Is it inositol 1,4,5-trisphosphate (IP_3) or diacylglycerol? What is the signal for the TRPC4/5 subgroup, especially under conditions where these channels appear to be agonist-activated and not store operated? Unfortunately, for many of the TRP channels, we probably know more about what doesn't activate them than what does. Finally, there are still a few orphan TRPs, with as yet no suggested physiological function, although the number is getting smaller. Perhaps we will be hearing about functions for some of these TRPs during this meeting.

The goal before us is not to solve all of these problems in the short time allotted to this meeting. Amidst the controversies in this relatively young field, we must first find common ground by at least agreeing upon our points of disagreement! Then, going forward, we hope to reach some consensus about what we know, what we don't know, and what we need to find out in the years ahead.

Molecular genetics of *Drosophila* TRP channels

Craig Montell

Departments of Biological Chemistry and Neuroscience, The Johns Hopkins University School of Medicine, Baltimore, MD 21205, USA

Abstract. The *Drosophila* TRP channel has served as a genetic model for understanding the functions, modes of activation, subunit assembly and protein interactions of TRP proteins in an *in vivo* context. During the last few years, it has become clear that TRP associates with a macromolecular complex, the signalplex, which includes TRP and many other signalling proteins critical for phototransduction. The central protein in this assembly is the PDZ protein INAD. Association of target proteins with INAD is required for rapid termination of the photoresponse and for proper localization of signalling proteins, such as TRP. In addition, TRP is reciprocally required for localizing INAD, indicating that TRP functions as both an ion channel and a molecular anchor. The TRP superfamily is comprised of six subfamilies, each of which includes homologues in *Drosophila*. Due to the similarities between mammalian and *Drosophila* TRP channels, and the tractability of genetic approaches in the fruitfly, investigations are underway to characterize the roles of additional *Drosophila* TRP proteins. Thus, the fruitfly continues to provide a valuable *in vivo* model to characterize the functions, interacting partners and activation mechanisms of TRP-related proteins.

2004 Mammalian TRP channels as molecular targets. Wiley, Chichester (Novartis Foundation Symposium 258) p 3–17

Ca^{2+} influx plays a pivotal role in regulating a diversity of processes ranging from the immune response to synaptic transmission, the insulin response, pain reception and phototransduction (reviewed in Berridge 1993). While voltage- and ligand-gated channels mediate Ca^{2+} influx in response to a variety of signals (reviewed in Berridge et al 2000), such channels do not account for many receptor-operated Ca^{2+} influx pathways identified in a variety of cell types. In particular the Ca^{2+} influx channels that are coupled *in vivo* to the stimulation of phospholipase C (PLC) have been enigmatic.

The two primary signalling pathways that activate PLC are initiated by receptor tyrosine kinases and G protein-coupled receptors (GPCRs) (reviewed in Berridge et al 2000). A widely held view of G protein-coupled signalling cascades is that they operate through stochastic interactions between activated receptors and

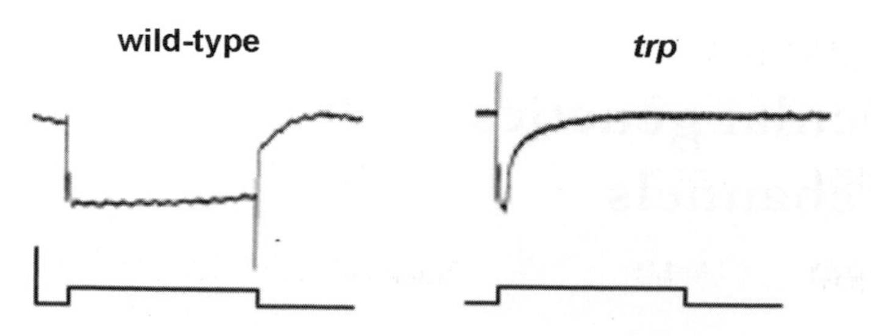

FIG. 1. ERG phenotype of *trp* mutant flies. The flies were dark-adapted for two minutes and exposed to a five second pulse of white light, indicated by the event marker. The vertical line to the left of the event marker below the wild-type ERG represents 5 mV.

downstream effector molecules. However, such a model does not account for the rapid kinetics of some cascades or for the specificity in G protein-coupled signalling. This latter phenomenon is an issue since many cells express multiple GPCRs, several of which may have the capacity to couple with the same heterotrimeric G protein. Yet activation of different GPCRs produces distinct biological effects.

To characterize the mechanisms underlying G protein-coupled Ca^{2+} influx, we have been focusing primarily on the phototransduction cascade in the fruitfly, *Drosophila melanogaster* (reviewed in Montell 1999). This cascade, which is a PLC dependent pathway, is completed in ~20 ms and represents the fastest known G protein-coupled cascade. The cascade culminates with Na^+ and Ca^{2+} influx. The Ca^{2+} not only contributes to depolarization of the photoreceptor cells, but in regulating adaptation and termination of the photoresponse.

Many mutations that disrupt the phototransduction cascade have been identified on the basis of screening for defects in the electroretinogram recording (reviewed in Pak 1991). Among these mutants is *trp*, which displays a transient response to light (Fig. 1). The *trp* mutation was a candidate for disrupting one class of light sensitive cation channel as the movement of the pigment granules, which is a Ca^{2+}-dependent phenomenon was transient in the mutant (Lo & Pak 1981). Moreover, light-dependent Ca^{2+} influx was reduced about 10-fold in *trp* flies (Hardie & Minke 1992). At the time the *trp* gene was cloned, it did not have significant homology to known proteins, but had a predicted topology of multiple transmembrane domains (Fig. 2A), consistent with other members of the superfamily of voltage- and ligand-gated ion channels (Montell & Rubin 1989).

Functional expression of TRP in heterologous systems provided strong evidence that it represented a new class of cation channel (Vaca et al 1994, Xu et al 1997). However, the mechanisms linking stimulation of PLC to activation of TRP *in vivo* remain controversial. The TRP protein does not appear to be activated in

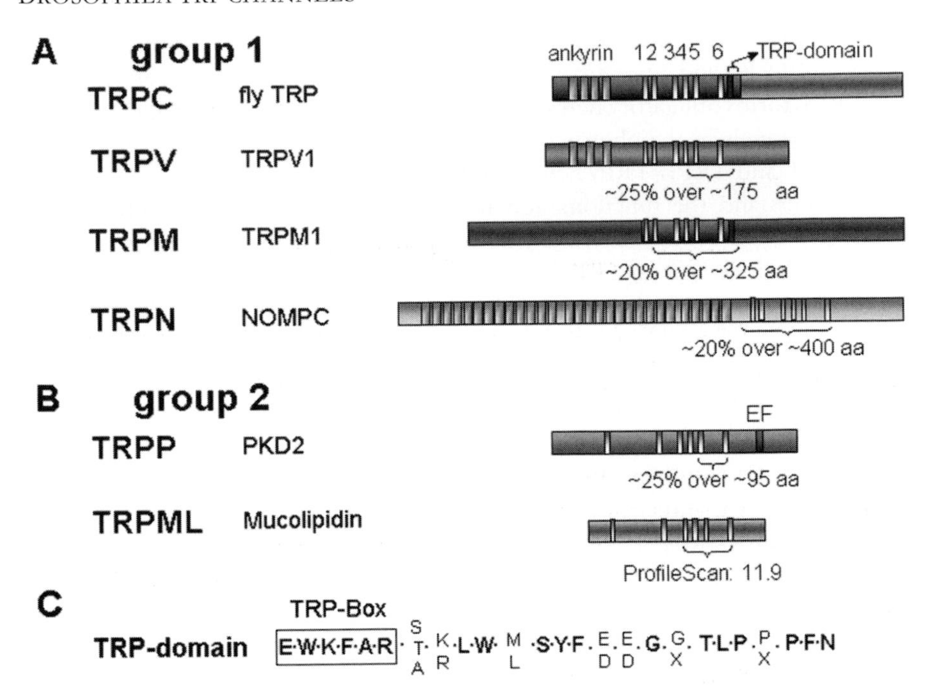

FIG. 2. The TRP superfamily. (A) The four TRP subfamilies, with the strongest homology to *Drosophila* TRP, are referred to as the group 1 TRPs (the percent identities to the TRPCs are indicated). The domain organization of the founding members of each subfamily is shown. There are two founding TRPV members, TRPV1 and osm-9; however, only TRPV1 is depicted. (B) Two TRP subfamilies, which are distantly related to *Drosophila* TRP, are referred to as the group 2 TRPs. The founding members of the two subfamilies are depicted. Mucolipin has no significant primary amino acid homology to *Drosophila* TRP. However, according to the ProfileScan algorithm (*http://hits.isb-sib.ch/cgi-bin/PFSCAN*), it is most related to TRP proteins. (C) Consensus sequence of the TRP domain. This 25 amino-acid motif is present in TRPC and TRPM proteins, C-terminal to the transmembrane segments.

photoreceptor cells through a mechanism involving release of Ca^{2+} from internal stores, as genetic elimination of either of the known Ca^{2+} release channels, the inositol 1,4,5-trisphosphate (IP_3) receptor and the ryanodine receptor, has no effect on the light-dependent Ca^{2+} influx (reviewed in Montell 1999). Evidence has been presented that TRP and the related channel, TRPL, may be activated by polyunsaturated fatty acids or diacylglycerol or through hydrolysis of phosphatidylinositol 4,5-bisphosphate (PIP_2) (Chyb et al 1999, Estacion et al 2001); however, this issue has not been resolved.

In addition to TRP, there are two other highly related channels, TRPL and TRPγ, which are expressed in photoreceptor cells (Phillips et al 1992, Xu et al 2000). These three channels form homo- and heteromeric assemblies, both *in vitro* and *in vivo* (Xu et al 1997, 2000). TRP is approximately 10-fold more abundant than TRPL and TRPγ, and is present primarily as a homomultimer. However, it appears that TRPL forms obligatory heteromultimers with either TRP or TRPγ. The three primary channels in the photoreceptor cells appear to be TRP homomultimers, TRP/TRPL and TRPL/TRPγ heteromultimers.

To understand the functions and mechanisms of activation of TRP, many investigators have been interested in identifying proteins that interact either directly or indirectly with TRP. The first TRP interacting protein identified was INAD (Huber et al 1996b, Shieh & Zhu 1996), a protein with five PDZ protein interaction modules. The observation that TRP binds to INAD raised a number of questions concerning the functions of the TRP/INAD interaction and whether other proteins bind directly to this complex. Our studies relevant to these questions are described below.

Given that activation of TRP is strictly dependent on the stimulation of PLC, it seemed plausible that there are proteins related to TRP, which may be responsible for the PLC-dependent Ca^{2+} influx pathways characterized in mammalian cells. In support of this proposal, was the identification of mammalian homologues of TRP (Wes et al 1995, Zhu et al 1995). Moreover, there are seven such proteins, referred to as TRPC channels, each of which includes three or four N-terminal ankyrin repeats, six transmembrane segments and a highly conserved 25 amino-acid segment, the TRP domain (Figs 2A and C, reviewed in Montell 2001) C-terminal to the transmembrane domains. Another common feature of TRPC channels is that they are stimulated through PLC signalling pathways (reviewed in Montell 2001). However, most of the work has been performed in *in vitro* systems. Therefore, there are many remaining questions concerning the roles, modes of activation, subunit composition and complexes formed by mammalian TRPC channels *in vivo*.

In addition to the TRPC channels, there are three other TRP subfamilies, TRPV, TRPM and TRPN, which encode over 20 channels (Fig. 2A, the group 1 TRPs; reviewed in Montell 2001). Each of these proteins also appears to contain six transmembrane segments, though other features are distinct from TRPC proteins. The TRPM members contain a TRP domain, but are devoid of ankyrin repeats, while TRPV proteins include three or four ankyrin repeats, but no TRP domain. TRPN members are unusual in containing 8–29 N-terminal ankyrin repeats. Some of the TRPV, TRPM and TRPN proteins are activated through mechanisms that are not dependent on PLC. These include changes in thermal conditions and osmolarity.

There are two additional TRP subfamilies, TRPP and TRPML, which include proteins distantly related to the TRPCs (Fig. 2B, the group 2 TRPs, reviewed in Montell 2001). The founding members of the TRPP and TRPML subfamilies, PKD2 and Mucolipin respectively, are mutated in many cases of polycystic kidney disease (PKD) and mucolipidosis (reviewed in Montell 2001, Sutters

2003, Slaugenhaupt 2002). PKD2 is of particular interest since PKD is a common disease affecting ~1/1000 individuals worldwide. PKD2 interacts directly with another multiple transmembrane domain protein, PKD1, and mutations in these two proteins account for nearly all cases of PKD (reviewed in Sutters & Germino 2003).

PKD1 and PKD2 are widely expressed and presumably function in a variety of tissues. In the worm, proteins related to PKD1 and PKD2 function in male mating behaviour (Barr et al 2001) and a sea urchin protein related to PKD1, REJ, has been implicated in the acrosomal reaction (Moy et al 1996). This latter observation raises the possibility that PKD1 and PKD2 may participate in male fertility. Two human homologues of PKD1 as well as PKD2 and the related protein, PKD2L2 are all expressed in the testis. Nevertheless, the roles and modes of activation of PKD2 are poorly understood. Proteins homologous to each of the six TRP subfamilies, including TRPP, are present in the fruitfly (reviewed in Montell 2003). Due to the relevance of PKD2 for human health and disease, we have recently begun to characterize the function of *Drosophila* PKD2.

Results and discussion

The composition and functions of the TRP signalplex

In vitro studies indicate that ion channel/PDZ protein interactions can affect channel clustering, at least in tissue culture (Kim et al 1995, Kornau et al 1995). To address whether the TRP/INAD interaction is required for the subcellular localization of TRP *in vivo*, we took advantage of an *inaD* allele, *inaD*^{P215}, which contains a point mutation in PDZ3 that disrupts binding to TRP (Shieh & Zhu 1996). In wild-type flies, TRP is specifically localized to the microvillar portion of the photoreceptor cells, the rhabdomeres, which is the site of phototransduction. In *inaD*^{P215} the localization of TRP is severely disrupted (Chevesich et al 1997). The TRP/INAD interaction is not required for initial targeting of TRP to the rhabdomeres, but rather for retaining TRP in the rhabdomeres (Li & Montell 2000, Tsunoda et al 2001).

TRP binds to only one of the five PDZ domains in INAD, raising the possibility that INAD associates with a variety of other signalling proteins. Consistent with this proposal, INAD binds directly to rhodopsin (Xu et al 1998), PLC (Huber et al 1996a, Chevesich et al 1997), protein kinase C (PKC) (Huber et al 1996a, Xu et al 1998), TRPL (Xu et al 1998), the NINAC myosin III (Wes et al 1999) and calmodulin (CaM) (Xu et al 1998) (Fig. 3). All of these proteins may be bound to a single complex as rhodopsin and PLC co-immunoprecipitate with TRP from wild-type fly heads, but not from *inaD*^{P215} (Chevesich et al 1997). With the exception of CaM, all of these proteins bind to PDZ domains. Moreover, some

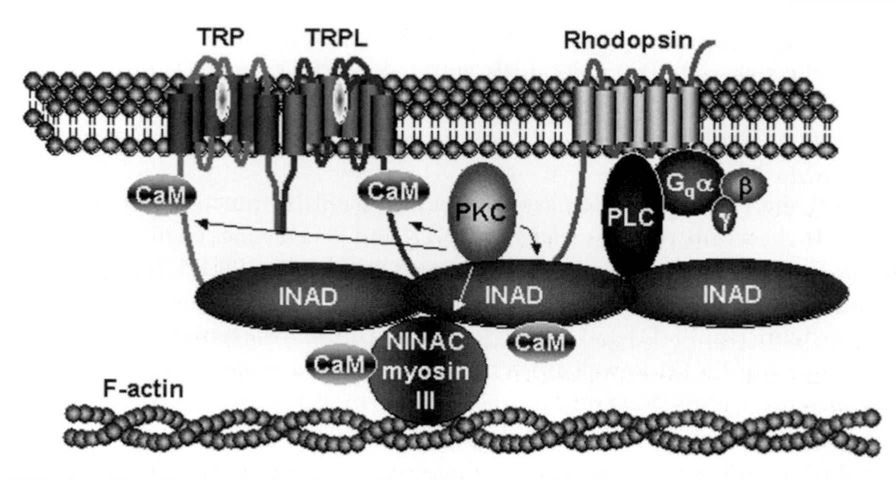

FIG. 3. The *Drosophila* signalplex. Six of the seven INAD target proteins bind to PDZ domains: TRP, TRPL, Rhodopsin, PLC, PKC and the NINAC myosin III. CaM binds to the linker region between PDZ1 and PDZ2. Members of the signalplex that bind CaM are indicated. INAD binding proteins, which are phosphorylated by PKC, are indicated by the arrows. The $G_q\alpha$ does not appear to bind directly to INAD.

of these proteins bind to the same and/or multiple PDZ domains. Since only one target protein associates with the hydrophobic pocket in a PDZ domain at any time, the question arises as to how INAD interacts simultaneously with all of its binding partners. We found that INAD also forms homophilic interactions and can self-assemble into a polymer *in vitro* (Xu et al 1998). This latter interaction appears to occur through portions of PDZ3 and PDZ4 distinct from those required for target binding. Therefore, the homophilic interactions of INAD appear to provide the necessary capacity to link all of the INAD target proteins in a single complex.

A key question concerns the functions of the INAD signalling complex (signalplex). One function appears to be for proper localization as three target proteins, TRP, PLC and PKC, require association with INAD for localization in the rhabdomeres (Chevesich et al 1997, Tsunoda et al 1997). In addition, association of the NINAC myosin III and PKC with INAD are required for rapid termination of the photoresponse. Since termination is a Ca^{2+}-regulated phenomenon, these proteins may need to associate with INAD to be situated near the TRP channel so that they can rapidly respond to the light dependent Ca^{2+} flux. PKC phosphorylates several targets in the signalplex, including NINAC (Li et al 1998), INAD and TRP (Huber et al 1996b, Liu et al 2000). CaM is also necessary for fast termination kinetics and at least four members of the signalplex bind to this Ca^{2+} regulatory protein: NINAC (Porter et al 1993), TRP (Chevesich et al 1997), TRPL (Phillips et al 1992) and INAD (Xu et al 1998).

Surprisingly, mutation of the INAD binding site in TRP has no impact on the kinetics of activation (Li & Montell 2000). However, we cannot exclude the possibility that indirect interactions between TRP and INAD participate in rapid signalling. Nevertheless, association of TRP with INAD has a role, in addition to a requirement for retention of TRP in the rhabdomeres. TRP and INAD have a reciprocal requirement for localization (Li & Montell 2000, Tsunoda et al 2001). A mutation that deletes the INAD binding site in TRP results in a defect in rhabdomere retention of INAD. Consequently, TRP, PLC and PKC also become mislocalized. These data indicate that TRP functions both as a molecular anchor and as a Ca^{2+} channel. Moreover, the finding that TRP is an anchor protein appears to account for the observation that TRP is a very abundant protein, which is present at similar levels to INAD.

Recent *in vitro* studies of several mammalian TRPC proteins indicate that they may also be coupled to supramolecular complexes (reviewed in Montell 2001). One candidate scaffold protein for TRPC4 and TRPC5 is the PDZ protein, NHERF. It will be of interest to determine whether TRP-containing macromolecular assemblies also occur *in vivo* and to address the function of such complexes. In addition, the analyses of *Drosophila* TRP raise the possibility that mammalian TRPs are not simply cation channels, but are multifunctional proteins, which carry out additional roles, such as anchoring other signalling proteins to specific cellular domains.

Requirement for *Drosophila* dPKD2 (Amo) for male fertility

To explore the functions of other members of the TRP superfamily in *Drosophila*, we recently initiated a phenotypic analysis of dPKD2 (Watnick et al 2003). Comparison of the human PKD2 deduced amino acid sequence with a conceptual translation of the full set of *Drosophila* proteins indicated that there is one PKD2 homologue (CG6504; Poisson probability score, $P(N)$, is $\sim 10^{-64}$; 968 amino acids), as well as three more distantly related proteins (CG9472, CG13762, CG16793; P(N) are $\sim 10^{-8}$ to 10^{-16}). Therefore, we set out to determine the function of CG6504, which we refer to as Amo (Almost there).

We found that *amo* was primarily expressed in male bodies. Very little *amo* RNA was detected either in male or female heads or in female bodies. Furthermore, the Amo protein was detected in the testis and in mature sperm. These data suggested that Amo functions in sperm and may be required for male fertility. However, this proposal could not be tested immediately since there were no existing mutations that disrupted the *amo* locus.

To generate a loss-of-function mutation in *amo*, we employed the recently described technique for creating mutations in *Drosophila* by homologous recombination (Rong & Golic 2000). This approach was successful as we found that the Amo protein was eliminated in the knockout flies. The *amo* flies displayed normal viability and there were no discernible morphological defects. Consistent with the Amo expression pattern, fertility of the *amo* males was severely reduced. In contrast, to these results, female fertility was indistinguishable from that of wild-type.

A reduction in male fertility could be due to any one of many possible defects. These include a decrease in sperm number, altered sperm motility or morphology, a perturbation in sperm storage or a defect in mating. Inadequate courtship behaviour could reflect an alteration in hearing as this behaviour is stimulated in part by a sound pattern produced by beating wings. We found that the quantity, morphology and motility of *amo* sperm were similar to wild-type. Furthermore, *amo* males display normal hearing and mating. The sperm are transferred to the females; however, there appears to be a defect in the quantity of sperm in the storage organs, the seminal receptacle and spermatheca. These data indicate that the reduced fertility in *amo* males results from a specific defect in sperm, leading to less efficient storage.

Currently, very little is known concerning the mechanisms of sperm storage. In particular the signals that are required for targeting the sperm to the storage organs have not been described. It is intriguing to speculate that Amo is a cation channel that participates in this process. Moreover, the identification of Amo provides the potential for using genetic approaches for identifying other loci that functionally interact with Amo and contribute to targeting of the sperm to the storage organs.

Conclusions

In view of the similarities of the *Drosophila* and mammalian TRP proteins and the availability of many genetic tools, the fruitfly provides a valuable model organism to identify the functions and molecular components of TRP signalling pathways. In particular, studies of *Drosophila* TRP show that it is activated *in vivo* through a PLC-dependent pathway, forms homo- and heteromeric channels *in vivo* and functions in a supermolecular signalling complex. Moreover, TRP is a multifunctional protein that plays roles as a molecular anchor, in addition to its more established role as a cation channel.

In vitro studies of mammalian TRPC proteins indicate that they are also activated through PLC-dependent pathways, are capable of forming heteromultimeric channels and are linked to supramolecular complexes, though in most cases, these observations have not been confirmed *in vivo*. Moreover, in a manner reminiscent of *Drosophila* TRP, many mammalian TRP proteins function in sensory physiology (reviewed in Montell et al 2002). Given these striking parallels between *Drosophila* TRP and the mammalian counterparts, it seems plausible that further characterization of the roles and modes of activation of

Drosophila TRPs will contribute to new and unexpected insights that will apply to mammalian TRP channels.

Acknowledgments

The work in the author's laboratory is funded by the NEI (EY08117 and EY10852) and the NIDDK (DK57325).

References

Barr MM, DeModena J, Braun D, Nguyen CQ, Hall DH, Sternberg PW 2001 The *Caenorhabditis elegans* autosomal dominant polycystic kidney disease gene homologs *lov-1* and *pkd-2* act in the same pathway. Curr Biol 11:1341–1346

Berridge MJ 1993 Inositol trisphosphate and calcium signalling. Nature 361:315–325

Berridge MJ, Lipp P, Bootman MD 2000 The versatility and universality of calcium signalling. Nat Rev Mol Cell Biol 1:11–21

Chevesich J, Kreuz AJ, Montell C 1997 Requirement for the PDZ domain protein, INAD, for localization of the TRP store-operated channel to a signaling complex. Neuron 18:95–105

Chyb S, Raghu P, Hardie RC 1999 Polyunsaturated fatty acids activate the *Drosophila* light-sensitive channels TRP and TRPL. Nature 397:255–259

Estacion M, Sinkins WG, Schilling WP 2001 Regulation of Drosophila transient receptor potential-like (TrpL) channels by phospholipase C-dependent mechanisms. J Physiol 530: 1–19

Hardie RC, Minke B 1992 The *trp* gene is essential for a light-activated Ca^{2+} channel in Drosophila photoreceptors. Neuron 8:643–651

Huber A, Sander P, Gobert A, Bähner M, Hermann R, Paulsen R 1996a The transient receptor potential protein (Trp), a putative store-operated Ca^{2+} channel essential for phosphoinositide-mediated photoreception, forms a signaling complex with NorpA, InaC and InaD. EMBO J 15:7036–7045

Huber A, Sander P, Paulsen R 1996b Phosphorylation of the *InaD* gene product, a photoreceptor membrane protein required for recovery of visual excitation. J Biol Chem 271:11710–11717

Kim E, Niethammmer M, Rothschild A, Jan YN, Sheng M 1995 Clustering of Shaker-type K^+ channels by interaction with a family of membrane-associated guanylate kinases. Nature 378:85–88

Kornau H-C, Schenker LT, Kennedy MB, Seeburg PH 1995 Domain interaction between NMDA receptor subunits and the postsynaptic density protein PSD-95. Science 269:1737–1740

Li HS, Montell C 2000 TRP and the PDZ protein, INAD, form the core complex required for retention of the signalplex in *Drosophila* photoreceptor cells. J Cell Biol 150:1411–1422

Li HS, Porter JA, Montell C 1998 Requirement for the NINAC kinase/myosin for stable termination of the visual cascade. J Neurosci 18:9601–9606

Liu M, Parker LL, Wadzinski BE, Shieh BH 2000 Reversible phosphorylation of the signal transduction complex in *Drosophila* photoreceptors. J Biol Chem 275:12194–12199

Lo M-VC, Pak WL 1981 Light-induced pigment granule migration in the retinular cells of *Drosophila melanogaster*. J Gen Physiol 77:155–175

Montell C 1999 Drosophila visual transduction. Ann Rev Cell Dev Biol 15:231–268

Montell C 2001 Physiology, phylogeny and functions of the TRP superfamily of cation channels. Sci STKE 90:RE1 *http://stke.sciencemag.org/cgi/content/full/OC_sigtrans;2001/90/re1*

Montell C 2003 The venerable inveterate invertebrate TRP channels. Cell Calcium 33:407–409
Montell C, Rubin GM 1989 Molecular characterization of the Drosophila *trp* locus: a putative integral membrane protein required for phototransduction. Neuron 2:1313–1323
Montell C, Birnbaumer L, Flockerzi V 2002 The TRP channels, a remarkably functional family. Cell 108:595–598
Moy GW, Mendoza LM, Schulz JR, Swanson WJ, Glabe CG, Vacquier VD 1996 The sea urchin sperm receptor for egg jelly is a modular protein with extensive homology to the human polycystic kidney disease protein, PKD1. J Cell Biol 133:809–817
Pak WL 1991 Molecular genetic studies of photoreceptor function using Drosophila mutants. In: Chader GJ, Farber D (eds) Molecular biology of the retina: basic and clinically relevant studies. Wiley-Liss, New York, p 1–32
Phillips AM, Bull A, Kelly LE 1992 Identification of a Drosophila gene encoding a calmodulin-binding protein with homology to the *trp* phototransduction gene. Neuron 8:631–642
Porter JA, Yu M, Doberstein SK, Pollard TS, Montell C 1993 Dependence of calmodulin localization in the retina on the *ninaC* unconventional myosin. Science 262:1038–1042
Rong YS, Golic KG 2000 Gene targeting by homologous recombination in Drosophila. Science 288:2013–2018
Shieh B-H, Zhu M-Y 1996 Regulation of the TRP Ca^{2+} channel by INAD in Drosophila photoreceptors. Neuron 16:991–998
Slaugenhaupt SA 2002 The molecular basis of mucolipidosis type IV. Curr Mol Med 2:445–450
Sutters M, Germino GG 2003 Autosomal dominant polycystic kidney disease: molecular genetics and pathophysiology. J Lab Clin Med 141:91–101
Tsunoda S, Sierralta J, Sun Y et al 1997 A multivalent PDZ-domain protein assembles signalling complexes in a G-protein-coupled cascade. Nature 388:243–249
Tsunoda S, Sun Y, Suzuki E, Zuker C 2001 Independent anchoring and assembly mechanisms of INAD signaling complexes in Drosophila photoreceptors. J Neurosci 21:150–158
Vaca L, Sinkins WG, Hu Y, Kunze DL, Schilling WP 1994 Activation of recombinant *trp* by thapsigargin in Sf9 insect cells. Am J Physiol 266:C1501–C1505
Watnick TJ, Jin Y, Matunis E, Kernan MJ, Montell C 2003 A flagellar polycystin-2 homolog required for male fertility in *Drosophila*. Curr Biol 13:2179–2184
Wes PD, Chevesich J, Jeromin A, Rosenberg C, Stetten G, Montell C 1995 TRPC1, a human homolog of a *Drosophila* store-operated channel. Proc Natl Acad Sci USA 92:9652–9656
Wes PD, Xu X-ZS, Li H-S, Chien F, Doberstein SK, Montell C 1999 Termination of phototransduction requires binding of the NINAC myosin III and the PDZ protein INAD. Nat Neurosci 2:447–453
Xu X-ZS, Choudhury A, Li X, Montell C 1998 Coordination of an array of signaling proteins through homo- and heteromeric interactions between PDZ domains and target proteins. J Cell Biol 142:545–555
Xu X-ZS, Li H-S, Guggino WB, Montell C 1997 Coassembly of TRP and TRPL produces a distinct store-operated conductance. Cell 89:1155–1164
Xu XZ, Chien F, Butler A, Salkoff L, Montell C 2000 TRPg, a *Drosophila* TRP-related subunit, forms a regulated cation channel with TRPL. Neuron 26:647–657
Zhu X, Chu PB, Peyton M, Birnbaumer L 1995 Molecular cloning of a widely expressed human homologue for the *Drosophila trp* gene. FEBS Lett 373:193–198

DISCUSSION

Muallem: Why doesn't this mutation cause a problem in human fertility? The effects seem to be restricted to the kidney. Is it because it is a specific mutation?

Montell: There is one paper describing fertility problems in men with polycystic kidney disease (Okada et al 1999); however, this study is not very definitive.

Putney: Is the mutation in the human disease a true null?

Montell: If you have a mutation in just one of the two alleles, that is all you need to get the disease because it emerges from loss of heterozygosity. In the case of flies, we are analysing the phenotype of a null mutant.

Westwick: Going back to the signalplex, which you mentioned at the beginning of your presentation, every time you get PLCβ activation, you are also going to get PI3K γ activation, which is G protein coupled. Is there any evidence for PI3K mutants or downstream products involving your signalplex?

Montell: We have been doing a lot of work lately on the role of PIP_3 in fly photoreceptors. A fascinating phenomenon in photoreceptor cells is that arrestin undergoes light-dependent shuttling, and this movement is regulated by PIP_3. If fly photoreceptor cells are maintained in the dark, most of the arrestin resides in the cell bodies. However, after exposure to light for about 5 minutes, most of the arrestin is concentrated in the rhabdomeres. A similar phenomenon is seen in mammalian rods and cones. If the animals are kept in the dark, most of the arrestin is in the inner segment. Upon light exposure, the majority of the arrestin moves to the outer segment. There are two questions: what is regulating this light-dependent translocation, and what is its function? We have found that fly arrestin binds directly to PIP_3, and if we mutate the PIP_3 binding site in arrestin there is a defect in the light-dependent translocation, which I have just described. We also found that these light-dependent translocations are important for long-term adaptation: not the rapid adaptation that occurs on a very fast time scale. These same defects in the light-dependent movement of arrestin are seen in a PTEN mutant and in a variety of mutants affecting the PI cycle, such as CDS and RDGB. If we overexpress PTEN we also see a defect. There is no effect on activation in a PTEN mutant. You asked whether PI3K binds to the signalplex. We have no evidence that it does.

Nilius: For PKD2, do you have any evidence that you need this interaction with PKD1? It has been published that only co-assembly of PKD1 and PKD2 forms functional channels (Hanoaka et al 2000).

Montell: No, but that is because we haven't looked yet.

Nilius: In your view, might PKD1 be an ion channel?

Montell: It would be interesting if PKD1 by itself did have some channel activity. What is known about PKD1 is that the two termini are on opposite sides of the membrane, with the N-terminus outside. So this would be different from most ion channels.

Delmas: There is no evidence that PKD1 can form a channel by itself. It is thought to form a non-selective cation channel only when it is co-expressed with polycystin-2 (Hanaoka et al 2000).

Montell: I have heard that there is one group that has unpublished evidence that it is functioning as an ion channel. We haven't looked in our lab. It is an open question.

Putney: I know you haven't looked at the fine structure where PKD is located in the sperm, but do you recall what the fine structure of *Drosophila* sperm looks like compared with human sperm, which has almost no intracellular membranes at all? I was surprised to see that it looks like it is distributed throughout the cell in *Drosophila* sperm as if it might be in endoplasmic reticulum (ER). Usually there isn't any ER in sperm.

Montell: In the primary spermatocytes it appears that the protein is distributed throughout the cell except the nucleus. In the mature sperm it appears that Amo is restricted to the tip of the tail.

Penner: With regard to the signalplex, what do you think is the function for bringing them together? What is the most important aspect of it? Is it the activation that requires the colocalization, or is it more the negative regulatory elements that need to be brought into the signalplex in order to provide this function?

Montell: A number of years ago I would have guessed that it had some role in activation. Maybe there is some subtle effect on activation. Certainly, as a first approximation, activation is relatively normal. We have evidence that it has a role in retaining proteins in the rhabdomeres. There was a nice paper from the Minke and Selingers labs (Cook et al 2000), which illustrated the importance of keeping proteins in the proper stoichiometries. If proteins aren't retained in the rhabdomeres their concentration goes down. In the case of the NINAC myosin III, if it is not bound directly to INAD we see a defect in termination of the photoresponse.

Penner: What is the activation mechanism?

Montell: I know Roger Hardie is going to be talking about his work dealing with polyunsaturated fatty acids and DAG. There is also interesting *in vitro* work from Bill Schilling suggesting that PIP_2 hydrolysis might be involved. We have some other preliminary results that suggest a different mechanism, but it is in the early stages.

Gill: You listed PKC as one of the members of the complex. I think you said it is involved in phosphorylating three different things including TRP.

Montell: It is involved in phosphorylating TRP. Len Kelly's lab showed it is involved in phosphorylating TRPL (Warr & Kelly 1996). It is also involved in phosphorylating NINAC and INAD.

Gill: Since DAG is potentially an important mediator, and PKC is an important part of this, functionally is there a role for PKC? What happens if you modify PKC?

Montell: There is an eye-enriched PKC. If this protein is eliminated, there is no effect on activation, though there are effects on termination. There is another PKC that is expressed in photoreceptor cells that is not eye-enriched. So even though the eye-enriched PKC doesn't have any activation defect when you get rid of it, we can't exclude that PKC has no role in activation.

Gill: Which one is in the complex, then?

Montell: The eye-enriched PKC. However, we don't have the reagents that would allow us to look at whether or not the PKC that isn't eye-enriched is in the complex. We don't have the appropriate antibodies.

Gill: The idea that TRPs are adaptor proteins as well as functional channels is an intriguing one. So you are thinking that if a TRP channel is eliminated this leads to disorganization of the structure. If you get rid of more than one do you have more dissociation? Are they all involved in anchoring?

Montell: No. There are four members of the signalplex that form what we call the core complex: TRP, INAD, PLC and PKC. We think they are all constitutively bound and present in stoichiometric concentrations. All the other proteins I described as INAD binding proteins do not interact in a stoichiometric fashion. TRPL, for example, is clearly not always bound to INAD. It can't be, because it also undergoes light-dependent translocations. Rhodopsin is also more abundant than INAD, so only a subset of the rhodopsin pool could be bound to INAD at any given time. There is no evidence that TRPL has any scaffolding role.

Gill: And yet they are in a complex most of the time.

Montell: I wouldn't conclude that most of the TRPL and rhodopsin pools are bound to INAD most of the time. The interactions of these proteins with INAD may be dynamic and transient.

Zhu: With regard to the PKD mutations, do flies have kidneys?

Montell: They have an open circulatory system. The closest thing might be the Malpighian tubules. They comprise a filtering system, but I wouldn't call the Malpighian tubules an equivalent of the kidneys. I would emphasize that some of the mammalian members of this family are not expressed in the kidney. Furthermore, one of the polycystin-2s, PKD2L2, is enriched in the testes. PKD1 and PKD2, which are associated with polycystic kidney disease, are also expressed in a variety of tissues, in addition to the kidneys.

Zhu: The human patients are all heterozygotes. In the flies do heterozygotes have any phenotype?

Montell: Heterozygous *amo* flies do not have a phenotype. Humans don't have any phenotype either until they have loss of heterozygosity.

Zhu: The idea behind the signalplex is presumably to facilitate fast activation kinetics, but from what you were saying there is no obvious effect.

Montell: The signalplex has an effect on the termination.

Zhu: How would you relate whatever you see in the photoreceptor to the heterologous system where you expressed the TRP and INAD, e.g. in S2 cells. Would the channel have a very fast activation?

Montell: No. In photoreceptor cells, TRP is activated within about 20 milliseconds. TRP activation is much slower in *in vitro* systems. However, in these heterologous systems TRP is not activated in a light-dependent manner.

Zhu: With the PIP_3 binding to arrestin and the resulting translocation, where do you think PIP_3 is? Is it membrane bound or free in the cytosol?

Montell: PIP_3 is a membrane-bound lipid. We don't know the relative concentrations of PIP_3 in the cell bodies and rhabdomeres. However, we hope to develop a reagent to look at this.

Taylor: I have a question about the notion that TRPs are required to assemble into a complex and the primary role of this might be to control rapid inactivation. I can't remember what the time course of an activation is in TRP knockouts, where you are relying on TRPL to drive the photoresponse. Is that very slow inactivation?

Montell: The normal inactivation—the termination of the photoresponse—is in the order of a few hundred milliseconds. It is not nearly as fast as the activation. If we disrupt the direct activation of TRP with INAD, we do not see any defect in inactivation or termination. However, a caveat is that we know that TRP is still interacting indirectly with the signalplex through interactions with other proteins. Thus, it is really not truly dissociated from the signalling complex. In mutants that do not express any TRP, the time-course of activation is relatively normal. During bright light stimulation, the light response is not sustained in *trp* mutant flies.

Groschner: Coming back to the molecular anchor function of TRP species, you told us that TRP is the molecular anchor and TRPL is not involved that much. If you consider two functional components, anchoring and ion conduction, you could imagine two structures: one is more for the scaffolding and not that much for the channel function, and there are other species in the signalplex without that much of a scaffold function.

Montell: TRP is about 10-fold more abundant than TRPL or TRPγ and it is the only one of the three channels that is present in stoichiometric proportions with INAD. It makes sense that TRPL does not serve a scaffold function as it is expressed at low levels and INAD undergoes light-dependent movements. However, TRP is not simply serving a scaffold function as it is also critically important for mediating light-dependent cation influx.

References

Cook B, Bar-Yaacov M, Cohen Ben-Ami H et al 2000 Phospholipase C and termination of G-protein mediated signalling in vivo. Nat Cell Biol 2:296–301

Hanaoka K, Qian F, Boletta A et al 2000 Co-assembly of polycystin-1 and -2 produces unique cation-permeable currents. Nature 408:990–994

Okada H, Fujioka H, Tatsumi N et al 1999 Assisted reproduction for infertile patients with 9+0 immotile spermatozoa associated with autosomal dominant polycystic kidney disease. Hum Reprod 14:110–113 [Erratum in Hum Reprod 14:1166]

Warr CG, Kelly LE 1996 Identification and characterization of two distinct calmodulin-binding sites in the TRPL ion-channel protein of Drosophila melanogaster. Biochem J 314: 497–503

Mammalian TRPC channel subunit assembly

William P. Schilling and Monu Goel

Rammelkamp Center for Education and Research, MetroHealth Medical Center, and the Department of Physiology and Biophysics, Case Western Reserve University School of Medicine Cleveland, OH 44109-1998, USA

Abstract. TRPC genes encode a family of ion channel proteins that appear to be responsible for Ca^{2+} influx following stimulation of membrane receptors linked to phospholipase C. TRPC channels are thought to be tetrameric, and there is growing evidence to suggest heteromultimeric channel assembly. However, the channel subunit composition *in vivo* and the rules governing subunit assembly remain largely unknown. Like the *Drosophila* TRP channels, the mammalian TRPCs may reside in large signalling complexes localized to subcellular microdomains by interaction with specific PDZ-containing scaffolding proteins. Selective localization within cellular signalling networks may play an important role in the mode of channel activation following receptor stimulation. Evidence for heteromultimeric TRPC channel assembly gleaned from overexpression studies will be reviewed and recent evidence for the selective association of native TRPC channel subunits in rat brain will be discussed.

2004 Mammalian TRP channels as molecular targets. Wiley, Chichester (Novartis Foundation Symposium 258) p 18–43

Transient receptor potential (TRP) channels, the topic of this symposium, were originally identified as critical players in *Drosophila* phototransduction. Twenty-one mammalian TRP family members have been identified to date. Based on primary structure, the TRP family has been subdivided into three major subgroups, TRPC, TRPV and TRPM (Montell et al 2002). There are seven mammalian TRP siblings, designated TRPC1–TRPC7, that exhibit 32–47% overall amino acid identity compared to the *Drosophila* isoforms. The TRPC family can be further divided into two major subgroups: TRPC1, C4, C5, and TRPC3, C6, C7. *TRPC2* is a pseudogene in humans (Wes et al 1995), but appears to play an important role in pheromone signalling and the acrosome reaction in rodents (Liman et al 1999, Jungnickel et al 2001). Like the *Drosophila* versions, the primary TRPC channels appear to be regulated by PLC-dependent mechanisms and thus are thought to play an important role in receptor-mediated Ca^{2+} signalling. Heterologous expression studies (Zitt et al 1996, Philipp et al 1996,

1998, Warnat et al 1999, Vazquez et al 2001), antisense experiments (Wu et al 2000, Liu et al 2000, Philipp et al 2000, Brough et al 2001), genetic disruption (Freichel et al 2001, Mori et al 2002), and adenovirus-mediated *invivo* over-expression (Singh et al 2001) suggest that the mammalian TRPC proteins are components of the elusive store-operated channels (SOCs) that underlie receptor-mediated Ca^{2+} influx in many cell types, although results in the literature are often conflicting in this regard. The molecular steps between activation of PLC and/or Ca^{2+} store depletion and subsequent channel activation remains unknown, but channels composed of TRPC3, TRPC6, and TRPC7 can be regulated by diacylglycerol (Okada et al 1999, Hofmann et al 1999).

Based on results from Northern blot analysis and RT-PCR, members of the TRPC channel family appear to be expressed in virtually all tissues and cell types (Garcia & Schilling 1997, Hofmann & Gudermann 2000, Riccio et al 2002). In many tissues and cultured cell lines, multiple TRPC family members are expressed (Fig. 1). Since the TRPC channels are thought to be tetramers, heteromultimeric channel assembly and functional expression of multiple TRPC channel subtypes in a single cell is possible. In this regard, epitope-tagged TRPC1 and TRPC3 co-immunoprecipitate when both are heterologously expressed in 293T cells (Xu et al 1997). Furthermore, co-expression of TRPC1 and TRPC3 in HEK293 cells gives rise to membrane currents with unique properties consistent with heteromultimeric assembly (Lintschinger et al 2002). Strubing et al (2001) found that novel channels were detected following heterologous expression of TRPC1 and TRPC5 in HEK293 cells, and that TRPC1 co-immunoprecipitated with TRPC5 and TRPC4 from rat brain lysates. Using a variety of different approaches including FRET, immunoprecipitation, and electrophysiological assays, Hofmann et al (2002) provide compelling evidence that TRPC1, TRPC4 and TRPC5 co-assemble, and that TRPC3, TRPC6 and TRPC7 co-assemble when heterologously expressed, but that no detectable cross-association occurs between the two major subgroups. Whether or not similar selective interactions occur among native TRPC channel subtypes remains an open question.

To determine whether TRPC channels can co-assemble we expressed individual channel pairs in Sf9 insect cells using recombinant baculovirus (Goel et al 2002). Cell lysates were subjected to immunoprecipitation (IP) using affinity-purified, rabbit polyclonal antibodies specific for each channel subtype. As seen in Fig. 2, TRPC4 and TRPC5 were found by Western blot in immunoprecipitates obtained using the TRPC1 antibody. Likewise, TRPC1 and TRPC5, co-IPed with TRPC4, and TRPC1 and TRPC4 co-IPed with TRPC5. However, antibodies to TRPC1, TRPC4 or TRPC5 failed to immunoprecipitate TRPC3, TRPC6 or TRP7 when individually co-expressed in a pair wise fashion in Sf9 cells. In contrast, TRPC6 and TRPC7 co-IPed with antibodies specific for TRPC3. Likewise, TRPC3 and TRPC7 co-IPed with TRPC6, and TRPC3 and TRPC6 co-IPed with antibodies

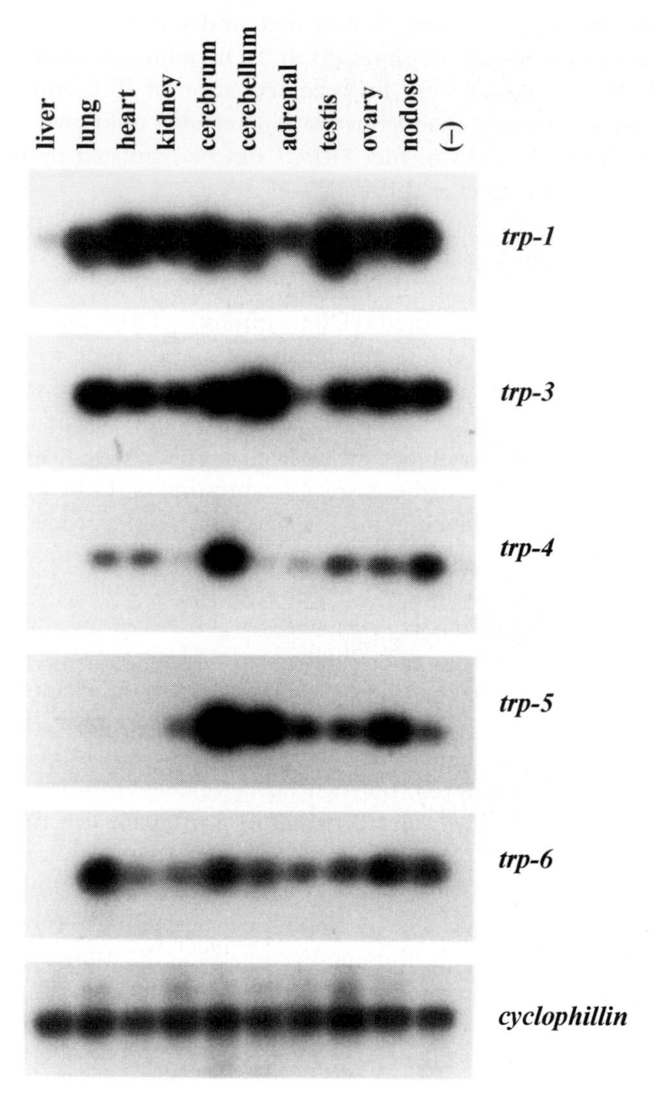
1A
liver
lung
heart
kidney
cerebrum
cerebellum
adrenal
testis
ovary
nodose
(–)
trp-1
trp-3
trp-4
trp-5
trp-6
cyclophillin

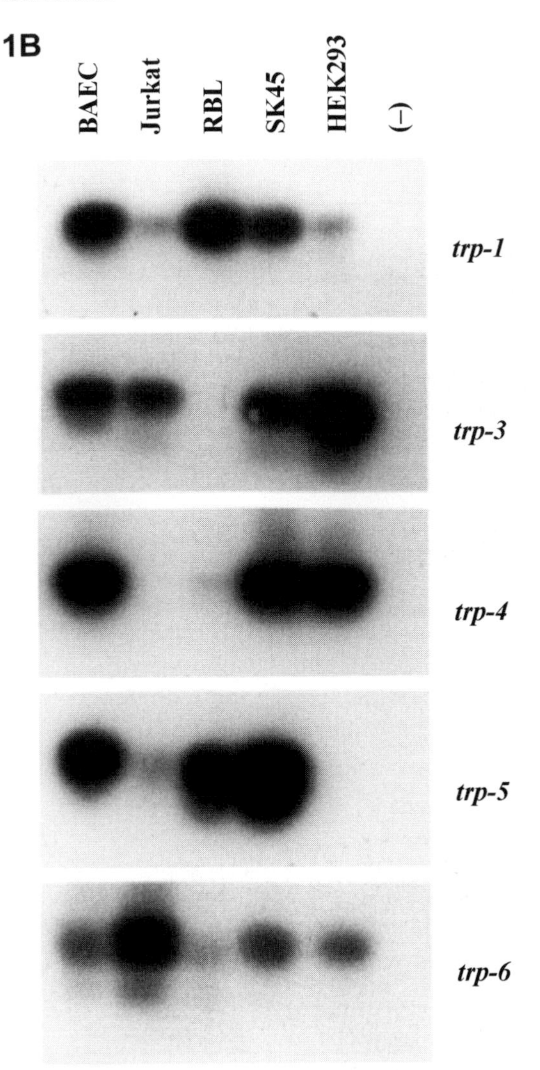

FIG. 1. Expression of mammalian TRPC homologues in adult rat tissues and selected cell lines. TRPC genes were amplified from the indicated tissues or cell lines by RT-PCR, blotted and probed with 5′ end-labelled synthetic oligonucleotides internal to the primers used for PCR. The column labelled (–) reflects PCR reactions performed without template. (From Garcia & Schilling 1997.)

specific for TRPC7 (Fig. 3), but antibody preparations specific for TRPC3, TRPC6 or TRPC7 failed to immunoprecipitate TRPC1, TRPC4 or TRPC5 when individually coexpressed. Thus, as suggested by the study of Hofmann et al (2002), TRPC1, TRPC4 and TRPC5 co-associate, and TRPC3, TRPC6 and

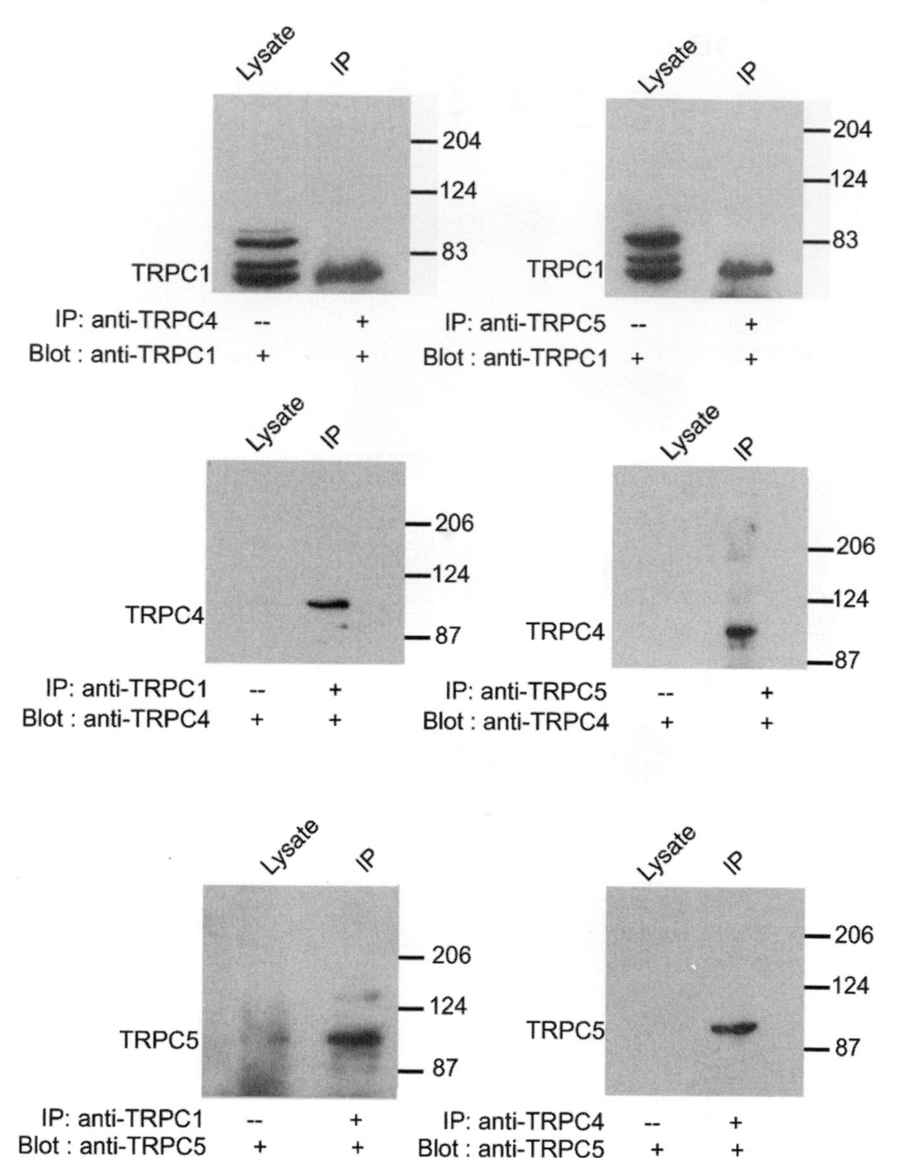

FIG. 2. TRPC1, TRPC4 and TRPC5 co-immunoprecipitate from Sf9 cell lysates. Sf9 cells were co-infected with recombinant baculovirus for expression of individual TRPC channel pairs; (*upper panels*) TRPC1 and TRPC4; (*middle panels*) TRPC4 and TRPC5; (*bottom panels*) TRPC1 and TRPC5. Each panel shows a Western blot using the indicated anti-TRPC antibody; *lane 1*, proteins from total cell lysates; *lane 2*, proteins immunoprecipitated with the indicated anti-TRPC channel antibody. (From Goel et al 2002.)

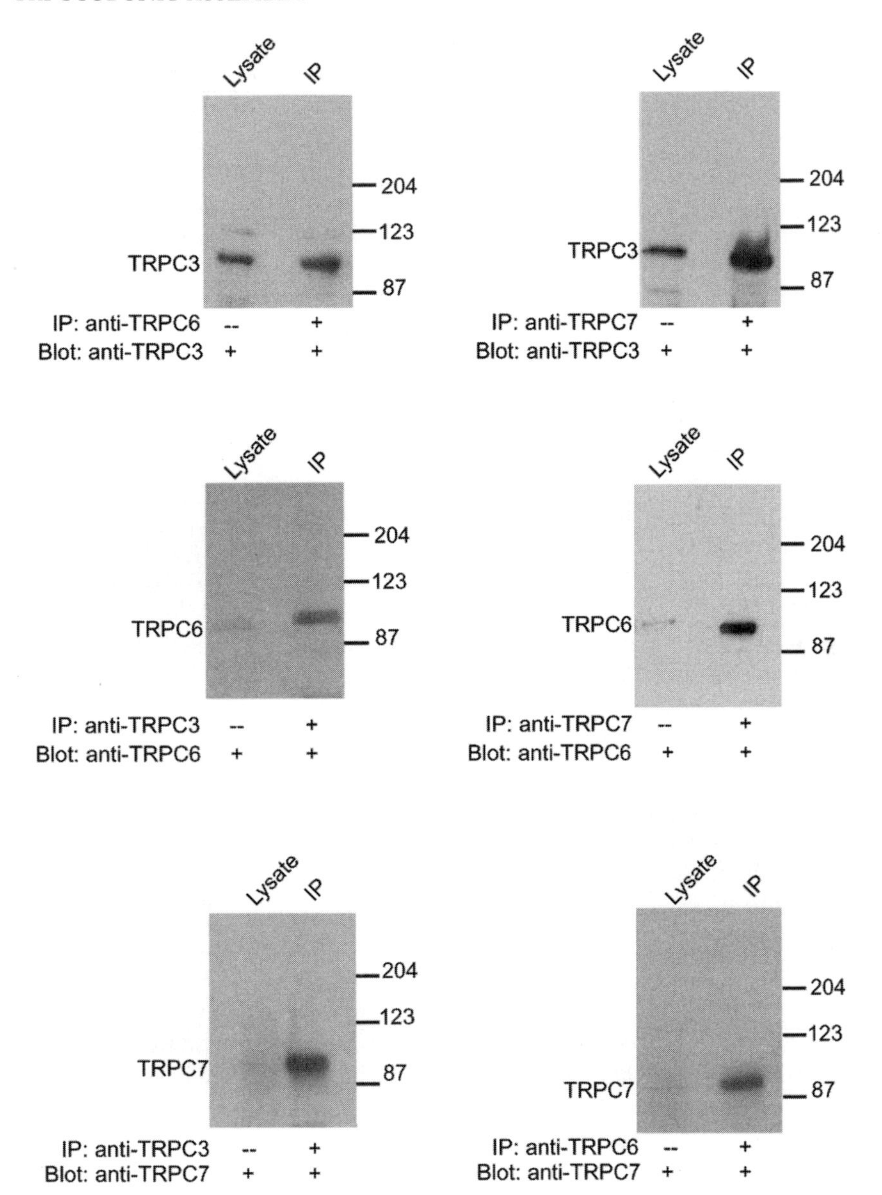

FIG. 3. TRPC3, TRPC6 and TRPC7 co-immunoprecipitate from Sf9 cell lysates. Sf9 cells were co-infected with recombinant baculovirus for expression of individual TRPC channel pairs; (*upper panels*) TRPC3 and TRPC6; (*middle panels*) TRPC6 and TRPC7; (*bottom panels*) TRPC3 and TRPC7. Each panel shows a Western blot using the indicated anti-TRPC antibody; *lane 1*, proteins from total cell lysates; *lane 2*, proteins immunoprecipitated with the indicated anti-TRPC channel antibody. (From Goel et al 2002.)

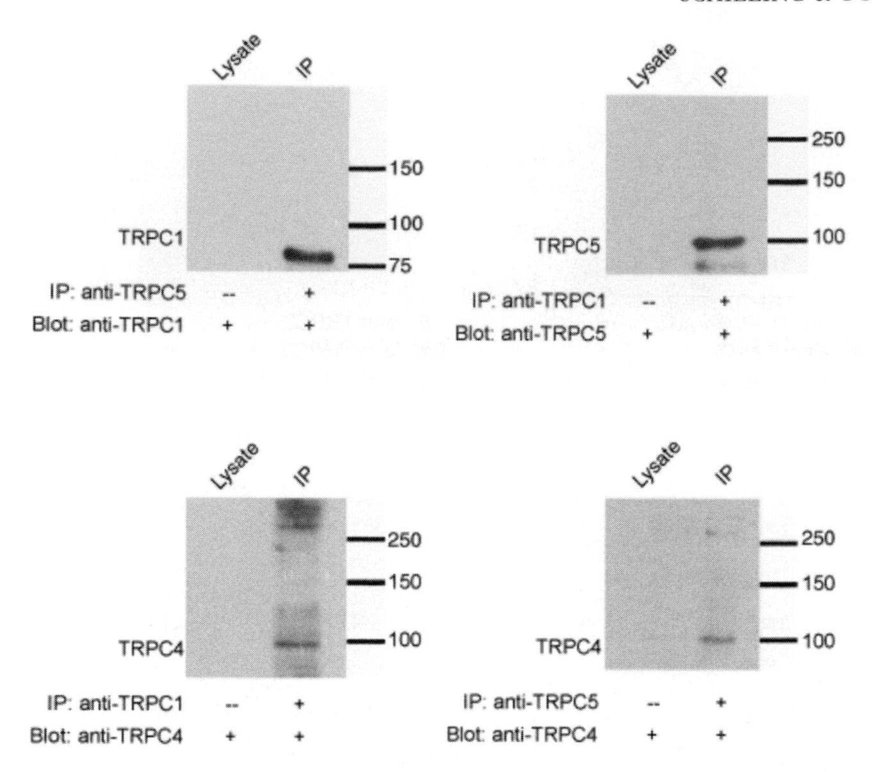

FIG. 4. TRPC1, TRPC4, and TRPC5 co-immunoprecipitate from rat cortex synaptosomes. Proteins from synaptosomal preparations isolated from rat cortex were immunoprecipitated and subjected to Western blot analysis. Each panel shows a Western blot using the indicated anti-TRPC antibody; *lane 1*, proteins from total synaptosomal lysates; *lane 2*, proteins immunoprecipitated with the indicated anti-TRPC channel antibody. (From Goel et al 2002.)

TRPC7 co-associate, but no cross-association between the two major TRPC channel subgroups is observed at least when heterologously expressed.

To determine if the TRPC channel subunit specificity observed in heterologous expressing systems applies to channels *in vivo*, similar immunoprecipitation experiments were performed from rat brain lysates. Specifically, co-IPs were performed from synapotosomal preparations freshly isolated from both rat cerebral cortex and cerebellum. As seen in Fig. 4, TRPC1, TRPC4 and TRPC5 co-IPed from rat synaptosomes. Likewise, TRPC3, TRPC6 and TRPC7 co-IPed from synaptosomal preparations (Fig. 5), but no cross association between the two major TRPC channel subgroups was observed. These results provide support for the hypothesis that TRPC channels within subgroups can form heteromultimers, but that heteromultimers between the subgroups do not occur either *in vitro* or

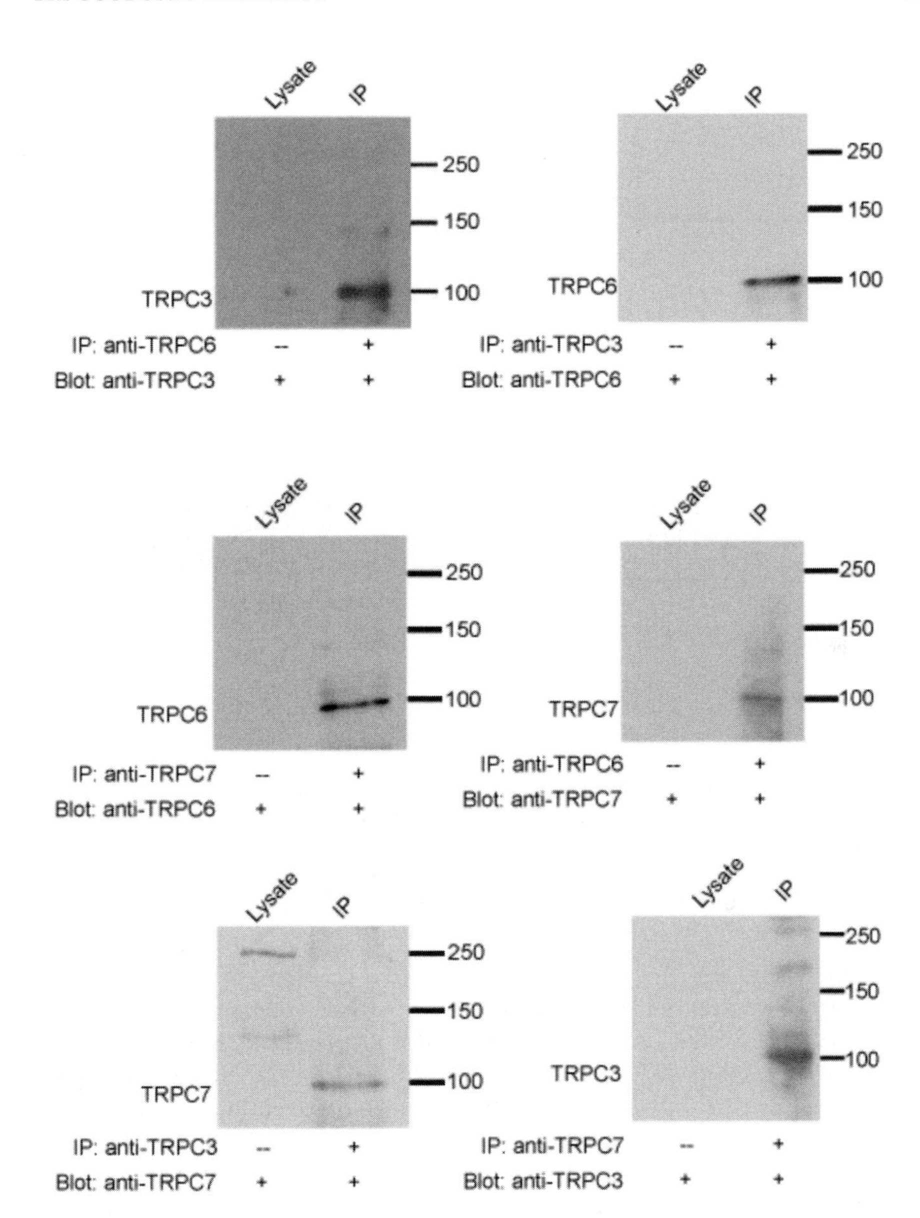

FIG. 5. TRPC3, TRPC6, and TRPC7 co-immunoprecipitate from rat cerebellar synaptosomes. Proteins from synaptosomal preparations isolated from rat cerebellum were immunoprecipitated and subjected to Western blot analysis. Each panel shows a Western blot using the indicated anti-TRPC antibody; *lane 1*, proteins from total synaptosomal lysates; *lane 2*, proteins immunoprecipitated with the indicated anti-TRPC channel antibody. (From Goel et al 2002.)

```
hTRPC3  LIKRYVLKA.QVDKENDEVNEGELKEIKQDISSLRYELLADKSQATEELAILIHKLSEKLNPS
hTRPC7  LIKRYVLKA.QVDRENDEVNEGELKEIKQDISSLRYELLEEKSQATGELADLIQQLSEKFGKN
hTRPC6  LIKRYVLQA.QIDKESDEVNEGELKEIKQDISSLRYELLEEKSQNTEDLAELIRELGEKLSME
hTRPC4  LVKRYVAAMIRDAKTEEGLTEENFKELKQDISSFRFEVLGLL..RGSKLSTIQSANASKESSN
hTRPC5  LVKRYVAAMIRNSKTHEGLTEENFKELKQDISSFRYEVLDLL..GNRK....HPRSFSTSSTE
hTRPC1  LVHRYLTSMRQKMQSTDQATVENLNELRQDLSKFRNEIRDLLGFRTSKYAMFYPRN.......
```

FIG. 6. Alignment of the coiled-coil regions of the TRPC channel proteins. This region, which is near the C-terminus, is highly conserved in each of the TRPC channel proteins. The residues in bold reflect the coiled-coil domain structure.

in vivo. The actual subunit composition, the stoichiometry of subunits that tetramerize to form a functioning channel, and the consequences of subunit arrangement on channel function and regulation remain to be determined. Likewise, the structural features of the TRPC channel proteins responsible for selective subunit association remain unknown. In this regard, recent studies suggest that coiled-coil regions in either the N- or C-terminal domains may play a role defining subunit interactions (Engelke et al 2002, Jenke et al 2003). As seen in Fig. 6, the coiled-coil region of TRPC1, TRPC4, and TRPC5 is conserved, but offset from the coiled-coil region of TRPC3, TRPC6 and TRPC7. This is consistent with the hypothesis that this region plays a role in selective subunit association, but this possibility awaits further investigation.

Studies in *Drosophila* photoreceptor cells have shown that TRP channels are part of a signalling complex (i.e. a signalplex) that is held together in part by the scaffolding protein, INAD (Shieh & Niemeyer 1995, Huber et al 1996, Chevesich et al 1997, Tsunoda et al 1997). INAD is composed of five tandem PDZ domains which act as protein binding modules. INAD interacts with itself and with several proteins involved in phototransduction creating clusters of signalling complexes (Xu et al 1998). Recent studies by our group have shown that the immunophilin, FKBP-59 binds to both TRPL and INAD *in vitro* and *in vivo* (Goel et al 2001) and that FKBP-59 can be displaced from TRPL by the immunosuppressant agent, FK506. Furthermore, heterologous expression studies have shown that FKBP-59 produces attenuation of TRPL channel activity in both fura-2 assays and in single channel recordings. These results suggest that FKBP-59 is a new member of the *Drosophila* signalplex. Interestingly, FKBP-12, another member of the immunophilin family, does not interact with TRPL either *in vitro* or *in vivo*. In contrast, recent studies have shown that *Drosophila* TRP interacts with FKBP-12, but not FKBP-59. Thus, the interaction of FKBPs with *Drosophila* TRP family members examined to date appears to be specific and selective.

Immunophilins are peptidyl-prolyl *cis/trans* isomerases which recognize specific XP dipeptides in target proteins (Marks 1996). In particular, leucyl-prolyl (LP) dipeptides are excellent substrates for the immunophilins. Site-directed mutagenesis revealed that FKBP-59 interacts with a proline-rich region in the C-terminal region of TRPL, 32 amino acids downstream of S6, the last membrane spanning segment. The proline-rich region, which in TRPL contains two LP dipeptides, is highly conserved in the other members of the *Drosophila* TRP family and in the mammalian TRPC orthologues. To determine whether immunophilin can interact with the mammalian TRPC channels, we cloned mammalian FKBP-52 and FKBP-12 and individually co-expressed them with each of the TRPC channel proteins in Sf9 insect cells using recombinant baculovirus. Results from immunoprecipitation experiments showed that TRPC1, TRPC4 and TRPC5 interact with FKBP-52, whereas TRPC3, TRPC6 and TRPC7 interact with FKBP-12. Similar results were obtained in rat brain. This suggests that FKBPs may play an important role in TRPC channel structure, function or regulation and future studies will address these issues.

One of the proposed mechanisms by which information concerning the repletion status of the endoplasmic reticulum (ER) is transmitted to the plasmalemmal channels is the 'conformational coupling' model (Irvine 1992). This model, which is analogous to electro-mechanical coupling that occurs between the ryanodine receptor and the voltage-gated Ca^{2+} channels in skeletal muscle, proposes that close physical association between proteins of the ER and the plasmalemma allows for information transfer via protein conformational change (Berridge 1997). Specifically, it has been suggested that the cytosolic domain of the inositol 1,4,5-trisphosphate receptor (IP_3R) binds to cytoplasmic regions of the TRPC channels. The binding of IP_3 and/or the depletion of Ca^{2+} from the ER induces a conformational change in the IP_3R that 'signals' the opening or activation of the TRPC channels (Kiselyov et al 1998, 2000). Delmas et al (2002) showed that TRPC1 channels overexpressed in superior cervical ganglion neurons, can be activated by thapsigargin and appear to be localized to a specific microdomain that includes a plasmalemmal bradykinin receptor and the IP_3R. Furthermore, these investigators showed that TRPC6 channels overexpressed in the same cell type, were activated by diacylglycerol, but not by store-depletion. TRPC6 channels were not localized to specific domains, but rather were randomly distributed on the cell surface. Thus, TRPC channel activation may depend on localization to specific subcellular microdomains, but how this localization is accomplished and maintained is unknown. Like the *Drosophila* TRP channels, it seems likely that PDZ-containing proteins will form the scaffolding for signalplex formation and localization in mammalian cells expressing the TRPC channels. In some mammalian cell types, the TRPC channel signalplex appears to be localized to caveoli (Lockwich et al 2000). The interaction between TRPC channels and the IP_3R, the localization of TRPC channels to caveoli, and the localization of TRPC channels to signalling microdomains will be discussed in subsequent chapters. However, to determine whether the subgroup-specific interactions extend to association with PDZ-containing proteins, we examined by immunoprecipitation the interaction of TRPC channels with INAD (Goel et al 2002). TRPC1, TRPC4 and TRPC5 co-IPed with INAD when heterologously co-expressed in Sf9 cells, but TRPC3, TRPC6 and TRPC7 did not. This result suggests that a specific mammalian INAD-like protein may play a role in targeting or subcellular localization of TRPC channels.

In summary, there is increasing evidence that TRPC channel proteins form heteromultimeric channel assemblies and that these channels are localized to specific subcellular microdomains. Such localization appears to be important for the mechanism by which the channels are activated by receptor stimulation. Our recent studies have shown that TRPC1, TRPC4 and TRPC5 may form heteromultimers that are linked to an INAD-like, PDZ-containing protein by the immunophilin, FKBP-52. In contrast, TRPC3, TRPC6 and TRPC7 may form non-tethered heteromultimers that interact with the immunophilin FKBP-12. Acute regulation of TRPC channel function by the drugs, FK506 and rapamycin, may constitute a novel mechanism of immunosuppressant action.

References

Berridge MJ 1997 Elementary and global aspects of calcium signalling. J Physiol 499:291–306

Brough G H, Wu S, Cioffi D et al 2001 Contribution of endogenously expressed Trp1 to a Ca^{2+}-selective, store-operated Ca^{2+} entry pathway. FASEB J 15:1727–1738

Chevesich J, Kreuz AJ, Montell C 1997 Requirement for the PDZ domain protein, INAD, for localization of the TRP store-operated channel to a signalling complex. Neuron 18:95–105

Delmas P, Wanaverbecq N, Abogadie FC, Mistry M, Brown DA 2002 Signaling microdomains define the specificity of receptor-mediated $InsP_3$ pathways in neurons. Neuron 34:209–220

Engelke M, Friedrich O, Budde P et al 2002 Structural domains required for channel function of the mouse transient receptor potential protein homologue TRP1B. FEBS Lett 523:193–199

Freichel M, Suh SH, Pfeifer A et al 2001 Lack of an endothelial store-operated Ca^{2+} current impairs agonist-dependent vasorelaxation in TRP4−/− mice. Nat Cell Biol 3:121–127

Garcia RL, Schilling WP 1997 Differential expression of mammalian *Trp* homologues across tissues and cell lines. Biochem Biophys Res Commun 239:279–283

Goel M, Garcia R, Estacion M, Schilling P 2001 Regulation of *Drosophila* TRPL channels by immunophilin FKBP59. J Biol Chem 276:38762–38773

Goel M, Sinkins WG, Schilling WP 2002 Selective association of TRPC channel subunits in rat brain synaptosomes. J Biol Chem 277:48303–48310

Hofmann T, Gudermann T 2000 Transient receptor potential channels as molecular substrates of receptor-mediated cation entry. J Mol Med 78:14–25

Hofmann T, Obukhov A G, Schaefer M, Harteneck C, Gudermann T, Schultz G 1999 Direct activation of human TRPC6 and TRPC3 channels by diacylglycerol. Nature 397:259–263

Hofmann T, Schaefer M, Schultz G, Gudermann T 2002 Subunit composition of mammalian transient receptor potential channels in living cells. Proc Natl Acad Sci USA 99:7461–7466

Huber A, Sander P, Gobert A, Bähner M, Hermann R, Paulsen R 1996 The transient receptor potential protein (Trp), a putative store-operated Ca^{2+} channel essential for phosphoinositide-mediated photoreception, forms a signaling complex with NorpA, InaC and InaD. EMBO J 15:7036–7045

Irvine RF 1992 Inositol phosphates and Ca^{2+} entry: toward a proliferation or a simplification. FASEB J 6:3085–3091

Jenke M, Sanchez A, Monje F, Stuhmer W, Weseloh RM, Pardo LA 2003 C-terminal domains implicated in the functional surface expression of potassium channels. EMBO J 22:395–403

Jungnickel M K, Marrero H, Birnbaumer L, Lemos JR, Florman HM 2001 Trp2 regulates entry of Ca^{2+} into mouse sperm triggered by egg ZP3. Nat Cell Biol 3:499–502

Kiselyov K, Xu X, Mozhayeva G et al 1998 Functional interaction between InsP3 receptors and store-operated Htrp3 channels. Nature 396:478–482

Kiselyov KI, Shin DM, Wang Y, Pessah IN, Allen PD, Muallem S 2000 Gating of store-operated channels by conformational coupling to ryanodine receptors. Mol Cell 6:421–431

Liman ER, Corey DP, Dulac C 1999 TRP2: a candidate transduction channel for mammalian pheromone sensory signaling. Proc Natl Acad Sci USA 96:5791–5796

Lintschinger B, Balzer-Geldsetzer M, Baskaran T et al 2002 Coassembly of Trp1 and Trp3 proteins generates diacylglycerol- and Ca^{2+}-sensitive cation channels. J Biol Chem 275: 27799–27805

Liu X, Wang W, Singh BB et al 2000 Trp1, a candidate protein for the store-operated Ca^{2+} influx mechanism in salivary gland cells. J Biol Chem 275:3403–3411

Lockwich TP, Liu X, Singh BB, Jadlowiec J, Weiland S, Ambudkar IS 2000 Assembly of Trp1 in a signaling complex associated with caveolin-scaffolding lipid raft domains. J Biol Chem 275:11934–11942

Marks AR 1996 Cellular functions of immunophilins. Physiol Rev 76:631–649

Montell C, Birnbaumer L, Flockerzi V et al 2002 A unified nomenclature for the superfamily of TRP cation channels. Mol Cell 9:229–231

Mori Y, Wakamori M, Miyakawa T et al 2002 Transient receptor potential 1 regulates capacitative Ca^{2+} entry and Ca^{2+} release from endoplasmic reticulum in B lymphocytes. J Exp Med 195:1–10

Okada T, Inoue R, Yamazaki K et al 1999 Molecular and functional characterization of a novel mouse transient receptor potential protein homologue TRP7. J Biol Chem 274:27359–27370

Philipp S, Cavalié A, Freichel M et al 1996 A mammalian capacitative calcium entry channel homologous to *Drosophila* TRP and TRPL. EMBO J 15:6166–6171

Philipp S, Hambrecht J, Braslavski L et al 1998 A novel capacitative calcium entry channel expressed in excitable cells. EMBO J 17:4274–4282

Philipp S, Trost C, Warnat J et al 2000 TRP4 (CCE1) protein is part of native calcium release-activated Ca^{2+}-like channels in adrenal cells. J Biol Chem 275:23965–23972

Riccio A, Medhurst AD, Mattei C et al 2002 mRNA distribution analysis of human TRPC family in CNS and peripheral tissues. Brain Res Mol Brain Res 109:95–104

Shieh BH, Niemeyer B 1995 A novel protein encoded by the InaD gene regulates recovery of visual transduction in Drosophila. Neuron 14:201–210

Singh BB, Zheng C, Liu X et al 2001 Trp1-dependent enhancement of salivary gland fluid secretion: role of store-operated calcium entry. FASEB J 15:1652–1654

Strubing C, Krapivinsky G, Krapivinsky L, Clapham DE 2001 TRPC1 and TRPC5 form a novel cation channel in mammalian brain. Neuron 29:645–655

Tsunoda S, Sierralta J, Sun Y et al 1997 A multivalent PDZ-domain protein assembles signalling complexes in a G-protein-coupled cascade. Nature 388:243–249

Vazquez G, Lievremont J-P, Bird GSJ, Putney JW Jr 2001 Human Trp3 forms both inositol trisphosphate receptor-dependent and receptor-independent store-operated cation channels in DT40 avian B lymphocytes. Proc Natl Acad Sci USA 98:11777–11782

Warnat J, Philipp S, Zimmer S, Flockerzi V, Cavalié A 1999 Phenotype of a recombinant store-operated channel: highly selective permeation of Ca^{2+}. J Physiol 518:631–638
Wes PD, Chevesich J, Jeromin A, Rosenberg C, Stetten G, Montell C 1995 TRPC1, a human homolog of a *Drosophila* store-operated channel. Proc Natl Acad Sci USA 92:9652–9656
Wu X, Babnigg G, Villereal ML 2000 Functional significance of human trp1 and trp3 in store-operated Ca^{2+} entry in HEK-293 cells. Am J Physiol Cell Physiol 278:C526–C536
Xu XZS, Li HS, Guggino WB, Montell C 1997 Coassembly of TRP and TRPL produces a distinct store-operated conductance. Cell 89:1155–1164
Xu XS, Choudhury A, Li X, Montell C 1998 Coordination of an array of signaling proteins through homo- and heteromeric interactions between PDZ domains and target proteins. J Cell Biol 142:545–555
Zitt C, Zobel A, Obukhov AG et al 1996 Cloning and functional expression of a human Ca^{2+}-permeable cation channel activated by calcium store depletion. Neuron 16:1189–1196

DISCUSSION

Nilius: What is the functional effect of this immunophilin binding on the TRPCs?

Schilling: We are just starting to look at this. In preliminary experiments, addition of FK506 to cells overexpressing TRPC6 caused inhibition of channel activity. We've observed this effect in both fura-2 assays and whole-cell recordings of TRPC6 currents.

Nilius: How did you activate the TRPC6 in these experiments?

Schilling: By stimulation of the muscarinic receptor with carbachol.

Li: What concentration of FK506 are you using?

Schilling: Typically 10–30 μM.

Putney: Did you compare the effect of FK506 on TRPC6 channels in which the putative binding site for FKBPs was mutated to knock-out FKBP binding?

Schilling: The functional studies with FK506 were done with intact full-length mouse TRPC6. We have created the non-binding mutants for *Drosophila* TRPL, but mutations in the FKBP binding domain of the mammalian TRPC channels have not been made.

Fleig: Do you think that the FKBP59 could be acting simply as a Ca^{2+} chelator?

Schilling: We have no evidence that FKBPs are acting as Ca^{2+} buffers. However, we have preliminary data indicating that calmodulin interacts with FKBP-59, but the functional implications of this interaction are unclear.

Fleig: Most voltage-gated ion channels have subunits other than α. In the case of TRP channels, they have been curiously devoid of such subunits so far. Is it just the interactions between different α subunits that change the biophysical properties of TRP channels? Is there any indication that there may be other subunits to be found?

Schilling: I don't know of anyone who has identified a specific subunit of a TRPC that is different from the individual TRPC channel proteins. Thus far, we have only identified immunophilins as potential TRPC channel accessory proteins. There may be others. Ultimately, our goal is to identify all proteins in the mammalian TRPC channel signalplex and determine specifically how they interact with the TRPC channel, either directly or perhaps indirectly through cytoskeletal elements or some other scaffolding protein. We can take clues from *Drosophila*: we might predict that PLC or some other protein in the signalling cascade will at least be in close proximity. It will be important to identify the PDZ-containing proteins that interact with each TRPC channel in mammalian cells. We can then ask specific questions about which particular regulatory proteins might be associated.

Nilius: What is your definition of a subunit?

Fleig: Different size, and has no pore-forming loop, thus presumably cannot substitute for an α subunit.

Ambudkar: I wanted to ask you about PDZ-containing scaffolding proteins that might interact with mammalian TRPCs. Your data, obtained with exogenously expressed INAD show that TRPC3, C6 and C7 do not interact with INAD. But you cannot rule out the possibility that there might be other endogenously expressed PDZ proteins that might interact with these TRPCs.

Schilling: There may well be another protein that interacts, but from the experiments we have done to date, it does not seem to be an INAD-like protein. We are currently evaluating the potential interaction of a number of PDZ-containing proteins, including PSD-95, GRIP, INAD-like protein and MUPP1 with the TRPC channel proteins.

Ambudkar: In the co-immunoprecipitations from the brain extracts, did you find any of these more commonly known PDZ proteins that are present in the brain?

Schilling: We haven't found any yet.

Penner: In terms of the multiple subunits composing various TRP channels, have you found any evidence that this changes their selectivity or activity?

Schilling: We have not started this line of investigation and the question is when to do these experiments. We can make guesses about the actual composition of the native channels, begin co-expression studies, make the appropriate concatamers and determine the functional consequences of the various subunit arrangements. Or we can wait until we know the actual subunit composition, the subunit stoichiometry, and the identity of accessory proteins present in the native signalling complex before asking questions concerning function.

Montell: If you consider the proteins that are known to bind to FKBP on the basis of co-IPs do any of these proteins interact with the TRPCs?

Schilling: We haven't looked specifically for other known FKBP interacting proteins, with the exception of one, the IP_3R. We know that FKBPs co-immunoprecipitate with IP_3R. Thus, the FKBPs may be part of a larger signalling complex that includes both the TRPC channels and IP_3R.

Authi: You showed some good evidence for TRPC1/4/5 associating and TRPC3/6/7 associating. Would you be happy in saying that TRPC1, for example, can form a homotetramer? Following on from that, are we happy that the homotetramer has sufficient pore-forming units that it can actually form a channel?

Schilling: We found functional, constitutively active, TRPC1 channels when overexpressed in Sf9 insect cells (Sinkins et al 1998). I would therefore say it can form homomultimers, and the channels do appear to be functional. However, the TRPC1 channels in this expression system were not regulated by store depletion with thapsigargin or by activation of G protein-coupled membrane receptors. Whether or not functional homotetramers of TRPC1 ever exist *in vivo* is unclear. We need to perform sequential immunoprecipitation experiments to determine the native subunit composition of the individual channels in the brain.

Penner: Have you tried TRPC1 overexpression in mammalian cells? It doesn't work for us.

Schilling: No. Perhaps it needs TRPC4 to help co-localize it to the plasma membrane as shown by Hofmann et al (2002). Have you tried that?

Penner: No.

Authi: We have been looking at platelets. We have detected a very small amount of TRPC1 that we can only see after immunoprecipitation, and we don't see any TRPC4 or TRPC5 at all in the cell. Every time you perform TRPC1 immunoprecipitation from brain homogenates, you get TRPC4 and 5 co-precipitating. Have you looked at any other cell where you can examine TRPC1 precipitation, and do you get any other TRPC protein with it?

Schilling: We plan to examine the TRPC subunit interactions in other tissues. Obviously, we started with brain because it seems to have the highest density of the TRPCs. In cultured cell lines (e.g. HEK cells and bovine aortic endothelial cells) we have had a difficult time finding evidence for endogenous TRPCs at the protein level. Therefore, we decided to start with the whole tissues and to then ask which specific cell type in each tissue expresses a specific TRPC channel.

Muallem: We know from some sparse functional data that it is possible to find combinations of TRPs that appear to have functional significance. In terms of PDZ binding, you can get a lot of binding to PDZ proteins but this doesn't mean that this is what happens *in vivo*.

Montell: I can underscore that point. With TRPV3 we obtained incredibly good interactions with PSD95 when we co-expressed the two proteins in tissue culture cells. But we can't get them to co-immunoprecipitate from brain tissue.

Schilling: It is important that the interaction of accessory proteins with the TRPC channels be examined at multiple levels. We must show that putative interacting proteins co-immunoprecipitate from native tissues with the channels, and we need to show by immunocytochemistry that they are present in the same cell and co-localize to the appropriate membrane. We then need to show that the interaction has a functional consequence. This can be accomplished in native tissues by examining the activity of known modulators. In the case of FKBP-12, it is known that FK506 displaces FKBP from its binding site on target proteins. Thus, FK506 at appropriate concentrations should have an effect on TRPC channel activity. Experiments of this type must be performed for every interacting protein that we identify by immunoprecipitation.

Putney: The issue here is that for the players you are looking at, you can carry everything you do *in vitro* to the brain except the PDZ domain protein, because we are not sure what the right candidate is in mammalian cells.

Montell: What about NHERF? Is there any evidence that NHERF is interacting with TRPC4 in native tissue?

Zhu: You can co-IP it from the brain with TRPC4. Also, NHERF co-IPed with PLCβ.

Muallem: This is a difficult protein. NHERF has only two PDZ domains, yet it is supposed to be associated with everything ever known to be able to transport something. One has to be a little bit careful. I don't know any activity that is associated with binding to PDZ domains that in one way or another hasn't been shown to be affected by binding to NHERF.

Zhu: The same thing could be said for INAD. It doesn't have an obvious effect except for a small effect on inactivation. We have to be careful not to try to deduce what the PDZ domain is doing from the limited amount of data.

Montell: The litmus test is whether you can make a point mutation in the target protein which prevents the co-IP *in vivo*.

Muallem: NHERF was knocked out and this had no effect on Ca^{2+} signalling. It has an incredibly mild effect on the function of the kidney. This is remarkable considering how much it was supposed to do. One has to be very careful about what the actual protein is doing *in vivo*.

Nilius: It is even trickier than this. For example, TRPV5 and TRPV6 have PDZ binding motifs, but what is bound is S100A10 which is not a PDZ domain protein (van de Graaf et al 2003). It is very tricky.

Schilling: We should also remember that the PDZ-containing protein may not be the scaffolding protein and that these interactions may be dynamic: they may be there only to bring specific proteins into play at particular times. In other words, there may be regulated movements of proteins into close proximity with the TRPC channels and this movement may be facilitated by a PDZ-containing protein. Some other protein may actually form the scaffolding. In this regard, the ankyrin repeats in the N-terminus of the TRPC proteins may interact with the cytoskeleton. This interaction may be important for localizing the TRPC channels to specific domains within the cell.

Gill: In fact, they may be there to prevent them interacting with certain things at specific times. They may have an inhibitory role.

Authi: Perhaps the heteromultimerization affects the location of the TRPC protein itself. There seems to be good circumstantial evidence that TRPC1 on its own seems to localize within internal structures. I am referring to some data that Hofmann and colleagues published in PNAS (Hofmann et al 2002) where if TRPC1 is expressed alone it locates within intracellular structures. Also, Mori's data tend to suggest that in DT40 cells in which TRPC1 had been knocked out, there is a reduction in Ca^{2+} release from intracellular stores (Mori et al 2002). This implies that if you just have TRPC1 expressed by itself, it seems to be more located in internal structures. The data of Hofmann et al (2002) suggest that if you co-express TRPC1 with either 4 or 5, it tends to go to the plasma membrane. Perhaps the multimerization of 1 with 4 or 5 affects the location of this protein.

Muallem: Is there any evidence to suggest how the FKBP proteins are working to regulate the channels?

Schilling: We do not know the functional implications of FKBP binding to mammalian TRPC channels. For *Drosophila* TRPL, FKBP59 appears to be involved in the Ca^{2+}-dependent regulation of the channel. We also have preliminary evidence that FK506 can change the sensitivity of *Drosophila* TRPL expressed in Sf9 cells to Ca^{2+}; it makes the channel much more Ca^{2+} sensitive.

Muallem: FKBPs have a very defined enzymatic activity.

Schilling: Yes, the FKBPs catalyse peptidyl-prolyl *cis/trans* isomerization reactions. At least for the ryanodine receptor, the enzymatic activity doesn't seem to be involved in their interaction with the channel. If you knockout the isomerase activity, FKBP-12 still interacts with the ryanodine receptor and modulates channel function (Timerman et al 1995).

Penner: It is important to consider that the TRPCs are likely to be non-specific ion channels. When dealing with non-specific ion channels we need to consider the membrane potential effect that they will create once they are activated. If you over-express a non-selective ion channel the membrane potential of those cells will depolarize, which, depending on the expression system, may fully set it to 0 mV or maybe slightly more negative. This depends on whether your expression system has Ca^{2+}-activated K^+ channels or other functions. Deducing the function of an ion channel on the basis of a Ca^{2+} signal is a fraught business. If your cell hyperpolarizes you are getting a larger $[Ca^{2+}]_i$ signal; if it depolarizes you are getting a smaller signal, without really changing the activity of the Ca^{2+} channels.

Schilling: The response in a Ca^{2+}/fura-2 assay provides a clue. We then examined the sensitivity of the TRPL channel to Ca^{2+}/CaM in excised inside-out membrane patches. There is an evolving story involving Ca^{2+}, FKBP-59 and the *Drosophila* TRPL channel, but we have not performed similar experiments with the mammalian TRPCs.

Scharenberg: You alluded to a potential protein folding role for FKBP-59. Have you done surface labelling and IP to see whether you can find native FKBP-59 in assembled plasma membrane channel complexes? In all the studies you show there are going to be a substantial fraction of intracellular channels.

Schilling: The next step will be to evaluate FKBP interaction with TRPC channels using synaptosomal plasma membranes. We will also use this preparation to evaluate TRPC subunit association and perhaps subunit stoichiometry.

Zhu: An important issue here is the cellular environment in relation to Ca^{2+} changes. There are differences among different cell lines we use for the expression of TRP channels. The HSG cells that Indu Ambudkar has been using show a clear hyperpolarization after stimulation with carbachol. In contrast, for non-transfected HEK cells the membrane potential changes very little (unpublished observations, M. X. Zhu). In Indu's study there is a clear demonstration that TRP1 functions as a SOC. One of the reasons for this might be that in that particular cellular environment the Ca^{2+}-activated K^+ channels hyperpolarize the membrane, leading to larger Ca^{2+} influx.

Montell: Are there mouse knockouts for FKBP-52 or 12, where you could look for defects in TRPC function?

Schilling: There are mouse knockouts for FKBP-12 and 12.6. The FKBP-12 knockout is embryonic and neonatal lethal. The mice die because of severe cardiomyopathy (Shou et al 1998). Disruption of the FKBP-12.6 gene causes cardiac hypertrophy in male mice (Xin et al 2002).

Barritt: How difficult would it be to take a more direct *in vivo* approach and try cross-linking? You start with the ability to detect TRP1 *in vivo* and then see what proteins are bound to it by cross-linking processes. The difficulty with this is that cross-linking will pick up many other interactions. Yet one of the issues we are talking around is how to define what is happening *in vivo* as compared with the *in vitro* situation.

Schilling: There are a number of experiments that can be done with cross-linking reagents. One approach to evaluate actual subunit composition, is to use cross-linking reagents followed by Western blot analysis to determine which TRPCs are in close proximity. The second approach would be to follow cross-linking with sucrose gradient centrifugation. A third approach is to perform sequential immunoprecipitation. Each approach has been used previously for a number of channel types. Cross-linking would play an important role in trying to understand which subunits are close to other subunits, but by lengthening the arm or spacer on the cross-linking reagent it may be possible to pull down novel binding partners.

Ambudkar: We have initiated experiments of this type. One approach we have taken is to biotinylate membrane preparations, solubilize, IP and then cross-link. We then cleave the cross-linker and use the biotin label to detect the proteins in the complex. We are able to obtain a reproducible profile of proteins. It is a tough experiment to do.

Barritt: Yes, but that is better than doing it in a whole cell extract, because then you are increasing the chance of cross-linking many proteins.

Ambudkar: Yes, that is correct. When cross-linking is done with whole cell extracts or in intact cells, we tend to get large aggregates of the protein which are fairly resistant to SDS.

Putney: Have you identified anything?

Ambudkar: Not yet.

Cox: Has anyone attempted to dimerize or trimerize FK506 connected with various length chains that probe the TRPL binding site you were referring to?

Schilling: Not that I am aware of. But the crystal structure of FKBP-12 has been solved with FK506 bound to it. There is quite a bit known about structure and function of the FKBP proteins.

Cox: You made a suggestion about multiple binding sites.

Schilling: Unlike FKBP-12 which has only a single FK506 binding site, FKBP-59 is thought to have up to three binding domains. We have cut FKBP-59 into three pieces corresponding to the individual FK506 binding domains to determine which domain actually interacts with INAD or the channel protein, and which domains have functional effects on channel activity.

Scharenberg: Actually, someone has made a dimer of FK506. They call it FK1012, and they use it as a chemical dimerizer.

Penner: Bill, is your interpretation of the FK506 data on the channel that it displaces FKBP, and then inhibits it? What is going on there?

Schilling: I don't know the answer to that, but it is possible that you can have FK506 binding to FKBP-52 without causing displacement from its target protein, because there are these multiple FK506 binding domains on FKBP-52. In fact, the first and second FK506 binding domains are highly conserved. It is possible that you could have three FK506 molecules bound to FKBP-52. These three FK506 binding domains on FKBP-52 may also independently interact with different proteins. Thus, one can envision FKBP-52 acting as a linker between the TRPC channel and a PDZ-containing protein.

Scharenberg: Have any FKBPs come up in the genetic screens of *Drosophila* phototransduction mutants?

Montell: No.

Putney: You made an off the cuff remark when you were talking about the *Drosophila* TRPL, INAD and FKBP interactions, that you saw the same thing in S2 cells. Do S2 cells express TRPL?

Schilling: We identified TRPL in S2 cell lysates by Western blot analysis. S2 cells also express FKBP and we were able to co-immunoprecipitate TRPL and FKBP-59 from S2 cell lysates.

Putney: What is the function of TRPL in S2 cells?

Schilling: I don't know.

Putney: This has been an issue that has come up before when we have talked about *Drosophila.* Do the TRPCs that *Drosophila* has function outside the photoreceptor? My feeling is that there is so much of it there that no one has really pushed the antibody work to see if it is anywhere else. So at the sensitivity at which you look at photoreceptors you don't see it anywhere else. At the sensitivity at which we can see TRPCs in the brain, we have a hard time finding them in the periphery.

Hardie: Julian Dow and Shireen Davies in Glasgow have Westerns and immunolocalization data indicating that TRP and TRPL are expressed in the Malpighian tubules (*Drosophila's* equivalent of a kidney) and *trp* and *trpl* mutants show defects in fluid secretion from these (S. Davies, J. Dow, personal communication).

Montell: The word 'specific' almost doesn't have a place in biology. One can refer to a protein as being enriched in a tissue. We have done microarray analyses to identify all of the eye-enriched proteins. TRP is enriched 150-fold in the eye versus other tissues. This is not specific; it is highly enriched. TRPL is about 100-fold enriched. In databases there are TRPL ESTs that are sequenced from embryos. These proteins are expressed in other places. John Carlson's lab has published a paper (Störtkuhl 1999) indicating that TRP has a role in olfactory adaptation.

Putney: The reason I got into this track is because it may be possible to get some clues about TRPC functions in mammalian cells by looking at the physiology outside of photoreceptors. All the TRPC mutants and multiple mutants are available in *Drosophila.* There is no obvious phenotype other than the visual one, but we could still do physiological studies on the various tissues.

Montell: There are other TRPs in flies. There are two TRPVs, one TRPM, several TRPNs, several polycystin 2-like proteins and one protein related to mucolipin.

Putney: But could we get clues about TRPC function by looking at other places than the photoreceptors?

Montell: These sorts of approaches sort of go against the Gestalt in *Drosophila*, where we primarily do forward genetics. For example, someone might be studying the passive immune response in flies and then by chance find that there is a TRP protein involved.

Hardie: One thing that is missing in *Drosophila* is a good electrophysiological I_{CRAC}. There probably is one there: for example, there is a thapsigargin-induced Ca^{2+} entry signal that can be measured in S2 cells, but S2 cells are not good for mutant analysis.

Montell: That's not exactly true. RNAi works great in S2 cells.

Gill: This is a very important point. If you have, for example, a TRP/TRPL double knockout in *Drosophila*, do they appear normal?

Hardie: They are actually pretty sick. At least, ours are. They are fertile and viable, but they are not that healthy.

Nilius: When you are trying to learn something about the function of TRPCs in mammalian tissue, I wouldn't go to *Drosophila*. I would look at conditional mouse knockouts.

Putney: The redundancy in the mammalian system may make that difficult. Those animals are being made now, and some exist, but it might be tough to interpret the results.

Nilius: It is difficult to translate a function in *Drosophila* to a mammalian setting.

Schilling: When Roger Hardie reported his results in S2 cells showing that thapsigargin stimulates Ca^{2+} entry (Yagodin et al 1998), we thought that there must be capacitative Ca^{2+} entry in *Drosophila* cells other than the eye. We started cloning and pulled out TRPγ. About four months later Craig Montell published on TRPγ expression in *Drosophila* eye (Xu et al 2000). But TRPγ is very similar to TRPC4. In fact, TRPC4 is 47% identical to TRPγ at the amino acid level, the highest identity of any TRPC with the *Drosophila* TRPs. The question is whether or not TRPγ might serve a SOC function in cell types other than the eye. Have you looked to see whether TRPγ is present in any significant quantities in any other tissue in *Drosophila*?

Montell: Looking at TRPγ expression isn't easy. Its expression is even lower than TRPL. It is very strange. We detect the protein only in the eye, but the RNA is not eye-enriched at all.

Zhu: Is INAD also enriched in the eye, as opposed to the periphery?

Montell: It is enriched 65-fold in the eye, but it is so abundant that the relatively lower expression in some other cell types might be rather significant. It is true that TRP and TRPL double mutant flies are pretty sick, so these channels must be doing something in addition to functioning in the eye. The only other thing we have checked is the auditory response, and this was normal.

Putney: Has anyone looked at their role in fly immunity?

Montell: Not to my knowledge. This would be interesting to look at.

Nilius: What would you consider is a good mammalian functional candidate for INAD?

Montell: There are many candidates. One might be GRP, which is a PDZ-containing protein that binds to glutamate receptors. I think it has seven PDZ domains. There is also PSD95 which has multiple PDZ domains and MUPP1, which is a huge protein with 13 PDZ domains.

Zhu: CIPP has 4 PDZ domains (Kurschner et al 1998). There is also NHERF2 (Hwang et al 2000). They recognize the same sequence as NHERF, so chances are they will also recognize TRPC4 and 5. There are just so many of them.

Nilius: Are there any functional data?

Zhu: There are no functional data with regard to how a PDZ domain regulates a mammalian TRP. I don't think there are many functional data, even in *Drosophila*.

Montell: It depends what you are calling function. Interactions between mammalian TRPs and PDZ proteins could have a role in localization and stability, rather than activation and deactivation. This is the case with the TRP/INAD interaction. Clearly, the interaction between INAD and TRP is critically important for retaining both these proteins in the rhabdomere and keeping them in the right stoichiometries. This is incredibly important for function. If you don't have TRP in the right place it is not functional. If you co-express INAD and TRP in tissue culture cells such as HEK cells, there is no effect.

Hardie: There are data that go against that. A recent paper reported that when INAD is co-expressed with *Callihphora* TRP it stops it working as a store-operated channel (Harteneck et al 2002).

Montell: At least in our hands there is no effect, and we wouldn't expect an effect on the basis of the *in vivo* data. *In vivo* when the direct TRP–INAD interaction is disrupted there isn't a profound effect on channel activity.

Gudermann: I think for TRPC2 the adaptor protein MUPP1 is an interesting candidate, but we have to be very careful what system we look at. It is highly expressed in the vomeronasal organ together with TRPC2. But when we were studying TRPC2 we had a hard time getting it functionally expressed in heterologous cell systems. From immunostaining and coprecipitation data, the adaptor protein MUPP1 and TRPC2 in the vomeronasal organ highlight an intriguing system indicating two components of a functional sensory signalling complex.

Muallem: We shouldn't restrict our thinking solely to pure PDZ-containing proteins. We can learn something from PSD95 and the post-synaptic density. Most of the proteins that interact with K^+ channels may have one PDZ motif, but they also have several other motifs. In PSD95 the whole complex is assembled around a completely different scaffold. INAD is the holy grail for the fly, but this might not be the case for the mammalian system.

Scharenberg: Analysis of any interaction domain in a protein should be carried out with caution, because if you overexpress it, you change the stoichiometry. It changes the physiology of the cell completely. You block interactions that may normally occur with other proteins. If you put mutations in a PDZ domain-containing protein, the same thing occurs. Any time you have a promiscuously interacting protein, it is an oversimplification to infer that only the interaction that you are interested in is being interrupted. It is a tricky business.

Zhu: With respect to NHERF, the reason I think it is so intriguing is that there are a number of similarities between it and INAD. For example, NHERF binds to ERM (ezrin–radixin–moesin). These proteins are bound to actin. We have evidence that TRPC4 is associated with actin by co-immunoprecipitation. This co-immunoprecipitation is dependent on the interaction of TRPC4 with NHERF. At the moment we don't know whether this interaction is helping the localization of TRPC4 with the plasma membrane, but clearly it doesn't have anything to do with the activation of the channel (unpublished results, M. X. Zhu).

Montell: Some of the best experiments that can be done with TRP/PDZ interactions in mammalian systems would be to get some of the mouse knockouts of these PDZ-containing proteins and see whether there are changes in the localization of candidate TRP proteins.

Putney: I have the impression from the literature that it is relatively easy to heterologously express 3, 6 and 7 and get them assembled into something that can be regulated through PLC and receptors and so forth. In contrast, it is very difficult with 1 and 4, but 5 seems to work. I wonder whether this is because 1 and 4 need a scaffolding protein that we are not putting in the cell, and 3, 6 and 7 don't.

Zhu: NHERF doesn't seem to help. It doesn't fix the problem.

Muallem: There are data that are a little different from that. NHERF seems to be quite important for getting TRPC4 to the plasma membrane. The data are very clean.

Zhu: This may be cell-type dependent.

Gill: A lot of people have had problems with TRPC1: when it is overexpressed it just stays in the ER and doesn't get to the plasma membrane. Indu Ambudkar, do you have any problems getting it to the plasma membrane?

Ambudkar: We can see the protein in the ER in some cells, and we have already published this (Wang et al 1999). So, I cannot say that the protein is in the plasma membrane in 100% of the cells. By FACS analysis, we see that >90% of the cells express the protein.

Penner: Your currents are still small.

Muallem: We can get TRPC1 to bind IP_3 receptors and Homer proteins. The currents we get are small, but they are measurable, at around 100 pA.

Ambudkar: There are several cell types where TRPC1 increases function. However, I'm not sure that the plasma membrane localization has been correlated with the functional effect in all these cases. Our studies show that TRPC1 can form active channels. It works very nicely in MDCK and HSG cells. In 293 cells we have had much more success with TRPC3 than TRPC1. These are all epithelial cell lines.

Authi: In the cell lines where TRPC1 goes to the plasma membrane, do they also have TRPC4 and 5?

Ambudkar: We have just recently obtained TRPC antibodies from Bill Schilling, so we hope to do these experiments soon. We have detected a number of TRPCs in various cell lines by RT-PCR, and so we assumed that we would find protein. But I am not sure that the story will turn out to be so simple. Where we expected proteins to be present, we have not been able to detect any. It is possible that we may have to enrich the protein preparation. We initially started working with crude membrane preparations, but now we are trying purified plasma membrane preparations.

Authi: With the purified membranes that we use, which are prepared using high voltage free flow electrophoresis, we can clearly see TRPC6 in the plasma membrane (of platelets). There is no doubting this at all. The very low expression of TRPC1 that we see appears to be only in the intracellular membranes.

Gill: We looked fairly hard at the TRPC1 knockouts in the DT40 cells. We couldn't see any real change in store-operated or receptor-induced entry. It seemed pretty much the same.

Penner: Did you look at currents?

Gill: No, we only did Ca^{2+} imaging.

Kunze: In carotid body oxygen-sensing glomus cells, the TRPC1 seems to be localized very close to the nucleus and doesn't extend to the plasma membrane. We don't know whether there is an intracellular function for the channel or, alternatively, the TRPC1 moves to the membrane under particular conditions. On the other hand, the neurons that innervate these cells also express TRPC1, and that TRPC1 appears to go out to the plasma membrane. The distribution and function may be cell dependent.

Delmas: We expressed TRPC1 in rat sympathetic neurons. We always got good expression of TRPC1 in the plasma membrane but still the current was very small.

Nilius: Reinhold Penner, do you think TRPC1 is a channel?

Penner: All I can say is that we've tried to express it in HEK cells, but it has never worked for us.

Putney: It is the same for us. In fairness, people who get TRPC1 to work don't get it to work well in HEK cells.

Zhu: It is a small effect. It works much better in a Ca^{2+}-free solution (Lintschinger et al 2000).

Ambudkar: We have not done much work with TRPC1 in HEK cells.

Putney: Is the channel regulated in HEK cells, or could you just see the channels in low Ca^{2+} conditions?

Groschner: In low Ca^{2+} you can see some conductance. It is constitutively active, and also regulated by changes in lipid mediators.

Penner: Which endogenous, native system would people choose to look at TRPC1 with? If I were to identify a large response in a native system, where would I look for TRPC1?

Zhu: Hippocampal pyramidal cells would be a good one. This is where you see the highest *in situ* signal.

Schilling: That's what I would say, also: neurons. But the channels might be TRPC1 and 4 or 5 heteromultimers.

Zhu: TRPC1, 4 and 5 are all very high in hippocampal neurons.

Authi: In terms of mRNA expression, I think smooth muscle cells have TRPC1 in reasonable amounts.

Benham: It is true that smooth muscle has TRPC1, 4 and 5, but smooth muscle generally expresses channels at a fairly low level, particularly vascular smooth muscle. This wouldn't be a cell type to choose if you are looking for a strong TRPC1 signal.

Barritt: We studied TRPC1 in liver cells (Brereton et al 2001). We first detected mRNA in those cells corresponding to TRPC1. From that time on I wondered whether TRPC1 realistically had a physiological role in liver cells. We haven't been able to test this further yet, or answer it. We spent some time looking for the protein, arguing that if we could detect this endogenously, we would have a reasonable foundation to go on to look at function (Chen & Barritt 2003). We are still in the position of not knowing in liver cells whether TRPC1 has a function. I still have doubts about this.

References

Brereton HM, Chen J, Rychkov G, Harland ML, Barritt GJ 2001 Maitotoxin activates an endogenous non-selective cation channel and is an effective initiator of the activation of the heterologously expressed hTRPC-1 (transient receptor potential) non-selective cation channel in H4-IIE liver cells. Biochim Biophys Acta 1540:107–126

Chen J, Barritt GJ 2003 Evidence that TRPC1 (transient receptor potential canonical 1) forms a Ca^{2+}-permeable channel linked to the regulation of cell volume in liver cells obtained using small interfering RNA targeted against TRPC1. Biochem J 373:327–336

Harteneck C, Kuchta SN, Huber A, Paulsen R, Schultz G 2002 The PDZ-scaffold protein INAD abolishes apparent store-dependent regulation of the light-activated cation channel TRP. FASEB J 16:1668–1670

Hofmann T, Schaefer M, Schultz G, Gudermann T 2002 Subunit composition of mammalian transient receptor potential channels in living cells. Proc Natl Acad Sci 99:7461–7466

Hwang JI, Heo K, Shin KJ et al 2000 Regulation of phospholipase C-beta 3 activity by Na^+/H^+ exchanger regulatory factor 2. J Biol Chem 275:16632–16637

Kurschner C, Mermelstein PG, Holden WT, Surmeier DJ 1998 CIPP, a novel multivalent PDZ domain protein, selectively interacts with Kir4.0 family members, NMDA receptor subunits, neurexins, and neuroligins. Mol Cell Neurosci 11:161–172

Lintschinger B, Balzer-Geldsetzer M, Baskaran T et al 2000 Coassembly of Trp1 and Trp3 proteins generates diacylglycerol- and Ca^{2+}-sensitive cation channels. J Biol Chem 275: 27799–27805

Mori Y, Wakamori M, Miyakawa T et al 2002 Transient receptor potential 1 regulates capacitative Ca^{2+} entry and Ca^{2+} release from endoplasmic reticulum in B lymphocytes. J Exp Med 195:673–681

Sinkins WG, Estacion M, Schilling WP 1998 Functional expression of TRPC1: a human homologue of the Drosophila TRP channel. Biochem J 331:331–339

Shou W, Aghdasi B, Armstrong DL et al 1998 Cardiac defects and altered ryanodine receptor function in mice lacking FKBP12. Nature 391:489–492

Störtkuhl KF, Hovemann BT, Carlson JR 1999 Olfactory adaptation depends on the Trp Ca^{2+} channel in Drosophila. J Neurosci 19:4839–4846

Timerman AP, Wiederrecht G, Marcy A, Fleischer S 1995 Characterization of an exchange reaction between soluble FKBP-12 and FKBP–ryanodine receptor complex. Modulation by FKBP mutants deficient in peptidyl-prolyl isomerase activity. J Biol Chem 270:2451–2459

van de Graaf SF, Hoenderop JG, Gkika D et al 2003 Functional expression of the epithelial Ca^{2+} channels (TRPV5 and TRPV6) requires association of the S100A10-annexin 2 complex. EMBO J 22:1478–1487

Wang W, O'Connell B, Dykeman R et al 1999 Cloning of Trp1β isoform from rat brain: immunodetection and localization of endogenous Trp1 protein. Am J Physiol Cell Physiol 276:C969–C979

Xin HB, Senbonmatsu T, Cheng DS et al 2002 Oestrogen protects FKBP12.6 null mice from cardiac hypertrophy. Nature 416:334–338

Xu XZ, Chien F, Butler A, Salkoff L, Montell C 2000 TRPgamma, a drosophila TRP-related subunit, forms a regulated cation channel with TRPL. Neuron 26:647–657

Yagodin S, Hardie RC, Lansdell SJ, Millar NS, Mason WT, Sattelle DB 1998 Thapsigargin and receptor-mediated activation of Drosophila TRPL channels stably expressed in a Drosophila S2 cell line. Cell Calcium 23:219–228

TRPC channel interactions with calmodulin and IP_3 receptors

Michael X. Zhu and Jisen Tang

Center for Molecular Neurobiology and Department of Neuroscience, The Ohio State University, 168 Rightmire Hall, 1060 Carmack Road, Columbus, Ohio 43210, USA

Abstract. Consistent with the conformational coupling mechanism, which suggests that store-operated channels are activated via physical interactions with intracellular calcium release channels, previous studies have demonstrated a functional coupling and a physical interaction between the transient receptor potential canonical type 3 (TRPC3) and inositol 1,4,5-trisphosphate receptor (IP_3R). The IP_3R–TRPC binding domains were determined using *in vitro* binding assays. This work aimed to study the effect of IP_3R–TRPC interaction on TRPC function. Pull-down experiments were used to study the binding of TRPC to IP_3R and to calmodulin. Patch clamp recordings in whole-cell and inside-out configurations were used to examine the effect of a TRPC-binding IP_3R fragment, Ca^{2+} and calmodulin on TRPC activity. We found that IP_3R and calmodulin compete for a common binding site at the TRPC C-terminus. TRPC channels are activated either by a peptide representing the TRPC-binding domain of IP_3R or by inactivation of calmodulin from the excised membrane patches. TRPC3 activity is inhibited by Ca^{2+} and calmodulin. Therefore, we have identified a critical IP_3R–TRPC interaction that is involved in the activation of TRPC-formed channels. We propose that IP_3Rs activate TRPC channels by displacing inhibitory calmodulin from a common calmodulin-IP_3R binding site located at the C-terminus of TRPC.

2004 Mammalian TRP channels as molecular targets. Wiley, Chichester (Novartis Foundation Symposium 258) p 44–62

Activation of phospholipase C triggers Ca^{2+} release from the endoplasmic reticulum (ER) through binding of inositol 1,4,5-trisphosphate (IP_3) to IP_3 receptors (IP_3Rs) located on the ER membrane. The depletion of Ca^{2+} from the ER store activates store-operated channels (SOCs) on the plasma membrane, allowing Ca^{2+} influx and replenishment of the depleted stores (Berridge 1995, Birnbaumer et al 1996). The molecular makeup and activation mechanism are the two major unanswered questions concerning SOCs. Transient receptor potential canonical (TRPC) proteins are the most likely molecular components of SOCs (Montell et al 2002). Mammals have seven TRPC isoforms, which can be divided into four subgroups as TRPC1, TRPC2, TRPC3/6/7 and TRPC4/5 (Zhu et al 1996). Distinct SOCs may be formed by multimerizations of either identical or different TRPC subunits, or TRPC and distantly related TRP homologues or other unidentified proteins, since TRPV6 and TRPM5 have also been suggested to form SOCs (Yue et al 2001, Perez et al 2002).

Several activation mechanisms have been proposed for SOCs. Among them, the conformational coupling hypothesis suggests that one or more ER-associated proteins may bind to SOCs and activate them through direct protein-protein interactions in a manner similar to the activation of skeletal muscle ryanodine receptor by dihydropyridine receptors (Irvine 1990, Berridge 1995). IP_3Rs seemed perfect for this role as they transverse the ER membrane, have fairly long cytoplasmic N-termini, and are already involved in releasing Ca^{2+} from the ER. Evidence supporting the role of IP_3Rs in conformational coupling includes that IP_3 activated SOCs in inside-out patches excised from several non-excitable cell types (Vaca & Kunze 1995, Kiselyov et al 1999, Zubov et al 1999). IP_3 also activated TRPC3 channels expressed in HEK293 cells and its effect was dependent on the IP_3Rs associated with the excised patches (Kiselyov et al 1998). To elucidate the mechanism by which IP_3Rs activate TRPC-formed channels, immunocoprecipitation experiments were used to show direct association between IP_3R and human TRPC3 or murine TRPC6 (Boulay et al 1999). Further experimentation using glutathione S transferase (GST) pull-down assays to test the binding between different regions of type 3 IP_3R (IP_3R3) and TRPC3 led to the identification of two TRPC-binding sites from the N-terminus of IP_3R3 and an IP_3R-binding site from the C-terminus of TRPC3 (Boulay et al 1999).

The purpose of this study was to examine the effects of IP_3R-TRPC interaction on TRPC function. We found that the IP_3R-binding site of TRPC also binds to calmodulin (CaM) in a Ca^{2+}-dependent manner. In *in vitro* binding studies, CaM and a TRPC-binding IP_3R fragment competed for binding to TRPC. Using patch clamp techniques, we showed that TRPC3 and TRPC4 were activated by a peptide representing the TRPC-binding domain of IP_3R3 and also by inactivation of CaM. The TRPC3 activity is sensitive to inhibition by Ca^{2+} and CaM. We conclude that a direct interaction between IP_3R and TRPC is involved in the activation of TRPC-formed channels and such activation is achieved through displacement of the inhibitory CaM from a common binding site located at the C-terminus of TRPC. This model of channel activation may be applicable to all TRPC-based channels since the common binding site is conserved among all seven TRPC isoforms.

Experimental procedures

DNA constructs for generating GST, enhanced blue fluorescence protein (EBFP), and maltose binding protein (MBP) fusion proteins containing fragments of IP_3R3 or TRPC were made as described (Zhang et al 2001a, Tang et al 2001). Methods for GST pull-down and CaM binding experiments have been described in Boulay et al

(1999) and Tang et al (2001). Stable HEK293 cell lines expressing a C-terminal haemagglutinin-tagged human TRPC3 (T3–9) and unmodified murine TRPC4α (T4–1, T4–60) were made and maintained according to Zhu et al (1998) and Zhang et al (2001b). Conditions for electrophysiological recordings of inside-out patches and the application of various treatments are described in Zhang et al (2001a) and Tang et al (2001).

For whole-cell recordings of T3–9, recording pipettes were pulled from micropipette glass and heat polished to 3–6 MΩ. Isolated cells were voltage-clamped at room temperature in the whole-cell mode using an Axopatch 200B Amplifier (Axon Instruments). Voltage commands were made using pCLAMP 8 and currents were recorded in a computer through a Digidata 1200 PC interface. Voltage ramps of 210 ms to +100 mV after a brief (25 ms) step to −120 mV from the holding potential of 0 mV were applied every 2 s. Data were acquired at the sampling rate of 2.5 kHz and filtered at 1 kHz. Solutions and other details are given in the figure legend. Junction potentials were corrected based on calculated values using pCLAMP 8.

Results and discussion

A CaM-binding site is colocalized with the IP_3R-binding site of TRPC

A previous study showed that the F2q fragment (Glu^{669}–Asp^{698}) of IP_3R3 bound to the T3C7 fragment (Met^{742}–Glu^{795}) of TRPC3 (Boulay et al 1999). Since at the equivalent regions of T3C7, TRPC1, C2 and C4 are less than 18% identical to TRPC3, it was difficult to predict whether IP_3Rs bind to the similar regions of other TRPCs. Studies have shown that channels formed by TRPC4/5 are different from those formed by TRPC3/6/7 in terms of mechanism of activation, ion selectivity, partnership for multimerization and interaction with scaffolding proteins (Philipp et al 1996, 1998, Hofmann et al 1999, 2002, Tang et al 2000, Goel et al 2002). Therefore, we tested the binding of murine TRPC4α to IP_3R3F2q with a hope that it might reveal features different from TRPC3.

First, we examined the interaction of ^{35}S-labelled TRPC4 N-terminus (aa1–364), transmembrane region (aa330–659) and C-terminus (aa659–750 and aa668–974) with GST-IP_3R3F2q using a GST pull-down assay. Because *Drosophila* TRP and TRP-like are CaM binding proteins (Phillips et al 1992, Chevesich et al 1997), we suspected that mammalian TRPCs should also bind to CaM. Thus, we also tested the interaction of these fragments with CaM-Sepharose. The results showed that the C-terminal fragments interacted with IP_3R3 and CaM. Further analyses showed that two separate C-terminal regions of TRPC4α, T4Cc (aa659–726) and T4CT2 (aa733–974) bound to both IP_3R3F2q and CaM (Tang et al 2001).

Since T4Cc is equivalent to T3C7, we first made smaller fragments from T4Cc to search for minimal sequences required for binding to IP_3R and CaM. These fragments were fused to either EBFP or MBP to increase the ^{35}S labelling and to facilitate the separation by SDS polyacrylamide gel electrophoresis. As shown in Fig. 1, there is a close correlation between IP_3R3F2q and CaM for binding to the TRPC4 fragments. Both proteins bound strongly to T4Cw (aa695–724). Further elimination of amino acids from either side of T4Cw reduced the binding to IP_3R as well as to CaM. Therefore, the binding sites for IP_3R3 and CaM within T4Cw are very close and inseparable using the deletion strategy. Examinations of equivalent regions from other isoforms revealed that the common CaM/IP_3R binding (CIRB) site is present in all TRPCs as well as in *Drosophila* TRP and TRP-like (Fig. 1C,D). Conservation of the CIRB site among the TRP proteins suggests that both IP_3R and CaM play important roles in regulating TRPC functions and they might do so either in a cooperative manner by facilitating each other's action or in a competitive manner by inhibiting each other's effect.

IP_3R and CaM compete for binding to the TRPC CIRB sites

To find out whether IP_3R and CaM bind to the CIRB site cooperatively or competitively, we examined the effect of CaM on the binding of ^{35}S-labelled MBP-T4Cw and MBP-T3C14 (aa761–795 of human TRPC3) to GST-IP_3R3F2q. With 70 μM free Ca^{2+} in the binding buffer, CaM dose-dependently inhibited the T4Cw-F2q and T3C14-F2q interactions with IC_{50} values of 2.1 and 1.2 μM, respectively. The inhibition at 20 μM CaM was 73% and 88% for T4Cw and T3C14, respectively (Fig. 2A). CaM also inhibited the binding of other TRPC CIRB sites to IP_3R3F2q to various degrees (Fig. 2B). Because the TRPC-CaM interaction is Ca^{2+} dependent (Zhang et a 2001a, Tang et al 2001), CaM only inhibited the TRPC-IP_3R interaction in the presence but not the absence of Ca^{2+} (Fig. 2B). These results indicate that at high Ca^{2+} concentrations, CaM competes with IP_3Rs for binding to the CIRB sites of all TRPC isoforms.

To test the effect of IP_3R on the TRPC-CaM interaction, we used an 18 amino acid peptide (F2v) containing Glu^{681}–Asp^{698} of IP_3R3, which represents the C-terminal portion of F2q that confers the IP_3R3 binding to TRPC (Tang et al 2001). The peptide dose-dependently inhibited the binding of CaM to the CIRB site of TRPC1–7 with IC_{50} values ranging from 1.7 to 90 μM (Fig. 2C). Interestingly, the peptide was about 10-fold more effective at blocking CaM binding to the CIRB site of TRPC3/6/7 than to that of other TRPC subgroups, suggesting that IP_3Rs are more capable of competing with CaM for binding to TRPC3/6/7 than to other TRPCs. These experiments were performed in the presence of 70 μM free Ca^{2+}, a favourable condition for CaM binding. Most

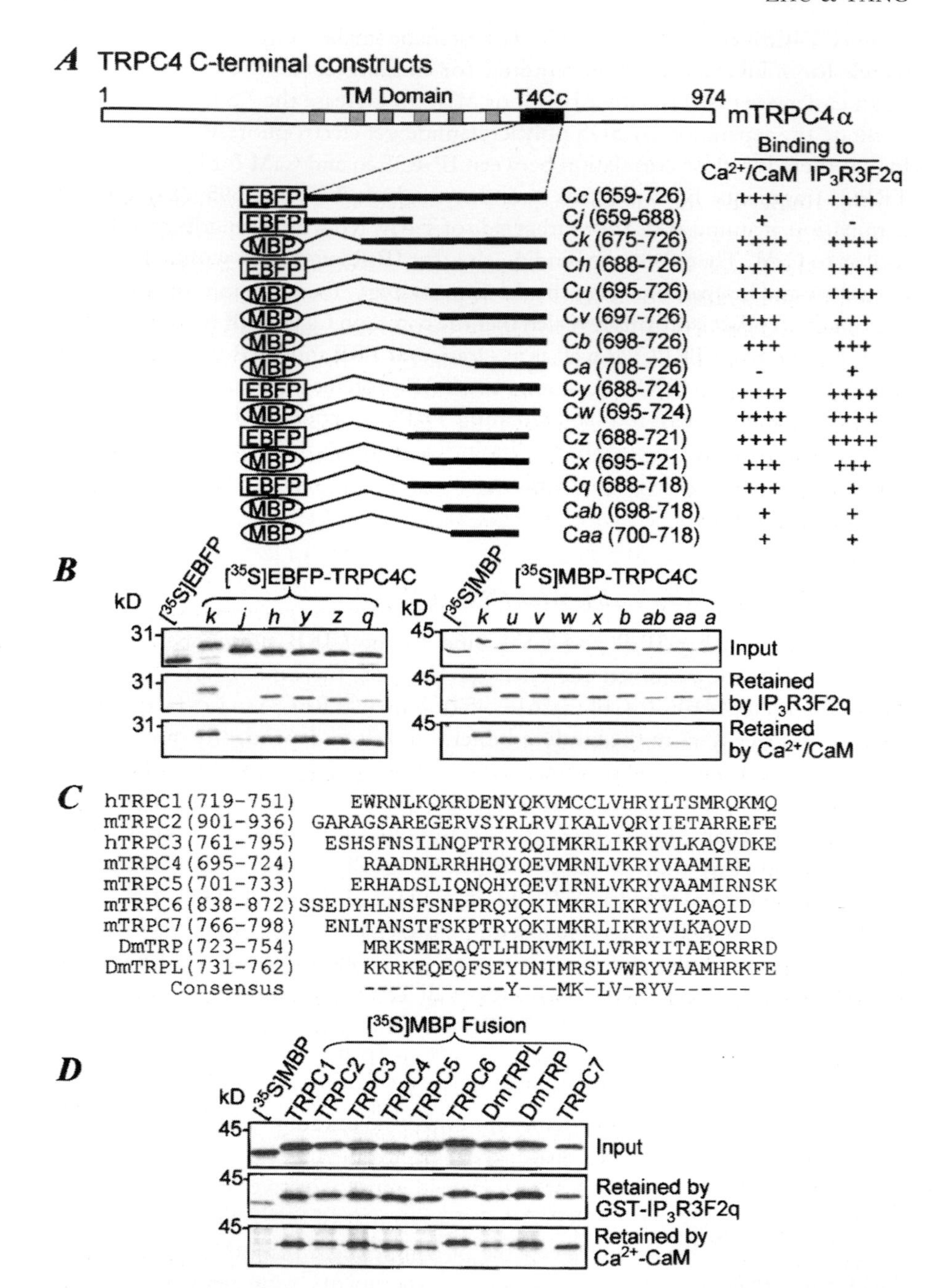

A TRPC4 C-terminal constructs
1
TM Domain
T4Cc
974
mTRPC4α
Binding to
Ca2+/CaM IP3R3F2q
Cc (659-726) ++++ ++++
Cj (659-688) + -
Ck (675-726) ++++ ++++
Ch (688-726) ++++ ++++
Cu (695-726) ++++ ++++
Cv (697-726) +++ +++
Cb (698-726) +++ +++
Ca (708-726) - +
Cy (688-724) ++++ ++++
Cw (695-724) ++++ ++++
Cz (688-721) ++++ ++++
Cx (695-721) +++ +++
Cq (688-718) +++ +
Cab (698-718) + +
Caa (700-718) + +
EBFP
MBP
B
[35S]EBFP
[35S]EBFP-TRPC4C
kD
k j h y z q
31
[35S]MBP
[35S]MBP-TRPC4C
k u v w x b ab aa a
45
Input
Retained by IP3R3F2q
Retained by Ca2+/CaM
C
hTRPC1(719-751) EWRNLKQKRDENYQKVMCCLVHRYLTSMRQKMQ
mTRPC2(901-936) GARAGSAREGERVSYRLRVIKALVQRYIETARREFE
hTRPC3(761-795) ESHSFNSILNQPTRYQQIMKRLIKRYVLKAQVDKE
mTRPC4(695-724) RAADNLRRHHQYQEVMRNLVKRYVAAMIRE
mTRPC5(701-733) ERHADSLIQNQHYQEVIRNLVKRYVAAMIRNSK
mTRPC6(838-872) SSEDYHLNSFSNPPRQYQKIMKRLIKRYVLQAQID
mTRPC7(766-798) ENLTANSTFSKPTRYQKIMKRLIKRYVLKAQVD
DmTRP(723-754) MRKSMERAQTLHDKVMKLLVRRYITAEQRRRD
DmTRPL(731-762) KKRKEQEQFSEYDNIMRSLVWRYVAAMHRKFE
Consensus -----------Y---MK-LV-RYV------
D
[35S]MBP Fusion
kD
[35S]MBP
TRPC1
TRPC2
TRPC3
TRPC4
TRPC5
TRPC6
DmTRPL
DmTRP
TRPC7
45
Input
Retained by GST-IP3R3F2q
Retained by Ca2+-CaM

likely at lower Ca^{2+} concentrations, IP_3Rs will be more effective at binding to the CIRB site of TRPC, preventing the CaM-TRPC interaction.

Ca^{2+} and CaM inhibit the activity of TRPC3

To determine the role of CaM on TRPC function, we performed whole-cell recordings using a stable TRPC3 cell line (T3–9). With 100 nM free Ca^{2+} in the pipette and 40 μM Ca^{2+} in the bath, application of 100 μM carbachol (CCh) caused a rapid increase of currents that displayed outward rectification at positive potentials and reversed at approximately 0 mV. The currents typically peaked at about 20–30 s and then gradually decreased (Fig. 3A). Control HEK293 cells showed no significant increase in current when stimulated with CCh, implicating that only TRPC3 activity is detected under these conditions. When Ca^{2+} in the pipette was increased to 500 nM, CCh-induced currents became much smaller (Fig. 3C), indicating that cytosolic Ca^{2+} has an inhibitory effect on agonist-induced TRPC3 activity. To determine whether the inhibition is mediated by CaM, we overexpressed the rat wild-type CaM in the T3–9 cells. As shown in Fig. 3B and C, with 100 nM Ca^{2+} in the pipette, CaM overexpression also inhibited CCh-induced TRPC3 activity. Voltage ramp measurements showed lower activities for CaM-transfected cells at both the negative and the positive potentials with no apparent change of the current-voltage relationship (Fig. 3B). Thus, increasing the cytosolic concentration of either Ca^{2+} or CaM inhibited agonist-stimulated TRPC3 activity.

A previous study had shown that in cell-attached patches of T3–9 cells, single channel activity of TRPC3 was activated by bath application of CCh. Excision of the patches to intracellular solution containing 100 nM free Ca^{2+} eliminated the activity, which was restored with application of IP_3 (Kiselyov et al 1998). We found that when excised to a Ca^{2+}-free solution containing 5 mM EGTA and no added Ca^{2+}, about one-third of the patches displayed TRPC3 activity (Fig. 4A–D), which is characterized by a relatively large single channel amplitude (3.2 pA at 60 mV) and a very brief mean open time (Fig. 4C). The open probability for the

---

FIG. 1. Colocalization of a common binding site for CaM and IP_3R3F2q on TRPC. (A) Compositions of EBFP or MBP fusion proteins containing subfragments of T4C*c*. Positions in the full-length murine TRPC4α are shown in the parentheses. Relative intensities of the fusion proteins retained by Ca^{2+}/CaM and GST-IP_3R3F2q are indicated by the + and − signs and are summarized from 2–4 experiments. (B) Binding results from representative experiments. (C) Amino acid compositions of mammalian TRPC1–7, *Drosophila* TRP (DmTRP) and TRP-like (DmTRPL) present in the MBP fusion proteins. 'm' and 'h' denote TRPC of murine and human origin, respectively. Positions for these sequences in the full-length proteins are shown in parentheses. (D) ^{35}S-labelled MBP-fusion proteins were tested for binding to GST-IP_3R3F2q and to Ca^{2+}/CaM. (Modified from Tang et al 2001.)

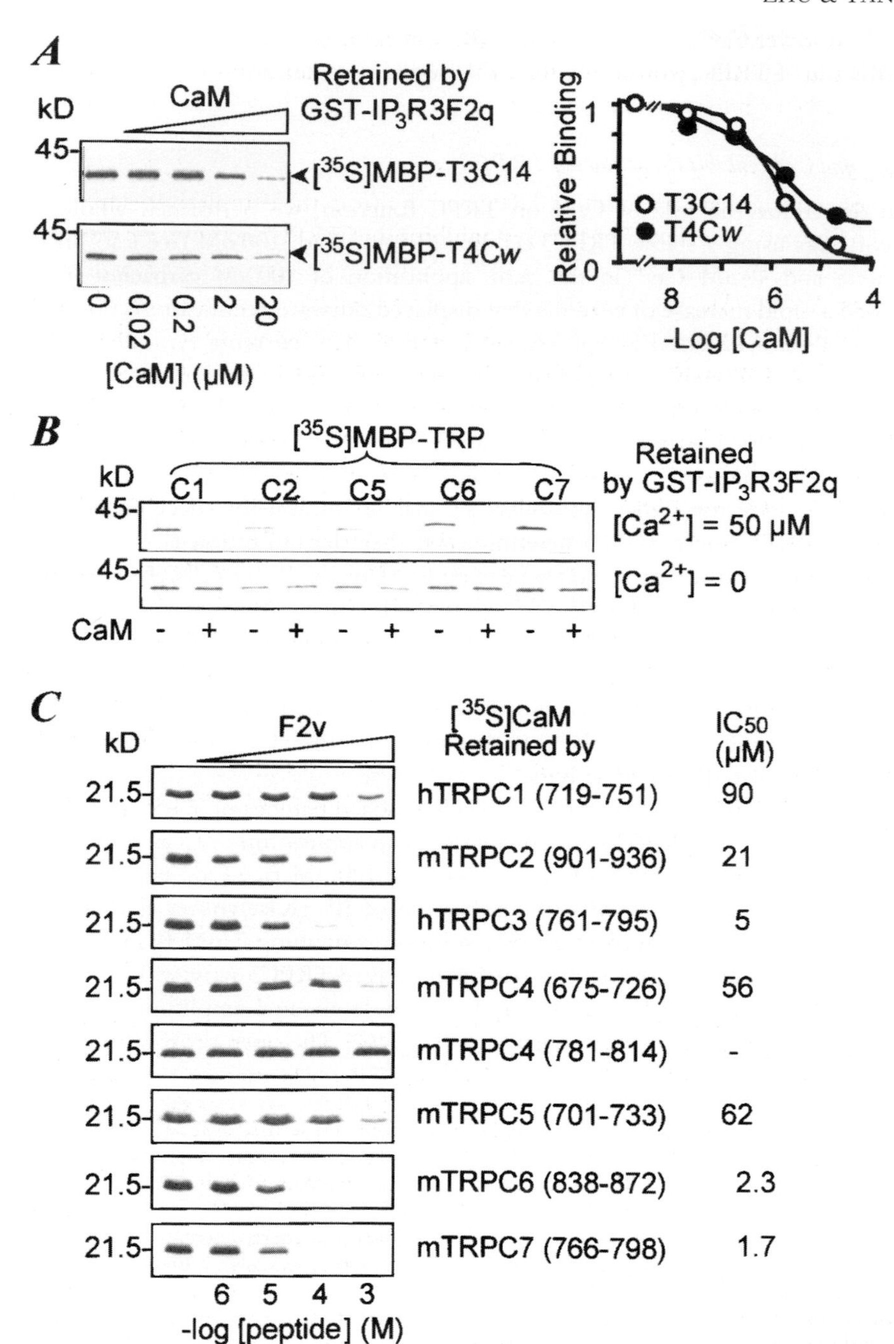
A
kD
CaM
Retained by
GST-IP3R3F2q
45-
[35S]MBP-T3C14
45-
[35S]MBP-T4Cw
0
0.02
0.2
2
20
[CaM] (μM)
Relative Binding
1
0
T3C14
T4Cw
8
6
4
-Log [CaM]
B
[35S]MBP-TRP
kD
C1
C2
C5
C6
C7
Retained
by GST-IP3R3F2q
45-
[Ca2+] = 50 μM
45-
[Ca2+] = 0
CaM - + - + - + - + - +
C
kD
F2v
[35S]CaM
Retained by
IC50
(μM)
21.5-
hTRPC1 (719-751)
90
mTRPC2 (901-936)
21
hTRPC3 (761-795)
5
mTRPC4 (675-726)
56
mTRPC4 (781-814)
-
mTRPC5 (701-733)
62
mTRPC6 (838-872)
2.3
mTRPC7 (766-798)
1.7
6 5 4 3
-log [peptide] (M)

outward current at 60 mV was more than 10 times higher than that for the inward current at −60 mV, consistent with the outward rectification of the whole-cell current. No similar current was seen in patches excised from control HEK293 cells. Application of 180 nM and 1.8 μM free Ca^{2+} to the cytoplasmic side of the patches inhibited the current by 83% and 93%, respectively (Fig. 4D,E), further demonstrating that Ca^{2+} inhibits TRPC activity from the cytoplasmic side.

To find out the role of CaM in Ca^{2+} inhibition of TRPC3 activity, we expressed in the T3–9 cells either the wild-type CaM or a CaM mutant, $CaM(EFmut)_4$, which carries Asp to Ala substitutions in all four of its EF hands (Xia et al 1998). $CaM(EFmut)_4$ interacted with TRPC3 weakly independent of Ca^{2+} (Zhang et al 2001a). The expression of $CaM(EFmut)_4$ caused a large increase in basal activity of patches excised to the Ca^{2+}-free solution (Fig. 4F,G). In contrast to cells expressing only TRPC3 that showed one or two open channels, cells co-expressing TRPC3 and $CaM(EFmut)_4$ displayed multiple open levels with flickers of brief openings, indicating that the CaM mutant had increased the opening frequency of TRPC3 without significantly affecting the mean open time. Surprisingly, raising Ca^{2+} to 1.8 μM did not inhibit the TRPC3 current but instead, increased the current by about threefold (Fig. 4F,G). On the other hand, no TRPC3 activity was detected in 10 excised patches from T3–9 cells overexpressing the wild-type CaM and increasing Ca^{2+} to 1.8 μM did not activate TRPC3 either (Fig. 4F,G). These results suggest that CaM is closely associated with TRPC3 under basal conditions and mediates Ca^{2+}-dependent channel inhibition. The CaM mutant displaced the endogenous CaM and thus resulted in spontaneous channel activation as well as its potentiation by Ca^{2+} through, maybe, a CaM-independent mechanism.

FIG. 2. Competition between CaM and IP_3R3 for binding to the TRPC CIRB site. (A) CaM inhibits the binding of CIRB site of TRPC4 (T4C*w*) and TRPC3 (T3C14) to IP_3R3. Left panel shows autoradiograms of ^{35}S-labelled MBP-T4C*w* or MBP-T3C14 retained by GST-IP_3R3F2q from representative experiments while the right panel shows averages of results of phosphorimaging analysis. (B) Representative experiments show that 20 μM CaM inhibits the binding of IP_3R3F2q to the CIRB site of TRPC1, C2, C5, C6 and C7 in a Ca^{2+}-dependent manner. 10 mM HEDTA and EGTA were used to buffer [Ca^{2+}] to 50 μM and 0 (< 10 nM), respectively. (C) peptide F2v inhibits binding of CaM to the CIRB site of TRPC1–7. Autoradiograms show ^{35}S-labelled CaM retained by GST fusion proteins containing the CIRB sequence of TRPC1–7 or the second CaM-binding site of TRPC4 (aa781–814), which binds to CaM but not IP_3R. Positions of TRPC sequences included in the GST fusion proteins are indicated in parentheses. The binding buffer contained 70 μM free Ca^{2+}. IC_{50} values were determined from results of phosphorimaging analysis of relative amount of [^{35}S] CaM retained by each TRPC fragment from two-three experiments. (Modified from Tang et al 2001.)

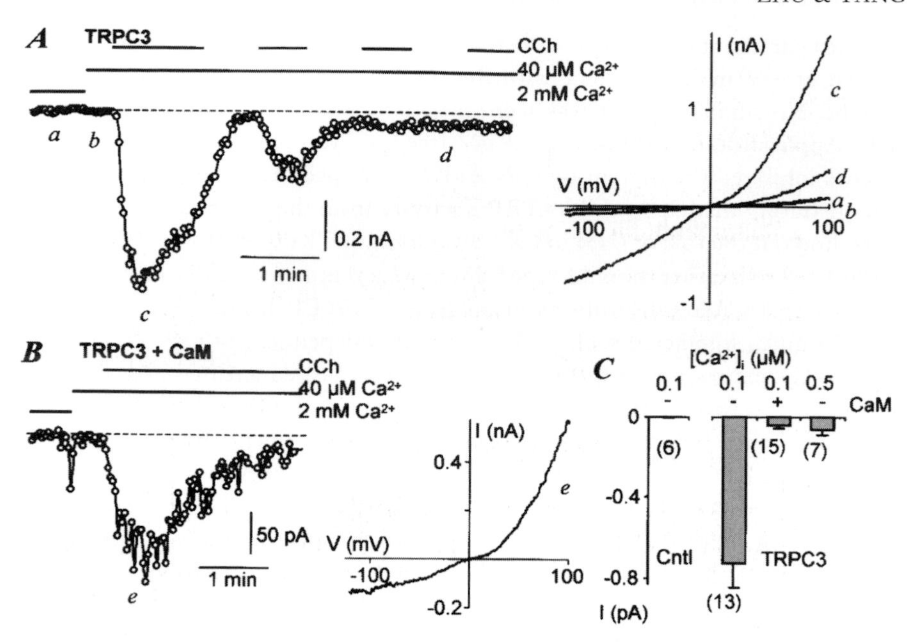

FIG. 3. Effect of Ca^{2+} and CaM on CCh-stimulated whole-cell currents in T3–9 cells. (A) Left, membrane currents at -95 mV in a T3–9 cell responding to the stimulation by 100 μM CCh. The internal solution contained (in mM), 140 caesium aspartate, 5 NaCl, 1 Na-ATP, 10 EGTA, 3.6 $CaCl_2$, 10 Hepes, pH 7.20 (free $[Ca^{2+}]=100$ nM). Whole-cell was made while the cell was in a bath solution containing (in mM), 160 NaCl, 2 $CaCl_2$, 10 Hepes, pH 7.40. After break-in, a solution of the same composition except that Ca^{2+} was reduced to 40 μM was perfused onto the cell. CCh was added as indicated by the upper bars. Right shows the current-voltage relationships obtained from voltage ramps applied at the time points indicated. (B) A similar experiment for a T3–9 cell over-expressing wild type CaM. (C) Average $\pm$SEM of currents at -95 mV at the peak of CCh stimulation for control cells (Cntl), cells expressing TRPC3 alone or TRPC3 + CaM. Intracellular $[Ca^{2+}]$ was either 100 or 500 nM as indicated. Numbers of cells used are indicated in parentheses.

TRPC channels are activated by CaM inactivation and by the IP_3R3 peptide

Two other methods were used to inactivate CaM from the inside-out patches. First, a CaM antagonist, calmidazolium (CMZ) at 10 μM, when perfused to the cytoplasmic side, activated TRPC3 (Table 1). The effect was reversed by the addition of 5 μM CaM ($\sim$84% inhibition, $n=5$). Second, a high affinity CaM-binding peptide, RS-20, which represents the CaM-binding site of chicken smooth muscle myosin light-chain kinase (Johnson et al 1996), also activated TRPC3 when applied to the cytoplasmic side of the excised patches (Table 1). In both cases, the single channel activity looked similar to that activated by co-expression with CaM(EFmut)$_4$. Control cells did not show any significant

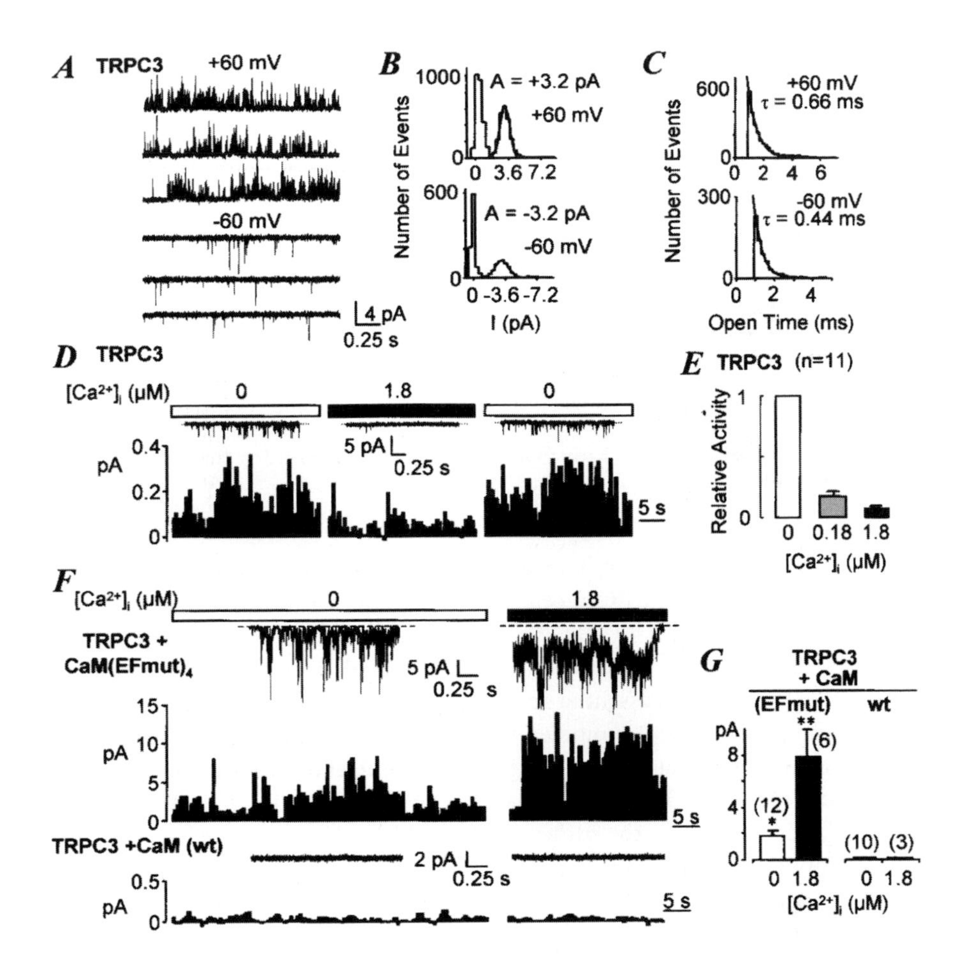

FIG. 4. Effects of Ca^{2+} and CaM mutant on TRPC3 activity in inside-out patches. (A–C) Single channel activities in patches excised from T3–9 cells to a Ca^{2+}-free solution. (A) Representative traces at transmembrane potential of +60 or −60 mV. (B,C) Amplitude (B) and open time (C) histograms generated from idealized traces encompassing 90 s recordings. (D) Effect of Ca^{2+} on the basal TRPC3 single channel activity. A patch held at −60 mV was perfused sequentially with intracellular solution containing 5 mM EGTA and no added Ca^{2+} (0 Ca^{2+}) and then a solution containing 1.8 μM free Ca^{2+}. Current recovered in 0 Ca^{2+}. Time plots show mean currents under each condition in 0.4 s bins. Representative traces are shown above the plots. Dashed lines indicate zero current levels. (E) Averages ± SEM ($n=11$) of inhibition of TRPC3 basal activity by Ca^{2+}. (F) Expression of CaM(EFmut)$_4$ (upper) but not wild-type CaM (lower) increased TRPC3 basal activity in excised patches. (G) average ± SEM of mean currents for numbers of patches shown in parentheses. Asterisks, $P < 0.01$ different from TRPC3 alone (*) or from 0 Ca^{2+} (**) by Student's t test. (Adapted from Zhang et al 2001a.)

TABLE 1 Ionic currents in inside-out patches from HEK293 cells expressing TRPC3 (T3–9) or its mutant, T3ΔC8

	Control	*T3–9*	*T3ΔC8*
Basal	0.03 ± 0.01 (17)	0.09 ± 0.03 (25)	0.03 ± 0.01 (15)
10 μM CMZ	0.12 ± 0.03 (5)	3.29 ± 0.57* (9)	0.09 ± 0.05 (8)
5 μM RS-20	n.d.	1.36 ± 0.51** (4)	n.d.
5 μM F2v	0.12 ± 0.07 (12)	2.11 ± 0.51* (15)	1.58 ± 0.46* (11)
5 μM F2vmut	n.d.	0.08 ± 0.04 (4)	n.d.

Currents in inside-out patches were recorded as in Fig. 4. CMZ and RS-20 were applied to the cytoplasmic side with an internal solution containing 18 μM free Ca^{2+} buffered with 5 mM EGTA. Peptide F2v and F2vmut were applied with an internal solution containing 5 mM EGTA and no Ca^{2+}. Data are average ± SEM of mean currents at −60 mV of 30 s to 150 s recordings from numbers of patches indicated in parentheses.
*, $P < 0.01$; **, $P < 0.05$ significantly different from the basal value as determined by Student's *t* test.
n.d. not determined.

increase in activity. However, increased currents were recorded from patches excised from two stable TRPC4α cell lines, T4-1 and T4-60, after application of either 1 or 10 μM CMZ (Table 2). These data suggest that under resting conditions, TRPC-formed channels are kept closed by CaM. Displacing the inhibitory CaM may be an essential and a final step for channel activation.

The IP_3R3 peptide, F2v, competed with CaM for binding to the CIRB site of TRPC (Fig. 2C). Application of 5 μM peptide F2v to the cytoplasmic side of the excised patches from T3–9 cells activated TRPC3 (Table 1). This activity was inhibited by 5 μM CaM in the presence of 18 μM Ca^{2+} by about 92% ($n = 3$). In contrast, application of 5 μM F2v mutant (F2vmut), which contains Gly in place

TABLE 2 Ionic currents in inside-out patches excised from HEK293 cells expressing TRPC4α

	Control	*T4-60*	*T4-1*
Basal	0.03 ± 0.03 (11)	0.01 ± 0.02 (32)	0.03 ± 0.01 (24)
1 μM CMZ	0.07 ± 0.03 (5)	0.15 ± 0.04* (11)	0.14 ± 0.03* (10)
10 μM CMZ	0.04 ± 0.04 (5)	0.57 ± 0.17* (6)	0.46 ± 0.10* (10)
5 μM F2v	0.06 ± 0.02 (5)	0.04 ± 0.04 (12)	0.08 ± 0.04** (5)
50 μM F2v	0.06 ± 0.05 (5)	0.52 ± 0.16* (14)	0.61 ± 0.19* (7)

Similar to Table 1. Data are average ± SEM of mean currents at −40 mV of 400 s recordings from numbers of patches indicated in parentheses.
*, $P < 0.01$; **, $P < 0.05$ significantly different from the basal value as determined by Student's *t* test.

of Trp^{693} and Trp^{696} of the F2v and is devoid of interaction with TRPC3 (Zhang et al 2001a), did not affect TRPC3 activity (Table 1). The current elicited by peptide F2v looked similar to that induced by co-expression with $CaM(EFmut)_4$ or inactivation of CaM using CMZ or RS-20. Consistent with the finding that peptide F2v is less effective at competing with CaM for binding to TRPC4 than TRPC3 (Fig. 2C), 5 μM peptide F2v did not significantly activate TRPC4. However, 50 μM peptide F2v caused a significant stimulation of TRPC4 activity in patches excised from both TRPC4 cell lines (Table 2). Most likely, peptide F2v activates the channel by displacing CaM bound to the CIRB site of the TRPC.

Like IP_3R3F2v, CMZ, RS-20, or $CaM(EFmut)_4$ may also displace the CaM bound to the CIRB site; however, this might not necessarily be true as these treatments would also affect CaM binding to other proteins. Deletion of aa776–796 from TRPC3 resulted in a mutant TRPC3 (T3ΔC8) that interacted with only IP_3R3F2q but not CaM (Zhang et al 2001a). Application of 10 μM CMZ to the cytoplasmic side of inside-out patches excised from cells expressing T3ΔC8 did not elicit any TRPC3-like current. However, to the same patches, 5 μM peptide F2v activated TRPC3-like currents (Table 1), indicating that T3ΔC8 forms functional channels responsive to activation by IP_3Rs but not by CaM inactivation. These results unequivocally show that CMZ and other treatments that remove or inactivate CaM activate TRPC channels by displacing CaM from the CIRB site of the TRPC.

Competition between CaM and IP_3Rs underlies the mechanism of conformational coupling

Our data show that CaM and IP_3Rs compete for binding to a common C-terminus site and thereby control the activity of TRPC channels. Under resting conditions, CaM keeps the channels closed by binding to the CIRB site. Binding of IP_3 and/or Ca^{2+} store depletion cause a conformational change that exposes the TRPC-binding domains from the IP_3Rs. The binding of IP_3Rs displaces CaM from the CIRB site and opens the TRPC channels. The Ca^{2+} coming through the channels increases the affinity of CaM to the CIRB site. Upon a decrease of the IP_3 signal, CaM will bind to channels again to cause inactivation (Fig. 5). This model enriches the conformational coupling hypothesis and explains how TRPC channels are regulated by IP_3Rs and CaM in response to the rise and fall of IP_3 and Ca^{2+} levels in the cytoplasm. However, it remains unclear how Ca^{2+} loss in the ER store would affect the conformation of IP_3Rs and expose the TRPC-binding domains. Nonetheless, a similar regulation was recently demonstrated for a native SOC in CHO cells, in which peptide F2v activated whereas CaM inhibited the SOC in whole-cell and single channel measurements (Vaca & Sampieri 2002). Therefore,

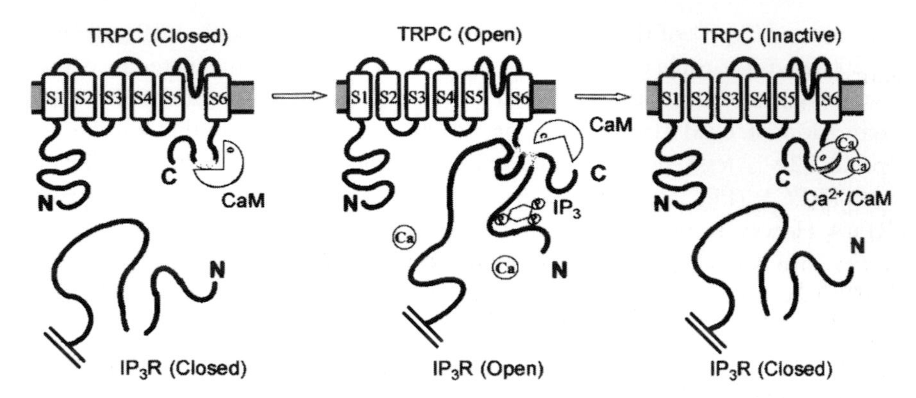

FIG. 5. A model of TRPC channel regulation by IP_3R and CaM. The simplified model assumes three states for TRPC: closed, open and inactive. Gating is controlled by the switch of the binding partner at the TRPC CIRB site. CaM is tethered to the CIRB site at rest (closed state). Activation of IP_3R by IP_3 following phospholipase C stimulation exposes the TRPC-binding domains from IP_3R, which in turn displaces CaM from the CIRB site. The removal of CaM leads to channel opening. Ca^{2+} influx through the channel enhances the affinity of CaM to the CIRB site, which will displace IP_3R when the IP_3 level decreases and result in channel closing. This presumably is an inactive state for TRPC since the subsequent decrease of Ca^{2+} concentration will lead to a lower affinity for CaM binding to the CIRB site and the return of TRPC to the closed state. S1–S6 indicate transmembrane segments. Grey lines indicate the CIRB site of TRPC and the TRPC binding domains of IP_3R. For simplicity, the transmembrane regions together with the C-terminus of the IP_3R as well as ER membranes are omitted.

the model presented here is likely to be applicable to native SOCs and other G_q/phospholipase C-activated channels composed of TRPC proteins.

Competition between CaM and another regulatory protein or domain for binding to a common site is well known for CaM-binding proteins, such as plasma membrane Ca^{2+}-ATPase (Carafoli 1991), olfactory cyclic nucleotide-gated channel (Varnum & Zagotta 1997) and the *N*-methyl-D-aspartate glutamate receptor (Zhang et al 1998). Yet, the regulation by CaM for some TRPC isoforms is more complicated than what we have discussed so far. Additional CaM-binding sites have been found in TRPC1, C2, C4 and C5 (Singh et al 2002, Tang et al 2001). These sites share no homology with each other and are thus likely to have distinct functions. For TRPC1, the second CaM-binding site has been shown to be involved in the slow Ca^{2+}-dependent inactivation of a SOC in human submandibular gland cells (Singh et al 2002).

Acknowledgements

Supported by National Institutes of Health grant NS42183 to M.X.Z. J.T is a recipient of postdoctoral fellowship of the American Heart Association Ohio Valley Affiliate.

References

Berridge MJ 1995 Capacitative calcium entry. Biochem J 312:1–11

Birnbaumer L, Zhu X, Jiang M, et al 1996 On the molecular basis and regulation of cellular capacitative calcium entry: roles for Trp proteins. Proc Natl Acad Sci USA 93:15195–15202

Boulay G, Brown DM, Qin N, et al 1999 Modulation of Ca^{2+} entry by polypeptides of the inositol 1,4,5-trisphosphate receptor (IP_3R) that bind transient receptor potential (TRP): evidence for roles of TRP and IP3R in store depletion-activated Ca^{2+} entry. Proc Natl Acad Sci USA 96:14955–14960

Carafoli E 1991 Calcium pump of the plasma membrane. Physiol Rev 71:129–153

Chevesich J, Kreuz AJ, Montell C 1997 Requirement for the PDZ domain protein, INAD, for localization of the TRP store-operated channel to a signaling complex. Neuron 18: 95–105

Goel M, Sinkins WG, Schilling WP. 2002 Selective association of TRPC channel subunits in rat brain synaptosomes. J Biol Chem 277:48303–48310

Hofmann T, Obukhov AG, Schaefer M, Harteneck C, Gudermann T, Schultz G 1999 Direct activation of human TRPC6 and TRPC3 channels by diacylglycerol. Nature 397:259–263

Hofmann T, Schaefer M, Schultz G, Gudermann T 2002 Subunit composition of mammalian transient receptor potential channels in living cells. Proc Natl Acad Sci USA 99:7461–7466

Irvine RF 1990 'Quantal' Ca^{2+} release and the control of Ca^{2+} entry by inositol phosphates — a possible mechanism. FEBS Lett 263:5–9

Johnson JD, Snyder C, Walsh M, Flynn M 1996 Effects of myosin light chain kinase and peptides on Ca^{2+} exchange with the N- and C-terminal Ca^{2+} binding sites of calmodulin. J Biol Chem 271:761–767

Kiselyov K, Xu X, Mozhayeva G et al 1998 Functional interaction between $InsP_3$ receptors and store-operated Htrp3 channels. Nature 396:478–482

Kiselyov KI, Semyonova SB, Mamin AG, Mozhayeva GN 1999 Miniature Ca^{2+} channels in excised plasma-membrane patches: activation by IP_3. Pflugers Arch 437:305–314

Montell C, Birnbaumer L, Flockerzi V 2002 The TRP channels, a remarkably functional family. Cell 108:595–598

Perez CA, Huang L, Rong M et al 2002 A transient receptor potential channel expressed in taste receptor cells. Nat Neurosci 5:1169–1176

Phillips AM, Bull A, Kelly LE 1992 Identification of a Drosophila gene encoding a calmodulin-binding protein with homology to the *trp* phototransduction gene. Neuron 8:631–642

Philipp S, Cavalie A, Freichel M et al 1996 A mammalian capacitative calcium entry channel homologous to Drosophila TRP and TRPL. EMBO J 15:6166–6171

Philipp S, Hambrecht J, Braslavski L et al 1998 A novel capacitative calcium entry channel expressed in excitable cells. EMBO J 17:4274–4282

Singh BB, Liu X, Tang J, Zhu MX, Ambudkar IS 2002 Calmodulin regulates Ca^{2+}-dependent feedback inhibition of store-operated Ca^{2+} influx by interaction with a site in the C terminus of TrpC1. Mol Cell 9:739–750

Tang Y, Tang J, Chen Z et al 2000 Association of mammalian trp4 and phospholipase C isozymes with a PDZ domain-containing protein, NHERF. J Biol Chem 275:37559–37564

Tang J, Lin Y, Zhang Z, Tikunova S, Birnbaumer L, Zhu MX 2001 Identification of common binding sites for calmodulin and inositol 1,4,5-trisphosphate receptors on the carboxyl termini of trp channels. J Biol Chem 276:21303–21310

Vaca L, Kunze DL 1995 IP_3-activated Ca^{2+} channels in the plasma membrane of cultured vascular endothelial cells. Am J Physiol 269:C733–C738

Vaca L, Sampieri A 2002 Calmodulin modulates the delay period between release of calcium from internal stores and activation of calcium influx via endogenous TRP1 channels. J Biol Chem 277:42178–42187

Varnum MD, Zagotta WN 1997 Interdomain interactions underlying activation of cyclic nucleotide-gated channels. Science 278:110–113
Xia XM, Fakler B, Rivard A et al 1998 Mechanism of calcium gating in small-conductance calcium-activated potassium channels. Nature 395:503–507
Yue L, Peng JB, Hediger MA, Clapham DE 2001 CaT1 manifests the pore properties of the calcium-release-activated calcium channel. Nature 410:705–709
Zhang S, Ehlers MD, Bernhardt JP, Su CT, Huganir RL 1998 Calmodulin mediates calcium-dependent inactivation of N-methyl-D-aspartate receptors. Neuron 21:443–453
Zhang Z, Tang J, Tikunova S et al 2001a Activation of Trp3 by inositol 1,4,5-trisphosphate receptors through displacement of inhibitory calmodulin from a common binding domain. Proc Natl Acad Sci USA 98:3168–3173
Zhang Z, Tang Y, Zhu MX 2001b Increased inwardly rectifying potassium currents in HEK-293 cells expressing murine transient receptor potential 4. Biochem J 354:717–725
Zhu X, Jiang M, Peyton M et al 1996 trp, a novel mammalian gene family essential for agonist-activated capacitative Ca^{2+} entry. Cell 85:661–671
Zhu X, Jiang M, Birnbaumer L 1998 Receptor-activated Ca^{2+} influx via human Trp3 stably expressed in human embryonic kidney (HEK)293 cells. Evidence for a non-capacitative Ca^{2+} entry. J Biol Chem 273:133–142
Zubov AI, Kaznacheeva EV, Nikolaev AV et al 1999 Regulation of the miniature plasma membrane Ca^{2+} channel I_{min} by inositol 1,4,5-trisphosphate receptors. J Biol Chem 274:25983–25985

DISCUSSION

Barritt: I'd like to ask about a prediction that you may be able to make from your studies. Take your data on TRPC3, for example. I think these suggest that if CaM is present, it will repress TRPC3 function, and if the CaM is removed it will open up the TRPC3 function. Could you use this to predict whether TRPC3 is involved in a particular cell type? If the experimenter uses a manipulation to make a dominant negative mutant of CaM or the peptide, and treats the cell with that, then it should be possible to see a change in inflow. Is that a reasonable prediction from your results?

Zhu: Yes. Up to this point it is fantasy to start thinking about the relative concentration of CaM in different microdomains. We don't have information on that. Our data simply suggest that there is a likelihood that the gating of TRPC is controlled by CaM. As long as the CaM is associated with it, the TRPC is not going to be active. Anything that inactivates or removes the CaM will cause TRPC activation.

Barritt: You are also, perhaps, implying that this would be in the context of any other factors that might also be regulating the TRP.

Zhu: I don't have any problem thinking that other mechanisms will be involved. For example, TRPC3 is activated by diacylglycerol (DAG). Does activation by DAG involve the removal of CaM? I don't know.

Nilius: Consider a resting cell, which has a Ca^{2+} concentration of less than 100 nM. The TRPs will never bind CaM, because the Ca^{2+} concentration will be in the range of 50 nM. I understand that you may have a mechanism that is involved in feedback inhibition of the channel, but I don't see any clue or explanation for activation of the channel.

Zhu: I don't have any information about whether TRPC is actually associated with CaM under resting conditions where the Ca^{2+} concentration is 50 nM. But the CaM concentration can be extremely high, at least in certain regions of the cell. We don't know whether CaM has to be associated with the TRPC directly under resting conditions, or is simply being brought very close to the TRPC. The binding is Ca^{2+} dependent, but in our data there is still a weak binding even in the absence of Ca^{2+}.

Nilius: Other people need a Ca^{2+} domain, but you need a CaM domain, creating a high local concentration of CaM.

Zhu: I don't really know exactly how CaM is distributed within a cell. But there could be CaM-presenting proteins situated very close to TRPCs, which would allow for local regulation.

Muallem: There are fairly good data from expression of tagged or fluorescently labelled CaM in some specific cells. For example, in smooth muscle fibres you can see some punctate concentrations, but in general the total CaM concentration is very high in the cells. However, the free CaM is pretty low in the sub micromolar range.

Putney: This is an important point. If you feel that under resting conditions regardless of the Ca^{2+} concentration there is some weak affinity of CaM for the channel, and that affinity is increased when it binds Ca^{2+}, then that is a different situation. This suggests that the mutant CaM, which is only deficient in its ability to bind Ca^{2+}, really isn't going to do anything when you overexpress it.

Zhu: It is not responding to Ca^{2+}. That is how I look at it.

Putney: By itself it shouldn't have any effect. It shouldn't be an activator of the channel, then.

Zhu: The mutant CaM appears to have a much better affinity than the wild-type in the absence of Ca^{2+}. It may be that when we overexpress the mutant it then occupies the binding site of the channel and we lose the Ca^{2+} regulation.

Putney: In your model it wouldn't bind so strongly that it couldn't be displaced by IP_3R.

Zhu: I don't know whether it will be displaced by IP_3R at all. Since our channel is already active due to the overexpression of mutant CaM, it would not matter whether there is IP_3R or not.

Groschner: This situation is reminiscent of the Ca^{2+}/CaM-dependent activation of L-type channels. For the dominant negative CaM to work, you have to postulate two binding sites, one where you can displace the CaM independently of Ca^{2+}, and a second one.

Zhu: The question is whether both the binding sites are on the TRPC channels. It could be that the Ca^{2+}-dependent one is on the TRPC channel and the Ca^{2+}-independent one is on another protein tightly associated with TRPC. We don't know this. From what we have looked at so far, it doesn't seem that there is any additional CaM binding site on any member of the TRPC3 family (I mean TRPC3, 6 and 7).

Penner: In your experiments you should be able to show whether this effect is irreversible. If you excise a patch and have a quiescent channel with CaM bound, and you apply calmidazolium to displace it so that the channel activates, is this irreversible?

Zhu: Yes.

Penner: So once the CaM is gone, you can't close it any more.

Zhu: Whenever we apply CMZ the activation is irreversible. We have to apply CaM back to the perfusion solution in order to stop it.

Putney: It should even be irreversibly activated by IP_3R.

Zhu: That is also true. With the peptide we have applied we never saw any recovery of activity unless we added CaM.

Muallem: There seem to be some differences in binding sites among the TRPs. Can this be extended to regulation?

Zhu: Perhaps. At least from our binding studies it requires much higher concentration of the IP_3R to compete with CaM for TRPC1, 4 and 5. This would suggest that even under conditions where we use the physiological agonist we would need a much stronger activation of the IP_3R to activate these channels than is needed for channels formed by TRPC3, 6 or 7.

Muallem: Do you actually see that?

Zhu: We haven't been able to compare the function of different TRPCs under the same experimental setting. In the literature we found that it was much easier to detect the activity of TRPC3, 6 and 7 than that of TRPC1, 4 and 5.

Putney: It is not obvious to me how the concentration dependence of activation by that peptide translates into the likelihood that an IP_3R, which has to interact on a stoichiometry of 1:1, would be able to activate the channels.

Zhu: What we make out of this is the affinity difference.

Putney: What we don't understand is what happens to the IP_3R that allows it to bind when stores are depleted or IP_3 is raised.

Zhu: What kind of conformation of IP_3R is required for it to start binding to TRPC? There is something called the effective concentration. It probably depends on how close the IP_3R is to the TRPC. It is possible that for TRPC3, the distance between the binding region of the IP_3R and TRPC doesn't have to be as close as that between IP_3R and TRPC4 where the affinity is weaker.

Putney: It is possible that the conformation of that peptide doesn't accurately reflect its conformation in the context of a whole molecule.

Zhu: Definitely not. That is why there are additional binding sites. We don't know what these binding sites are doing.

Scharenberg: Might you see IP_3-dependent binding to the TRPs? Do you have IP_3 in your immunoprecipitations?

Zhu: This is something I have been trying to address. This requires a construct with both the IP_3R binding site and the TRPC binding site. In all the studies we have done so far they are separated. They are not in the same construct. We are trying to make a construct that contains both and then somehow find a condition mimicking the intracellular environment.

Schilling: Can you immunoprecipitate the IP_3R with TRPC3 from brain tissue?

Zhu: We haven't been able to do this. Lutz Birnbaumer's group has immunoprecipitated IP_3Rs by TRPC3 and TRPC6 after overexpressing them in HEK cells.

Schilling: Is this interaction displaced by IP_3?

Zhu: I don't think so.

Schilling: Since both the TRPC channels and the IP_3Rs are presumably tetramers, would the IP_3R simultaneously interact at four points on the TRPC channel?

Zhu: I don't really know how the subunits are arranged, but it wouldn't be an ideal arrangement if the four subunits from the IP_3R bind directly to the four subunits of the TRPC channel. One thing we don't really know is the structure of the IP_3R. There is no information on this. I also don't know where Ca^{2+} comes out from the IP_3R: when it opens does it come out from the middle, or the side?

Fleig: Do you know which Ca^{2+} source is most important, the one that is being released or the one that is going through the channel into the cells?

Zhu: In terms of mediating CaM inhibition, I would think it is the Ca^{2+} coming through the TRPC channel that is most important.

Taylor: The fact that so much of the CaM effect persists in the excised patches does imply a slow dissociation rate for the CaM from that site. Is it sufficient just to have competition between the IP_3 peptide and the CaM, or do you need an acceleration of dissociation rate, such that it is interacting somewhere else to actively promote dissociation?

Zhu: There is definitely a Ca^{2+} dependency. This would indicate that there is a requirement for lowering the affinity. If you apply the peptide in the presence of Ca^{2+}, then it wouldn't work. We have to use a Ca^{2+}-free solution in order to get that activation. On the other hand, when we used RS-20, a peptide representing the CaM-binding site of chicken smooth muscle myosin light chain kinase we had to include Ca^{2+}, because its binding to CaM is Ca^{2+} dependent.

Taylor: If you are just relying on the passive dissociation rate of CaM before the IP_3 peptide can jump in and prevent it rebinding, then it is not going to go any faster than that off-rate.

Zhu: We imagine that under physiological conditions, IP_3 receptor has to play an active role in displacing CaM. The competition study we did was in the presence of Ca^{2+}, in order to see the peptide competing with CaM for binding. Otherwise we wouldn't see CaM binding. Even in the presence of Ca^{2+}, assuming a very active IP_3R, you will still get displacement. It is not a passive process.

Taylor: You envisage that IP_3R would accelerate the dissociation.

Zhu: Yes.

Taylor: This requires that it binds to an additional site to the one that overlaps with CaM.

Putney: Has anyone measured the kinetics of TRPC activation by electrophysiology? From the fluorescence it is definitely in the order of a second or so. It happens pretty quickly.

Zhu: Unless we have a direct activator it is difficult. With the whole cell experiment we are using carbachol. It goes through the G protein pathway.

Putney: It is still pretty fast, compared with the off-rate of CaM in your patches.

Scharenberg: When you dialyse IP_3 in, do you see current activation in your system?

Zhu: No, we have never been able to use IP_3 to activate anything. I wonder whether I have bought the correct kind of IP_3!

Nilius: The mechanism you were describing seems to be very similar to the Ca^{2+}-dependent inactivation of TRPV6. The CaM binding site in TRPV6 is extremely unusual (it is not conserved within the TRPV6 species).

Zhu: There is a lot of similarity with this particular mode of competitive binding for many channels. The cyclic nucleotide-gated channels in olfactory cells are activated by cGMP or cAMP. However, there is a CaM binding site at the N-terminus which also binds to the C-terminus of the same channel. Whenever the C-terminus and N-terminus bind to each other the channel is activated, and then CaM is activated and competes with the C-terminus. The only difference there is that this is intracellular binding, whereas here the IP_3R and TRPC interaction is intercellular. A similar phenomenon exists in the NMDA receptor where the NR1 subunit has a CaM binding site where alpha actinin also binds. For the NMDA receptor, the CaM is also inhibitory. The same could be true for voltage-gated Ca^{2+} channels, except that we don't know whether the CaM binding site of a voltage-gated Ca^{2+} channel also binds to some other part of the channel itself or another protein.

Plasma membrane localization of TRPC channels: role of caveolar lipid rafts

Indu S. Ambudkar, So-ching Brazer, Xibao Liu, Timothy Lockwich and Brij Singh

Secretory Physiology Section, Gene Therapy and Therapeutics Branch, National Institute of Dental and Craniofacial Research, National Institutes of Health, Bethesda MD 20892, USA

Abstract. GPCR-mediated activation of the Ca^{2+} signalling cascade leads to stimulation of Ca^{2+} influx into non-excitable cells. Both store-dependent and independent channels likely contribute towards this Ca^{2+} influx. However, the identity of the channels and exact mechanism by which they are activated remains elusive. The TRPC family of proteins has been proposed as molecular components of these channels. Studies from our laboratory and others have shown that mammalian TRPC proteins are assembled in a multiprotein complex that includes various key Ca^{2+} signalling proteins. However, relatively little is known regarding the mechanisms involved in the assembly of the TRPC channel complex in the plasma membrane. We have reported that TRPC1 and TRPC3 signalling complexes are associated with caveolar lipid raft domains (LRDs) in the plasma membrane. Recently we have examined the role of caveolin-1 in the regulation of TRPC channels and store-operated Ca^{2+} entry (SOCE). Based on our studies, we suggest that (1) caveolin 1 has a potentially critical role in the localization of TRPC channels plasma membrane caveolar LRDs, and (2) the molecular architecture of caveolae can facilitate intramolecular interactions between TRPC channels and associated proteins that are involved in activation and/or inactivation of SOCE.

2004 Mammalian TRP channels as molecular targets. Wiley, Chichester (Novartis Foundation Symposium 258) p 63–74

Activation of cell surface receptors which are coupled to inositol lipid signalling, results in phosphatidylinositol 4,5-bisphosphate (PIP_2) hydrolysis, generation of diacylglycerol (DAG) and inositol 1,4,5-trisphosphate (IP_3), release of Ca^{2+} from internal Ca^{2+} stores, and activation of plasma membrane Ca^{2+} influx channels (Putney et al 2001). Two main types of plasma membrane channels have been associated with this Ca^{2+} signalling mechanism: store-independent channels that appear to be activated as a result of PIP_2 hydrolysis, likely by DAG, and store-operated Ca^{2+} channels (SOCs), which are activated primarily in response to depletion of Ca^{2+} from the internal Ca^{2+} stores, i.e. in the absence of detectable increase in IP_3. It has been suggested that the store-independent channels are either activated directly by PIP_2 hydrolysis (i.e. PIP_2 functions as a tonic inhibitor of the channel) or by DAG. While the mechanism involved in store-operated Ca^{2+} entry (SOCE) is not yet known, it has been proposed that channel activation might involve: interaction with IP_3R in the ER, a diffusible factor released or generated in response to internal Ca^{2+} store depletion and regulated recruitment of channels by fusion of intracellular vesicles. Presently, there are no data to conclusively identify the mechanism(s) underlying activation and gating of these plasma membrane Ca^{2+} channels.

There are sufficient data to demonstrate that members of the TRPC family are involved in agonist-stimulated Ca^{2+} signalling (Minke & Cook 2002, Zitt et al 2002). It is now clear that some TRPCs, e.g. TRPC1 and TRPC4, can be activated by conditions that induce depletion of the internal Ca^{2+} store in the absence of increased PIP_2. Others, like TRPC3, TRPC6 and TRPC7 are activated by agonists that stimulate PIP_2 hydrolysis and by DAG. It is presently unclear whether both these types of Ca^{2+} channels are simultaneously activated in cells following stimulation. Furthermore, the exact molecular mechanisms which couple the activation of these channels to a G protein-coupled receptor (GPCR)-stimulated Ca^{2+}-signalling cascade are not yet known. We hypothesized that the physiological function of the various TRPC channels might be resolved by identifying the proteins that interact with them and by determining the cellular milieu in which they function.

Ca^{2+} signalling complexes and microdomains

A large number of studies have demonstrated that agonist-stimulated Ca^{2+} influx occurs within specific spatially restricted microdomains (Isshiki & Anderson 1999, Isshiki et al 2002). We have shown that during Ca^{2+} influx-dependent refill of internal Ca^{2+} stores there is minimal diffusion of $[Ca^{2+}]$ in the sub-plasma membrane region due to rapid uptake into the ER by the SERCA pump (Liu et al 1998). Such studies suggest a close apposition between the ER and plasma membrane at the site of Ca^{2+} influx which is consistent with several of the models proposed for the activation of SOCE channels as well as non-store-operated Ca^{2+} channels, such as TRPC3. Thus, it has been suggested that the architecture of such a Ca^{2+} signalling microdomain can facilitate direct physical, or functional, coupling between the molecular components that are involved in the activation, or inactivation, of plasma membrane Ca^{2+} influx channels.

Biochemical and morphological data also support the suggestion that Ca^{2+} signalling occurs in functionally distinct microdomains. It has been recently recognized that Ca^{2+} signalling proteins are assembled in multiprotein complexes (Muallem & Wilkie 1999, Kiselyov et al 2003). A well studied prototype is the *Drosophila* TRP complex which is assembled by the scaffolding action of INAD, a multi-PDZ domain containing protein. INAD binds to a number of signalling proteins, such as TRP, TRPL, calmodulin (CaM), PLC, G protein and PKC (Montell 2001). We and others have previously reported that TRPC proteins are also assembled in multimeric protein complexes with key Ca^{2+} signalling proteins (Lockwich et al 2000, 2001, Tang et al 2000). Although several mammalian PDZ domain-containing proteins have been shown to act as scaffolds for receptor-associated signalling complexes in the plasma membrane, little is known about the organization of TRPC channels in mammalian cells or whether they have specific cellular locations. Thus, information regarding the assembly of the mammalian TRPC1-associated signalling complex is key to understanding the mechanism involved in SOCE.

Caveolae and their role in Ca^{2+} signalling

Functionally and biochemically distinct microdomains, formed by the lateral packing of glycosphingolipids and cholesterol within the membrane bilayer, have been identified in plasma membranes (Hooper 1999). These liquid-ordered domains called lipid raft domains (LRDs) are characterized by their insolubility in nonionic detergents under certain conditions, e.g. 1% Triton X-100 at 4 °C. Caveolae are a specialized form of LRD that are found as vesicles in the subplasma membrane region of cells or as invaginations of the plasma membrane (Liu et al 2002, Razani et al 2002). Caveolae have also been found as grape-like clusters, rosettes, fused as elongated tubes, or trans-cellular channels in certain types of tissues. Caveolin 1, a cholesterol binding protein, is enriched in caveolae and found in a variety of non-excitable cell types. Caveolin 3, another member of the caveolin gene family, is found only in muscle cells. Caveolin 2 is also as widely distributed as caveolin 1. Although it is known to form a complex with caveolin 1, its exact function is not yet known.

Caveolin 1 is a 22 kDa protein containing two membrane anchoring domains, a protein scaffolding domain, and lipid modification sites. It is known to interact with most proteins via its scaffolding domain. Caveolin 1 also forms highly stable homo-oligomers. These oligomers are believed to represent the assembly units of caveolae. Individual caveolin oligomers interact with each other via C-terminal domains. Such interactions form a network of caveolin in the plasma membrane region which is suggested to drive the invagination of caveolae. A number of cellular functions have been ascribed to caveolae. The most well studied of these are vesicular transport (transcytosis, endocytosis and potocytosis), cellular cholesterol homeostasis (including transport of newly synthesized cholesterol), and regulation of signal transduction mechanisms.

A large number of proteins have been found to be localized in caveolae and these have one or more binding sites for ca1veolin 1. These binding sites interact with the scaffolding domain in caveolin 1 and are rich in aromatic amino acids and a number of them also appear to have characteristic motifs. It is presently unclear whether caveolin binding actually regulates trafficking of some proteins to caveolae. Many of the signalling proteins are directly routed to caveolae via lipid modification of specific amino-acid sequences. Once they reach the caveolae, they interact with caveolin 1, which acts as a scaffold to retain them. The majority of the caveolae-associated proteins are involved in a variety of cellular signalling events. In fact, entire signalling modules like the PDGFR-Ras-ERK module have been localized to caveolae. Thus, it has been proposed that arrangement of the lipids and scaffolding proteins within LRD forms a platform for the assembly of a number of proteins into signalling complexes. This compartmentalization of the signalling molecules can increase the rate of interactions and enhance cross-talk networks. In addition to its role as a molecular scaffold, caveolin also modulates the activity of the proteins localized in caveolae. A well-studied example of this is the eNOS system, where caveolin binding to eNOS exerts a tonic inhibition. Activation of eNOS involves binding to calmodulin and release from caveolin 1. Similarly, PKCα and $G_{\alpha i}$ are also inhibited by caveolin 1.

Importantly, key protein and non-protein molecules involved in the Ca^{2+} signalling cascade, such as PIP_2, $G_{\alpha q/11}$, muscarinic receptor, PMCA and IP_3R-like protein, and Ca^{2+} signalling events such as receptor-mediated turnover of PIP_2 have also been localized to caveolar microdomains in the plasma membrane (Isshiki & Anderson 1999, Liu et al 2002). An interesting study showed that the agonist-stimulated Ca^{2+} signal in endothelial cells originates in specific areas of the plasma membrane that are enriched in caveolin 1 (Isshiki et al 1998). Furthermore, we and others have reported that intact lipid rafts are required for activation of SOCE (Lockwich et al 2000, Isshiki et al 2002, Kunzelmann-Marche et al 2002). Thus, it has been proposed that caveolae might regulate the spatial organization of Ca^{2+} signalling by contributing to the localization of the Ca^{2+} signalling complex as well as the site of Ca^{2+} entry. However, exactly how caveolae regulate SOCE is not yet known.

Localization of TRPC-associated signalling complex in caveolar lipid raft domains

Studies in our laboratory have provided strong evidence that TRPC1 is a component of the SOCE channel in HSG cells (Liu et al 2000, 2003). Further, we have demonstrated that TRPC1 is assembled in a multimeric complex of Ca^{2+} signalling proteins that is associated with caveolin-scaffolding LRD in the plasma membrane (Lockwich et al 2000). The components of this complex include IP_3R, caveolin 1, CaM, PMCA, and $G_{\alpha q/11}$. TRPC1 fulfilled several criteria that have been used to describe the association of proteins with such domains:

- insolubility following treatment of HSG cell membranes with Triton X-100 at 4 °C, even after inclusion of KI to disrupt cytoskeletal interactions
- presence in low density fractions of membranes treated with Triton X-100 +KI at 4 °C
- appearance in the high density fraction of membranes treated with Triton X-100 +KI at 37 °C
- increased solubilization with Triton X-100 following treatment of cells with CD (β methyl cyclodextran) to deplete plasma membrane cholesterol
- co-migration with caveolin 1 on density gradients and co-immunoprecipitation with caveolin 1. Consistent with our data, TRPC1 has also been shown to co-localize with caveolin 1 in sperm cells (Trevino et al 2001).

We have also reported that TRPC3, like TRPC1, is assembled in a multimolecular signalling complex containing PLCβ, G$\alpha_{q/11}$, IP$_3$R, SERCA, ezrin and caveolin 1 (Lockwich et al 2001). The proteins that were identified in the TRP3 complex are appropriate candidates for the regulation of TRP3, which reportedly functions as a Ca^{2+} influx channel activated in response to agonist-stimulation of PLC, either by diacylglycerol produced as a result of PIP$_2$ hydrolysis or by interactions with the IP$_3$R. Importantly, our data demonstrate that the TRP3 signalling complex is internalized by conditions such as cytoskeletal rearrangements, which have been known to induce internalization of caveolae. Further studies will be required to determine whether this internalization represents a mechanism of trafficking of the TRPC3 channel complex and whether it is involved in the regulation of agonist-stimulated Ca^{2+} influx. Recently, TRPC4 was also shown to be present in caveolae in interstitial cells of Cajal and this localization appeared to be critical for regulation of the pacemaker activity of these cells (Torihashi et al 2002). Thus, several TRPC channels appear to be localized in caveolar LRD. Important questions that arise from these findings are: (1) does caveolar localization of TRPC channels determine the regulation of their function, and (2) is the subcellular localization of the agonist-stimulated Ca^{2+} influx mechanism(s) physiologically relevant?

Role of caveolin 1 in plasma membrane localization of TRPC1

Our recent studies have addressed the role of caveolin 1 in TRPC1 function. We have identified two putative caveolin binding domains in the N- and C-terminal regions, respectively, of TRPC1. Interestingly, TRPC1 which lacked the N-terminal caveolin 1 binding domain failed to be trafficked to the plasma membrane. Further, expression of this protein exerted a dominant negative effect on the SOCE. In contrast, full-length TRPC1 was routed to the plasma membrane and induced a 1.5- to 2.0-fold increase in SOCE. We have reported earlier that truncation of the C-terminus itself, which includes the C-terminal caveolin 1 binding site, does not alter localization of TRPC1 (Singh et al 2000). A mutant caveolin 1, lacking its protein scaffolding and membrane anchoring domains, disrupted plasma membrane localization of both endogenous and exogenously expressed TRPC1 and suppressed SOCE. This effect was not due to a generalized disruption of plasma membrane lipid raft domains. Further, we have shown that TRPC1 from cells expressing mutant caveolin 1 was relatively more soluble in detergent than from cells expressing full length caveolin 1. Since TRPC1 does not appear to contain any consensus sequences for lipid modification, we suggest that direct interaction of the TRPC1 N-terminus with caveolin 1 plays a critical role in the trafficking of the protein to specific plasma caveolae where it is assembled into the SOCE channel complex.

Implications of TRPC1–caveolar interactions in the regulation of SOCE

Based on the presently available data, we suggest that caveolin 1 could contribute to regulation of SOCE in several ways. Firstly, as suggested by the studies discussed above, TRPC1 trafficking to the plasma membrane is mediated via interaction with caveolin 1. Thus, caveolin 1 has a potentially critical role in the assembly of TRPC1-containing SOCE channels in plasma membrane caveolar LRDs. We and others have also reported that disruption of LRDs disrupts activation of plasma membrane TRPC1 channels. Thus caveolar LRDs might facilitate and coordinate the signals that lead to activation of SOCE via two possible mechanisms. (1) Since Ca^{2+} signalling proteins that lead to the activation of SOCE are colocalized in the same microdomain, caveolae could mediate interactions between the TRPC channels and proposed regulatory proteins such as IP_3R or PLCγ (Kiselyov et al 1998, Patterson et al 2002). The invaginated morphology of caveolae would uniquely enable the plasma membrane in this region to have access to regulatory components located further inside the cells, such as the ER. (2) Since caveolae are also found as subplasma membrane vesicles, it is interesting to speculate that they might act as holding platforms for pre-assembled Ca^{2+} signalling complexes, or key components of such complexes, which upon stimulation of the cell are recruited to the plasma membrane via vesicle fusion. It is important to note that proteins involved in docking and membrane fusion are enriched in caveolae (Isshiki & Anderson 1999). Such regulation of SOCE would be consistent with the secretion–coupling model. Finally, caveolae could also function as regulators of SOCE inactivation. Caveolin 1 is known to act as a tonic inhibitor of a number of signalling proteins. Additionally, caveolae have been shown to undergo dynamic internalization and the internalized vesicles have been shown to fuse with the ER. Thus, during prolonged activation of Ca^{2+} influx, channels could be down-regulated via internalization. An interesting idea proposed by Isshiki and Andersen (1998) is that this process would allow the external Ca^{2+} to be trapped in the vesicles and delivered to the ER. Thus, recycling of plasma membrane Ca^{2+} channels would both limit the number of 'active' channels and provide a route for the refill of internal Ca^{2+} stores with external Ca^{2+}. Another possible mechanism for inactivation of SOCE would be the exit of TRPC1 from caveolae and internalization via clathrin-coated pits. Channels internalized this way would be routed to endosomes for degradation.

In conclusion, the presently available data suggest that (i) localization of TRPC1 in the plasma membrane and assembly in the SOCE channel complex is dependent on caveolin 1, and (ii) intactness of caveolar lipid raft domains is required for activation of SOCE. Additionally, caveolin could also more directly alter the function/and or regulation of TRPC1. Present studies in our laboratory are directed towards further understanding these potentially important roles of caveolin in SOCE and TRP channel function.

References

Hooper NM 1999 Detergent-insoluble glycosphingolipid/cholesterol-rich membrane domains, lipid rafts and caveolae. Mol Membr Biol 16:145–156

Isshiki M, Anderson RG 1999 Calcium signal transduction from caveolae. Cell Calcium 26: 201–208

Isshiki M, Ando J, Korenaga R et al 1998 Endothelial Ca^{2+} waves preferentially originate at specific loci in caveolin-rich cell edges. Proc Natl Acad Sci USA 95: 5009–5014

Isshiki M, Ying YS, Fujita T, Anderson RG 2002 A molecular sensor detects signal transduction from caveolae in living cells. J Biol Chem 277:43389–43398

Kiselyov K, Xu X, Mozhayeva G et al 1998 Functional interaction between InsP3 receptors and store-operated Htrp3 channels. Nature 396:478–482

Kiselyov K, Shin DM, Muallem S 2003 Signalling specificity in GPCR-dependent Ca^{2+} signalling. Cell Signal 15:243–253

Kunzelmann-Marche C, Freyssinet J, Martinez MC 2002 Loss of plasma membrane phospholipid asymmetry requires raft integrity. Role of TRP channels and ERK pathway. J Biol Chem 277:19876–19881

Liu X, O'Connell A, Ambudkar IS 1998 Ca^{2+}-dependent inactivation of a store-operated Ca^{2+} current in human submandibular gland cells. Role of a staurosporine-sensitive protein kinase and intracellular Ca^{2+} pump. J Biol Chem 273:33295–33304

Liu X, Wang W, Singh BB et al 2000 Trp1, a candidate protein for the store-operated Ca^{2+} influx mechanism in salivary gland cells. J Biol Chem 275:3403–3411 [Erratum in J Biol Chem 275:9890–9891]

Liu P, Rudick M, Anderson RG 2002 Multiple functions of caveolin-1. J Biol Chem 277: 41295–41298

Liu X, Singh BB, Ambudkar IS 2003 TRPC1 is required for functional store-operated Ca^{2+} channels: role of acidic amino acid residues in the S5-S6 region. J Biol Chem 278: 11337–11343

Lockwich TP, Liu X, Singh BB, Jadlowiec J, Weiland S, Ambudkar IS 2000 Assembly of Trp1 in a signaling complex associated with caveolin-scaffolding lipid raft domains. J Biol Chem 275: 11934–11942

Lockwich T, Singh B, Liu X, Ambudkar IS 2001 Stabilization of cortical actin induces internalization of transient receptor potential 3 (Trp3)-associated caveolar Ca^{2+} signaling complex and loss of Ca^{2+} influx without disruption of Trp3-inositol trisphosphate receptor association. J Biol Chem 276:42401–42408

Minke B, Cook B 2002 TRP channel proteins and signal transduction. Physiol Rev 82:429–472

Montell C 2001 Physiology, phylogeny, and functions of the Trp superfamily of cation channels. Sci STKE 90:RE1 *http://stke.sciencemag.org/cgi/content/full/OC_sigtrans;2001/90/re1*

Muallem S, Wilkie TM 1999 G protein-dependent Ca2+ signaling complexes in polarized cells. Cell Calcium 26:173–180

Patterson RL, van Rossum DB, Ford DL et al 2002 Phospholipase C-gamma is required for agonist-induced Ca^{2+} entry. Cell 111:529–541

Putney JW Jr, Broad LM, Braun FJ, Lievremont JP, Bird GS 2001 Mechanisms of capacitative calcium entry. J Cell Sci 114:2223–2229

Razani B, Woodman SE, Lisanti MP 2002 Caveolae: from cell biology to animal physiology. Pharamacol Rev 54: 431–467

Singh BB, Liu X, Ambudkar IS 2000 Expression of truncated transient receptor potential protein 1 α (Trp1α): evidence that the Trp1 C terminus modulates store-operated Ca^{2+} entry. J Biol Chem 275:36483–36486

Tang Y, Tang J, Chen Z et al 2000 Association of mammalian trp4 and phospholipase C isozymes with a PDZ domain-containing protein, NHERF. J Biol Chem 275:37559–37564

Torihashi S, Fujimoto T, Trost C, Nakayama S 2002 Calcium oscillation linked to pacemaking of interstitial cells of Cajal; requirement of calcium influx and localisation of TRP4 in caveolae. J Biol Chem 277:19191–19197

Trevino CL, Serrano CJ, Beltran C, Felix R, Darszon A 2001 Identification of mouse trp homologs and lipid rafts from spermatogenic cells and sperm. FEBS Lett 509:119–125

Zitt C, Halaszovich CR, Luckhoff A 2002 The TRP family of cation channels: probing and advancing the concepts on receptor-activated calcium entry. Prog Neurobiol 66:243–264

DISCUSSION

Muallem: Do any of the other proteins that sit in the complex have a caveolin binding site? When you disrupt caveolae, does this disrupt localization of everything or only TRPC1?

Ambudkar: Among the proteins I described, $G\alpha_{q/11}$ has a caveolin binding site. It is also acylated, which allows it to interact directly with the plasma membrane. This means that it can get to the plasma membrane rafts without caveolin. Once it gets there it binds to caveolin. We are currently trying to see whether delta caveolin 1 expression disrupts the whole signalling complex. We know that carbachol-stimulated internal Ca^{2+} release is not decreased, which suggests that the components involved in the proximal signalling events are able to get to the plasma membrane without caveolin 1.

Muallem: So TRPC1 is not the anchor as in *Drosophila*.

Montell: You showed that as a result of your caveolin deletion you no longer got plasma membrane localization of TRPC1. However, you still got very good colocalization in the cell body. Nevertheless, you concluded that you needed the binding for plasma membrane localization. Could you expand on this? It looked like they were still binding but weren't getting to the plasma membrane.

Ambudkar: Yes, we do see intracellular colocalization of TRPC1 and delta caveolin. What I think is happening is that the delta caveolin can interact with TRPC1, but it cannot get to the plasma membrane since it lacks its membrane-interacting domains.

Montell: Your conclusion was that TRPC1–caveolin binding was needed for it to get to the membrane.

Ambudkar: Our conclusion was that plasma membrane localization of TRPC1–caveolin complex relies on the ability of caveolin to get to the plasma membrane. We have not yet determined whether caveolin 1 affects its routing to the plasma membrane or its retention once it gets there.

Montell: It looked like you were disrupting the plasma membrane association, rather than the interaction between TRPC1 and caveolin specifically.

Ambudkar: That is correct. We have shown that delta caveolin 1 and TRPC1 can interact with each other. We have proposed that caveolin has to get to the plasma membrane in order for functional TRPC1 channels to assemble. Because we have removed its membrane-interacting domain, delta caveolin 1 can no longer traffic to the plasma membrane, thus TRPC channel assembly is disrupted.

Montell: In the case where you delete the caveolin binding site from TRPC1, is there any effect on the localization of caveolin? In other words, does TRPC1 have some role in anchoring caveolin?

Ambudkar: We do not have any conclusive data as yet.

Nilius: How do you reconcile the general importance of your data with the fact that the caveolin 1 knockout is more or less completely normal? If you are dealing with an important mechanism I would have expected to see more. Second, several cells do not have caveolin, but they have wonderful Ca^{2+} signalling.

Ambudkar: I wish I had an easy explanation, but I don't. The fact is, many cells contain caveolar domains and a number of Ca^{2+} signalling proteins are localized in these domains. Thus, I believe these domains are very interesting and potentially important in cellular signalling events. We have to also appreciate that compensatory mechanisms can occur in knockout mice. Thus, other protein or lipid components might allow formation of caveolar-like LRDs. It would be interesting to look at Ca^{2+} signalling mechanisms in some of these knockout mice.

Nilius: CaCo-2 cells and many tumour cells lack caveolin, but they still have store-operated Ca^{2+} entry

Ambudkar: It is possible that they might have other proteins that can substitute for caveolin 1.

Muallem: We have to be careful about knockouts. In endothelial cells caveolin is immensely important. In the caveolin knockout mouse there isn't any abnormality in endothelial cell function and the change in blood pressure is trivial. It is not easy to explain. But this doesn't mean that caveolae aren't important.

Ambudkar: The function of caveolae could be cell-specific. Whether caveolin is there or not, lipid rafts are present and there are other proteins that can bind to cholesterol or to the other lipid components and act as scaffolds.

Nilius: Don't you think that it is odd that an ion channel should sit in such a structure that would reduce access to it significantly?

Ambudkar: I don't think access of Ca^{2+} to the channel is a problem, because caveolae carry out potocytosis as well as transcytosis which involve entry of much larger molecules into the caveolae.

Groschner: I want to comment on Bernd Nilius' point about the channels sitting in invaginations. What we see is that HEK cells have very few invaginations but there are definitely planar raft-like structures. If we overexpress caveolin 1 we get a lot more invaginations. What happens when you overexpress caveolin 1 in your cells?

Ambudkar: From what we have done so far, it appears that the plasma membrane signal of TRPC1 is increased.

Scharenberg: With regard to your data on β methyl cyclodextran, you pretreat for 10 min with this. Does it do any Ca^{2+} signalling on its own? Have you ever just put this on cells to see what happens?

Ambudkar: Yes, we've checked this and it doesn't do anything. We did image during the pretreatment. We see slow recovery, which suggests that it does not directly act from the outside.

Putney: The problem with this drug is that you have to use a lot of it, between 5 and 10 mM. In one cell line, we used this for an entirely different reason: to block endocytosis of receptors. This did block endocytosis of receptors without any effect on capacitative Ca^{2+} entry measured by fura-2.

Taylor: I am worried about how conclusive the evidence is that caveolin plays a part in getting TRP1 to the membrane or keeping it there. If you overexpress the dominant negative caveolin, even if it is not the normal vehicle, it is now stuck in the ER and anything that binds to it is going to be stuck there with it. One might get this result whether caveolin normally plays a part or not. Then the other piece of evidence was that you took out the caveolin binding site from the TRP and found that it too was retained in the ER. Unless you can show that this isn't a misfolded protein, it could be that this is what happens if the protein simply misfolds.

Ambudkar: We have deleted other regions in TRPC1 and it can get to the plasma membrane. So, unless the caveolin-binding domain also somehow specifically affects protein folding/stability, the deletion should not be a problem.

Scharenberg: You have to be careful with deletions. We deleted a tiny bit of TRPM2, about 20 amino acids right at the N-terminus, and we didn't see surface transport.

Ambudkar: As I said, we have deleted other domains, one of which is upstream from the caveolin 1 binding region, and that did not affect plasma membrane localization.

Taylor: I am not sure how you address this. How do you check whether or not it is a protein misfolding problem? Nonetheless, it seems that another deletion doesn't answer the question: you may have deleted between domains.

Ambudkar: You are suggesting that since we have removed the caveolin binding domain, the protein doesn't fold properly and is trapped in the ER. To me this suggests that caveolin is important for TRPC1 trafficking and might have a chaperone-like function. When it is bound to caveolin it is properly folded and can be assembled into plasma membrane channels.

Taylor: That assumes that the only thing chopping a big chunk out of a protein does is to stop it binding to the thing you first thought of. I don't think this prevents that manipulation also causing a protein folding problem.

Ambudkar: As I said earlier, we have deleted other regions, in both the C- and N-terminus and the protein doesn't get stuck in the ER. If this caveolin-binding region is a specific region that alters the folding of the protein, then this is also an interesting result in itself. We are also addressing this by performing site-directed mutagenesis of amino acid residues in the caveolin 1 binding region.

Montell: You could look for a point mutation in TRPC1 that disrupts its interaction with caveolin. This is not to say that a point mutation couldn't disrupt folding, but it would be less likely. The caveat to this is that I have seen many cases in our own lab in which single amino acid changes cause instability of a protein; however, upon deleting a functional domain entirely, the protein is completely stable.

Penner: You have beautiful coronal staining of your channels. How deep is this area of the caveolae? When you do the experiment functionally, that is you stimulate cells so that most of the caveolae move up to the membrane, can you sharpen that corona? Is that feasible?

Muallem: Of course not. The resolution of the confocal microscope isn't sufficient.

Penner: How thick is it?

Ambudkar: It is a very narrow region, we have not yet measured it exactly.

Muallem: It is less than 0.4 microns. That is why we do FRET.

Ambudkar: We are now imaging GFP-labelled TRP3 in HEK-293 cells to see the movement of the protein when it is being trafficked to the plasma membrane. We can manipulate this by tetanus toxin treatment which will block the fusion event. In these experiments it is very difficult to differentiate protein that is present just below the plasma membrane from that inserted into the membrane. We are resolving this by using methods which determine surface expression of the protein.

Putney: Something I think is relevant here is that with TRP3, in our stable cell lines with GFP fusion, fluorescence is uniformly distributed around the entire cell. If this is in caveolae there must be a lot of them. With transient transfection, in contrast we get punctate labelling. There is a difference.

Muallem: One possibility is that you are overexpressing to such an extent that you are decorating the entire membrane. Again, though, fluorescence microscopy can only give you a resolution of up to 100 nm, which is not sufficient to answer these kinds of questions.

Scharenberg: The Tsien monomeric GFP labelled with myristolate gave a beautiful demonstration of the existence of rafts. This would be another nice way for Indu Ambudkar to address these issues, because this would produce a robust FRET signal.

Assembly and gating of TRPC channels in signalling microdomains

Patrick Delmas

Intégration des Informations Sensorielles, CNRS, UMR 6150, IFR Jean Roche, Faculté de Médecine, Bd. Pierre Dramard, 13916 Marseille, France

Abstract. A variety of plasmalemmal Ca^{2+}-permeable channels, many of which are assembled from TRPC channels and are regulated by elements of the phosphatidylinositol pathway, may fulfil the role of store-operated channels (SOCs) and receptor-operated channels (ROCs). Growing evidence suggests that TRPC channels are clustered into spatially restricted microdomains that are important interaction sites for signalling molecules and for the induction of selective cellular responses. For example, TRPC1, which is activated solely by the depletion of internal stores in neurons, is assembled in a Ca^{2+} signalling complex, composed of the bradykinin receptor, $G\alpha_q$ subunit, phospholipase C (PLC)β and inositol 1,4,5-trisphosphate receptor (IP_3R) whereas TRPC6, which is activated by phosphatidylinositol 4,5-bisphosphate (PIP_2) hydrolysis *per se*, is evenly distributed. Thus, differential targeting of TRPCs in microdomains allows different receptors to selectively recruit different Ca^{2+} entry pathways. TRPCs also co-assemble with members of the TRPP group, the polycystins. Because the polycystin proteins are thought to function as sensors of the extracellular environment, it can be hypothesized that TRPC channels are involved in a wide range of cellular functions other than those of SOCs and ROCs, including mechanotransduction.

2004 Mammalian TRP channels as molecular targets. Wiley, Chichester (Novartis Foundation Symposium 258) p 75–97

Ca^{2+} plays very important roles in regulating fundamental cell functions ranging from contraction and secretion to proliferation, differentiation and cell death. In virtually all cells, stimulation of phospholipase C (PLC)-coupled receptors is accompanied by an increase in intracellular Ca^{2+} concentration. These Ca^{2+} signals occur via Ca^{2+} release from internal stores, which, in some instances, causes secondary Ca^{2+} influx across the plasma membrane. This behaviour led to the formulation of the capacitative Ca^{2+} entry (CCE) hypothesis by Putney (1990), which was substantiated later by the identification of the Ca^{2+} release-activated Ca^{2+} current (I_{CRAC}) in the plasma membrane of various cell types (Parekh & Penner 1997). These channels are referred to as store-operated channels (SOCs), that is, they are activated by a reduction in Ca^{2+} in the endoplasmic reticulum (ER) (Putney et al 2001).

In the past few years, experimental evidence supported the view that the products of the *Drosophila* transient receptor potential (TRP) genes as well as a number of its seven mammalian homologues (TRPC) may fulfil this definition of SOCs. Other TRP homologues do not appear to be store-operated and are referred to as receptor-operated channels (ROCs) or non-capacitative channels. They are activated in response to agonist binding to PLC-coupled receptors but not — or not necessarily — by store emptying (Montell 2001). In all cases, signalling events downstream from receptor activation and PLC stimulation activate the TRP channels.

Despite the recent advances made in the characterization of TRPC channels, there is still no clear-cut biophysical correlate between TRPC-mediated responses in recombinant systems and SOCs in native cells. To make matters worse, recombinant studies aimed at characterizing functional properties of TRPC channels have generated a rather confusing picture (Vennekens et al 2002). These current gaps in our knowledge of the interrelationship between TRPCs and SOCs stress the need for further work on the properties of SOCs in native systems. They also suggest that TRPCs may associate with some, as yet unidentified, regulatory proteins, which may be responsible for the multi-functional properties of TRPC channels in different cell systems. Even more profoundly, one fundamental question remains as to whether TRPC can be gated by mechanisms distinct from PLC signalling and store depletion and thus serve functions other than those of SOCs and ROCs. A challenging impetus therefore remains to identify the physiological ligands and molecular cues that regulate TRPC channels and to define how these different factors adapt TRPC functions to the unique requirements of each tissue. Answering these questions should have important implications for our understanding of the exact physiological role of the different TRPC gene products.

Considering the coupling of TRPC channels to multiple PLC-coupled membrane receptors, one central question is how receptors expressed in the same cells can generate specific TRP signals? Evidence has grown to suggest that signalling specificity is determined by the localization of the signalling molecules in spatially discrete microdomains (Delmas & Brown 2002, Kiselyov et al 2003). Perhaps the best example of such structures is the INAD-assembled signalling complex of *Drosophila* photoreceptors, which organizes key components of the phototransduction cascade, including TRP channels, into a multiprotein signal transduction unit (Minke & Cook 2002). In the human, an INAD-like protein has been described and a protein called 'regulatory factor of the Na^+/H^+ exchanger' (NHERF) appears to be able to bring together PLCβ1 and TRPC4 (Tang et al 2000). Assembly of signalling molecules in microdomains greatly facilitates key reactions in the signalling cascade. Thus, TRPC1 and TRPC3, together with multiple Ca^{2+} signalling proteins, are present in caveolar microdomains, which provide a platform for the coordination of the molecular interactions possibly leading to the activation of SOCs (Lockwich et al 2000, 2001) (see companion paper by Ambudkar et al 2004, this volume). Also, many TRPC proteins seem to be able to interact directly with inositol 1,4,5-trisphosphate receptors (IP_3Rs) and ryanodine receptors (RyRs) located in the ER (Kiselyov et al 1999, 2000, 2001, Tang et al 2001), enabling bi-directional signalling between the plasma membrane and the ER and providing strong evidence that both IP_3R and RyR can act as transducers of the Ca^{2+} store status.

The influence of these microdomains on TRPC signalling and in generating receptor-specific Ca^{2+} signals is discussed below with emphasis on results obtained on TRPC channels expressed in sympathetic neurons (Delmas et al 2002a). Additionally, some preliminary evidence that TRP channels may co-assemble with TRP-related proteins of the polycystin family (TRPP) will be presented, suggesting that gating and functions of TRPC channels are perhaps even more complex that portrayed above.

Experimental procedures

DNA constructs

cDNAs for hTRPC1, hTRPC3 and mTRPC6 have been kindly provided by Dr Christian Harteneck and were subcloned into pcDNA3 vectors. hPKD1 and mPKD2 cDNAs were subcloned into pcDNA3.1 vectors and were described previously (Delmas et al 2002b).

Cell culture and DNA delivery

Sympathetic (SCG) neurons were isolated from Sprague-Dawley rats (2 weeks old) and cultured on glass coverslips as described (Delmas et al 2000). Plasmids were diluted to 100 μg/ml in Ca^{2+}-free solutions (pH 7.3) and were pressure-injected into the nucleus of neurons (Delmas et al 2002a,b). Cells were maintained 24 h in culture after injection.

Perforated whole cell recording and microvesicles

Whole-cell currents were recorded using the perforated patch method. Briefly, amphotericin B (0.1 mg/ml in DMSO) was dissolved in the intracellular solution consisting of (mM): K^+ acetate, 90; KCl, 30; $MgCl_2$, 3; Hepes, 40 (pH 7.3). The external solution consisted of (in mM): NaCl 130, KCl 3, $MgCl_2$ 1, HEPES 10, tetrodotoxin (TTX) 0.0005, $CaCl_2$ 0.1–1, D-glucose 11 (pH 7.3). Recordings were obtained with an Axopatch 200A amplifier and filtered at 2 kHz. Outside-out perforated microvesicles were obtained as described (Delmas et al 2002a). Briefly, patch pipettes were pulled from thin-walled borosilicate glass capillaries, polished and coated with Sylgard. They had resistance of 6–13 MΩ when filled with the internal solution described above. After partial permeabilization, the pipette was withdrawn from the cell to form an outside-out perforated vesicle of about 2 μm diameter. For each experiment presented below n was $\geqslant 4$.

Results and discussion

Role of signalling microdomains in the differential activation of TRPC1 and TRPC6 by PLC-coupled receptors

The gating of heterologously expressed TRPC channels in sympathetic neurons. Recording of membrane currents in the voltage-clamp mode of the patch clamp technique was used to study receptor activation of TRPC channels in sympathetic neurons. Thus, I will refer to 'agonist-induced TRPC currents' as a more general term for TRP-mediated Ca^{2+} entry, reflecting their general conducting properties rather than their strict Ca^{2+} transport capabilities.

Mechanisms of gating of mTRPC6, hTRPC1 and hTRPC3 were assessed 24 h after intranuclear cDNA delivery. In many expression systems, TRPC3 and TRPC6 are among the TRPC subunits activated by the lipid second messenger diacylglycerol (DAG) (Hofmann et al 1999, Ma et al 2000). In sympathetic neurons, mTRPC6 current was activated by OAG (50 μM), a cell-permeable analogue of DAG, but not by thapsigargin (TG, 500 nM), an inhibitor of the ER Ca^{2+}-ATPase (SERCA) that is responsible for store filling (Fig. 1, Delmas et al 2002a). Introducing IP_3 (15 μM) into the cellular cytoplasm through microinjection had no effect on mTRPC6. hTRPC3 could be activated by either OAG or microinjection of IP_3, in agreement with previous data suggesting that TRPC3 may be gated by DAG (Hofmann et al 1999) as well as by activation of IP_3Rs (Ma et al 2000). hTRPC1 activated in response to TG or IP_3 but not OAG (Fig. 1 and see Fig. 3) (Delmas et al 2002a).

Collectively, these data indicate that DAG is necessary and sufficient to activate mTRPC6 in sympathetic neurons. Thus, TRPC6 functions as a lipid-gated channel that fulfils the role of non-capacitative channel, in good agreement with data on recombinant TRPC6 in cell lines (Hofmann et al 1999) and on native TRPC6 channels in vascular smooth muscle cells (Inoue et al 2001). The picture with hTRPC3 is more complex. Its ability to sense both lipid mediators and IP_3R activation suggests it can act as a capacitative as well as a non-capacitative channel. Thus, TRPC3 is multi-functional, gated by DAG on the one hand and coupled to IP_3R, possibly via direct protein–protein interactions (Kiselyov et al

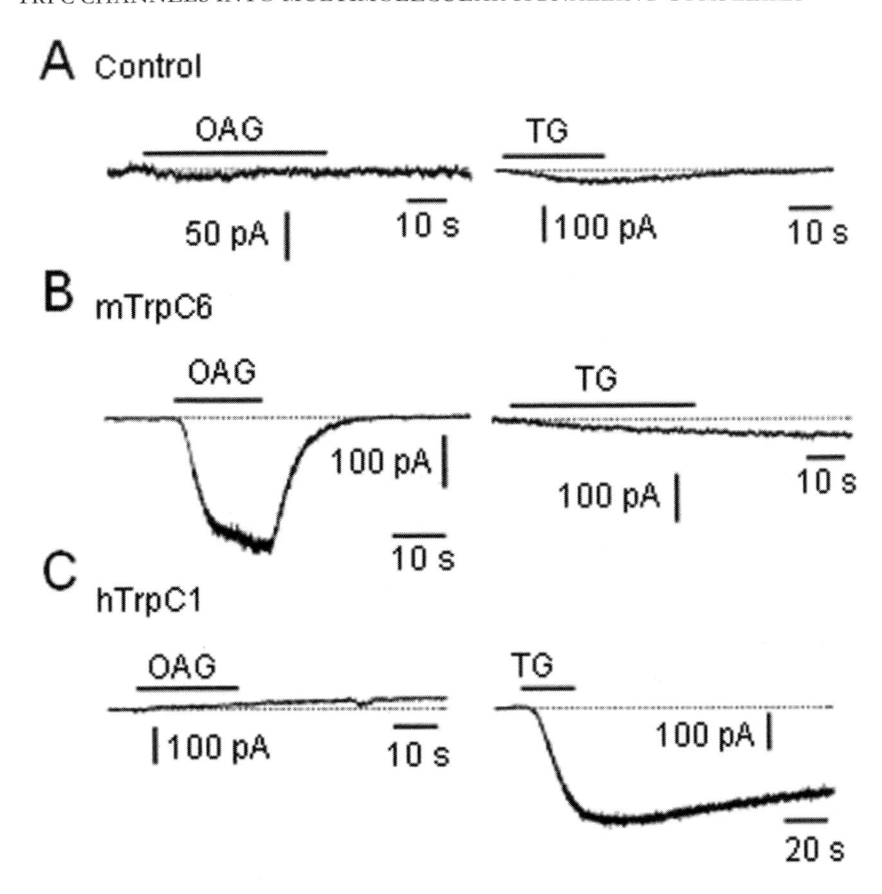

FIG. 1. Mechanisms of activation of hTRPC1 and mTRPC6 in sympathetic neurons. Effects of 1-oleoyl-2-acetyl-glycerol (OAG, 50 μM) and thapsigargin (TG, 500 nM) recorded in control neurons (A) and in neurons expressing either mTRPC6 (B) or hTRPC1 (C). Cells were voltage-clamped at -70 mV using the perforated patch method. Reprinted with permission from Delmas et al (2002a).

1999) on the other. Either coupling may occur independently as evidenced by studies in DT40 triple IP_3R knockout cells (Venkatachalam et al 2001).

In marked contrast, the interpretation of our results indicates that TRPC1 is a capacitatively coupled channel. With one notable exception that failed to detect store dependence (Lintschinger et al 2000), studies on recombinant TRPC1 (Zitt et al 1996) as well as those using antisense strategies in native cells (Xu & Beech 2001, Brough et al 2001, Liu et al 2000) provide strong support for these channels acting as store-operated channels.

Selective coupling of TRPC1 to IP_3R-associated bradykinin receptors. The activation of TRPCs by constitutively expressed bradykinin (B_2 subtype, B_2R) and muscarinic (M_1 subtype, M_1AChR) receptors was tested using saturating concentrations of the respective agonists, bradykinin (BK, 100 nM) and oxotremorine M (Oxo-M, 5 μM). In a variety of cells, stimulation of either of these two $G_{q/11}$-coupled receptors leads to PLCβ-mediated production of IP_3 and subsequent Ca^{2+} release through IP_3Rs. Both BK and Oxo-M activated mTRPC6 and hTRPC3 whereas BK only gated hTRPC1 (Fig. 2). Other receptors known to couple to $G_{i/o}$-type G-proteins such as the α_2-adrenergic and somatostatin receptors were inefficient in stimulating any of these TRPCs.

Activation of mTRPC6, hTRPC3 and hTRPC1 by B_2R and/or M_1AChR was blocked by application of U73122 (10 μM), an inhibitor of PLC, consistent with the requirement of PLCβ in TRPC gating. Effects of BK on hTRPC1 and hTRPC3 were prevented by the membrane-permeable inhibitors of IP_3R, xestospongin C (50 μM) and diphenylboric acid 2-amino-ethyl ester (2APB, 80 μM) (Fig. 2B). Injection of IP_3 occluded the activation of hTRPC1 but not mTRPC6 (Fig. 3). Thus, although both M_1AChR and B_2R stimulate PLCβ, activation of IP_3R-operated hTRPC1 was restricted to B_2R.

Overall, the picture that is emerging is that TRPC1 coupling to specific receptors reflects the functional compartmentalisation of the plasma membrane. Thus, in sympathetic neurons, the actin cytoskeleton provides a molecular scaffold upon which B_2R, $G_{q/11}$, PLCβ, IP_3R and most notably TRPC1 are assembled in a signalling complex (Delmas et al 2002a). Such interactions ensure efficient coupling of B_2R to IP_3R-operated TRPC1. Perhaps these plasma membrane domains are parented to caveolae, which have been shown to play a crucial role in TRP signalling mechanisms. As one example, submandibular gland cells target TRPC1 in caveolin-scaffolding domains where it interacts more or less directly with multiple Ca^{2+} signalling molecules including $G_{q/11}$ and IP_3R (Lockwich et al 2000). Although the M_1AChR complex also houses $G_{q/11}$ and downstream PLCβ machinery, it lacks IP_3R, making it unable to stimulate TRPC1. As these examples illustrate, neurons seem to target TRPC1 and TRPC6 to different signalling domains, depending on whether or not they accommodate IP_3Rs. Hence, the IP_3R-specific targeting of TRPCs allows different PLC-coupled receptors to selectively recruit different Ca^{2+} entry pathways.

Functional association of TRPC1 with polycystin signalling complexes. One question remains whether TRPC channels serve solely as SOCs and ROCs. Because of the ubiquity of TRPC channels in excitable and nonexcitable cells, it is conceivable that, in some particular cell types, their reported functions are epiphenomena, while their main role is altogether different. The interaction of TRPC1 with polycystin 2 (Tsiokas et al 1999), a member of the TRP-related group (TRPP) that

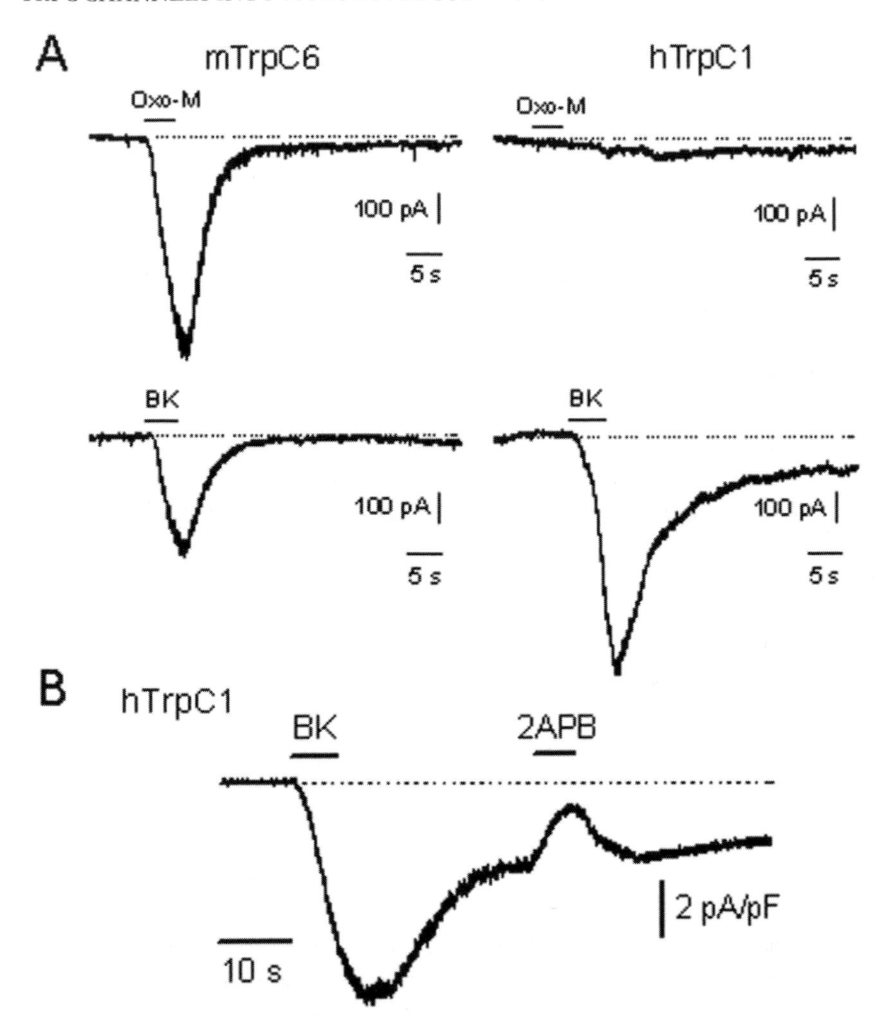

FIG. 2. Differential activation of hTRPC1 and mTRPC6 by M_1AChR and B_2R. (A) Shown are inward currents induced by oxotremorine M (Oxo-M, 5 μM, top traces) and bradykinin (BK, 100 nM, bottom traces) in cells expressing either mTRPC6 or hTRPC1 (as indicated). Holding potential, −80 mV. (B) Block of BK-evoked hTRPC1 currents by 2 APB (80 μM). Reprinted with permission from Delmas et al (2002a).

shares significant sequence homology (25%) to TRPC channels (Montell 2001), allows some predictions on less conformist roles for the TRPCs.

Polycystin 2 is a ~1000 amino acid protein with six putative transmembrane segments that forms a Ca^{2+}-permeable, non-selective cation channel (Hanaoka et al 2000, González-Perret et al 2001, Vassilev et al 2001). Although polycystin 2

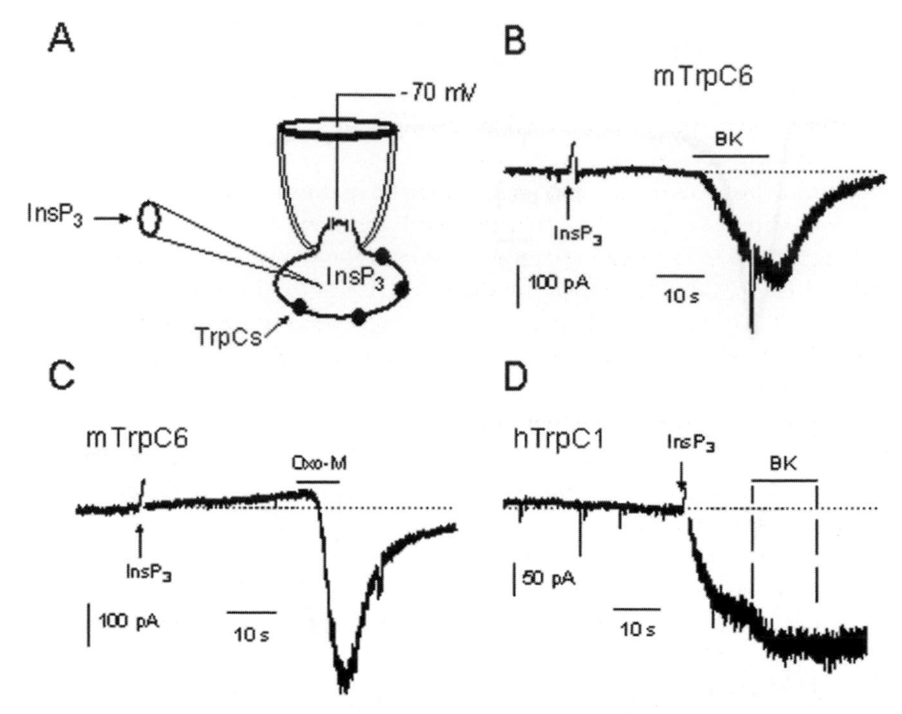

FIG. 3. The B_2R but not the M_1AChR couples to hTRPC1 via IP_3R. (A) Schematic diagram of intracellular microinjection of IP_3 during perforated patch clamp recording. (B–D) Cytoplasmic microinjection of IP_3 (0.5 pl at 100 μM, arrows) activated hTRPC1 but not mTRPC6. Note that injection of IP_3 occluded the activation of hTRPC1 by BK (100 nM, D) but not the activation of mTRPC6 by BK or Oxo-M (B,C). Reprinted with permission from Delmas et al (2002a).

shares structural similarities with TRPCs, it is lacking the ankyrin repeats typical of the TRP group (Mochizuki et al 1996). Its C terminus is thought to extend into the cytoplasm and contains a coiled-coil domain that can physically bind to polycystin 1, another member of the polycystin family (see below), and other distinct regions responsible for the interaction with TRPC1 and for the homodimerization of polycystin 2 (Qian et al 1997, Tsiokas et al 1997, 1999).

Polycystin 1 is an integral membrane protein of about 4300 amino acids thought to function as an atypical G protein-coupled receptor. It is composed of 11 transmembrane domains, a large N-terminal extracellular region containing a number of adhesive domains (Hughes et al 1995, Sandford et al 1997) and a C-terminal business end that can activate various G proteins (Delmas et al 2002b, Parnell et al 2002).

Both polycystin 2 and polycystin 1 are defective in autosomal dominant polycystic kidney disease (ADPKD) (Wu & Somlo 2000). Because mutations in either of these two proteins produce virtually identical clinical symptoms, irrespective of the causative proteins, it is thought that they function as constituents of a common signalling pathway or as interacting partners within a heteromeric polycystin complex. Backing up these observations, mutation analysis in *Caenorhabditis elegans* indicates that the homologues of polycystin 1 and polycystin 2 are involved in a single genetic pathway in sensory neurons consistent with a function in chemo- and mechano-transduction (Barr & Sternberg 1999). In a similar vein, both proteins act in the same pathway and contribute to fluid-flow sensation by the primary cilium in the renal epithelium (Nauli et al 2003). From these data and the conservation of polycystin pathways in multiple cell types, it can be hypothesized that polycystins act as sensors of the extracellular environment, initiating a variety of intracellular transduction signals in multiple cellular processes. Are TRPCs involved in any of these?

Despite intensive efforts aimed at delineating the activation mechanisms of polycystins, the molecular events that activate the polycystin complex are still unknown. The results presented below unravel a few.

To reconstitute polycystin complexes, sympathetic neurons were intranuclearly microinjected with cDNAs encoding polycystin 2 (PC2) and polycystin 1 (PC1) (Delmas et al 2002b). This allows the study of the structure/function relationships of polycystin complexes in living cells, as well as the effects of co-expressing TRPC1. To possibly mimic dynamic regulation of the polycystin complex, a polyclonal antibody (MR3) that binds to the N-terminal extracellular domain of PC1, was used as a putative ligand (Fig. 4). Local application of MR3 (1/100) to cells co-expressing PC1/PC2 produced cation currents that were blocked by amiloride and La^{3+}, two known blocking agents of PC2 channels in lipid bilayers (Fig. 4B) (González-Perret et al 2001). MR3-induced currents were also blocked by an antibody raised against PC2, indicating that PC2 comprised the channel pore. No currents were seen in cells expressing hPC1, mPC2 or TRPC1 alone. Gating of PC2 by MR3 did not require G proteins since MR3 responses remained unchallenged by procedures (pertussis toxin, GDP-β-S, anti-G$\alpha_{q/11}$ antibody) that typically block G protein activation. Collectively, these data suggest that binding of antibodies to the large extracellular domain of PC1 results in conformational change of the polycystin complex that ultimately leads to PC2 activation, a mechanistic sequence that fits well with the recently proposed role of polycystin complexes as mechano-fluid stress sensors in cilia embryonic kidney cells (Nauli et al 2003).

Co-expression of TRPC1 with PC1/PC2 complex resulted in *ca.* fourfold reduced MR3 currents (Fig. 4C). Also, in these cells, B_2R stimulation was unable to activate TRPC1, suggesting that TRPC1 is no longer part of the B_2R–IP_3R microdomain but rather associated with the polycystin complex (Fig. 4D). Single channel recording using cell-attached patches and outside-out microvesicles

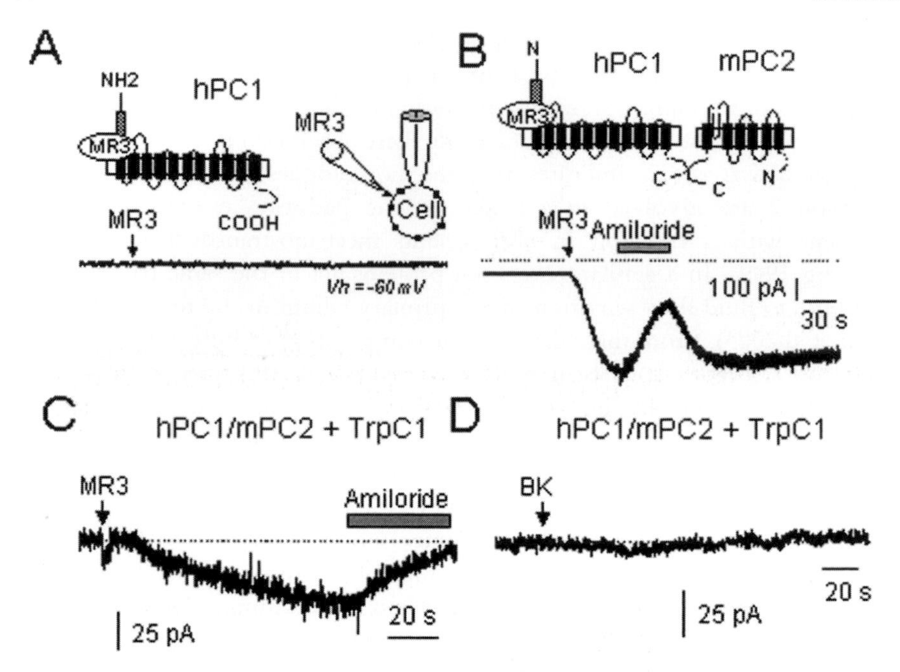

FIG. 4. Activation of PC1/PC2/TRPC1 ion channel complex by MR3 antibodies but not by bradykinin. (A,B) Upper panels: schematic illustration of the membrane topology for hPC1 and mPC2. Currents evoked by 10 s application of the anti-hPC1 antibody MR3 (1/100) in cells expressing hPC1 (A) or hPC1/mPC2 (B). Note that MR3 activated an inward current only in the hPC1/mPC2-expressing cell, which was blocked by bath application of amiloride (80 μM). Dashes lines indicate the null-current baseline at −60 mV. (C,D) Effects of MR3 and BK on cells expressing hPC1/mPC2/TRPC1.

provided further evidence for this. Recordings in cells expressing PC1/PC2 revealed inward channel currents with main chord conductances of 90–130 pS with Na^+ as charge carrier (Figs 5A,D). These channels were gated by MR3, suppressed by amiloride or La^{3+} and were not detected in mock cells or in cells expressing PC1 alone. Inward channel currents gated by MR3 in cells co-expressing TRPC1 along with PC1/PC2 had a main chord conductance of 25 pS, which is very much comparable to the single channel conductance of TRPC1 (13–16 pS) gated by bradykinin in cells expressing TRPC1 alone (Figs 5B–D).

Taken together, these data suggest that TRPC1 and PC1/PC2 may form functionally associated 'subunits' of a heteromultimeric complex that reconstitutes a novel surface Ca^{2+}-permeant cation channel. At this juncture, the precise molecular stoichiometry of this complex is uncertain. However, because of the propensity of TRPC1 to heterodimerize with PC2 (Tsiokas et al 1999), it is plausible that the complex consists of a PC2/TRPC1 channel heteromer bound to

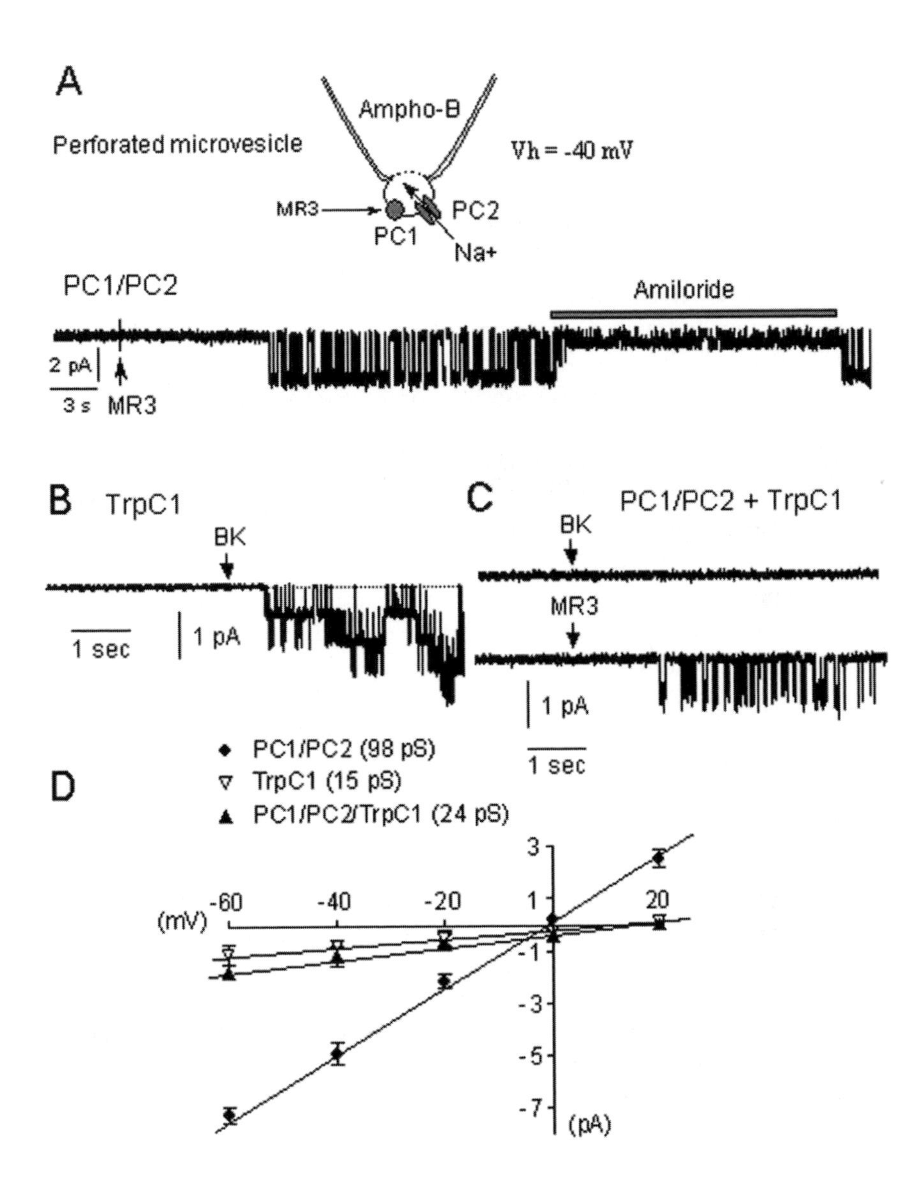

FIG. 5. Single channel properties of polycystin/TRPC1 ion channels in perforated microvesicles. Representative single-channel activities recorded in outside-out microvesicles excised from cells expressing mPC2/hPC1 (A), TRPC1 (B) or hPC1/mPC2/TRPC1 (C) in response to MR3 or BK (as indicated). Note the difference in single channel conductances (D). $[Na^+] = 143$ mM, $[Ca^{2+}] = 0.1$ mM.

a single PC1 protein via C-terminal tethering. Thus, the activation of the PC2/TRPC1 channel would proceed through a structural rearrangement of PC1.

Although further work is clearly needed to define the relative importance of these complexes both *in vitro* and *in vivo*, this proposed mechanism may be paradigmatic for the function of TRPCs and other TRP-related proteins in a variety of cell types. Ca^{2+}-dependent signalling by the polycystin/TRPC1 complex is reminiscent of that occurring during the acrosome reaction, a prerequisite for sperm-egg fusion in sea urchin sperm. In these cells, it is the receptor for egg jelly (suREJ1-3), a membrane glycoprotein sharing domain homology with PC1, that binds to the egg jelly and triggers the exocytotic acrosome reaction by increasing intracellular Ca^{2+} (Moy et al 1996, Mengerink et al 2002). Recently, it has been shown that TRPC2 plays an essential role in the sustained increase of Ca^{2+} influx responsible for the acrosome reaction (Jungnickel et al 2001). Analogously, NOMPC, a member of the TRP-related group (TRPN) that contains many ankyrin repeats, mediates mechanoreceptor currents in insect bristles (Walker et al 2000). A *Caenorhabditis elegans* homologue was also identified in mechanosensitive neurons of the nematode (Walker et al 2000).

Thus, on the basis of developments reported here, it is conceivable that TRPCs and related polycystin proteins may function as extracellular sensors that regulate Ca^{2+} transport. Because TRPCs are widely expressed and play multifunctional roles, it would be interesting to test whether the ability to respond to 'mechanical' stimuli is a general feature of TRPCs.

Conclusion

Functional studies in heterologous systems have provided conflicting results on the regulation of TRPC channels (Minke & Cook 2002). Although this could result from the intrinsic characteristics of TRPC channels, the detailed studies of TRPC signalling pathways provide alternative clues regarding these apparent discrepancies. Growing evidence indicates that TRPs are incorporated into multiprotein signalling microdomains, which create local conditions that enhance molecular interaction of the segregated transduction elements. Thus, the features of TRPC channels in a foreign cellular environment may reflect their abnormal interaction with host proteins. In native cells, the localization of TRPCs in different signalling complexes allows PLC-coupled receptors to selectively recruit different Ca^{2+} entry pathways, thereby coordinating the activities of multiple TRPCs. The new picture that is emerging from the functional interaction of TRPCs with polycystin proteins may also suggest that TRPC functions are not only related to SOCs and ROCs but also to mechanotransduction. Only in conjunction with studies on TRP-related proteins will we gain important clues on the multi-functionality of TRPCs.

Acknowledgments

This study was supported by the CNRS and the Wellcome Trust Programme grant 038171. I wish to thank Drs Christian Harteneck and Günter Schultz (Institut für Pharmakologie, Freie Universität Berlin) for TRPC cDNAs, and David A. Brown (Dpt of Pharmacology, UCL, London), Jing Zhou (Brigham and Women's Hospital and Harvard Medical School, Boston) and Marcel Crest (Intégration des Informations Sensorielles, CNRS, Marseille) for support and constructive discussion.

References

Ambudkar IS, Brazer SC, Liu X, Lockwich T, Singh B 2004 Plasma membrane localization of TRPC channels: role of caveolar lipid rafts. Wiley, Chichester (Novartis Found Symp 258) p 63–74

Barr MM, Sternberg PW 1999 A polycystic kidney-disease gene homologue required for male mating behaviour in C. elegans. Nature 401:386–389

Brough GH, Wu SW, Cioffi D et al 2001 Contribution of endogenously expressed Trp1 to a Ca^{2+}-selective, store-operated Ca^{2+} entry pathway. FASEB J 15:1727–1738

Delmas P, Brown DA 2002 Junctional signaling microdomains: Bridging the gap between the neuronal cell surface and Ca^{2+} stores. Neuron 36:787–790

Delmas P, Abogadie FC, Buckley NJ, Brown DA 2000 Calcium channel gating and modulation by transmitters depend on cellular compartmentalization. Nat Neurosci 3:670–678

Delmas P, Wanaverbecq N, Abogadie FC, Mistry M, Brown DA 2002a Signaling microdomains define the specificity of receptor-mediated $InsP_3$ pathways in neurons. Neuron 34:209–220

Delmas P, Nomura H, Li X et al 2002b Constitutive activation of G-proteins by polycystin-1 is antagonized by polycystin-2. J Biol Chem 277:11276–11283

González-Perret S, Kim K, Ibarra C et al 2001 Polycystin-2, the protein mutated in autosomal dominant polycystic kidney disease (ADPKD), is a Ca^{2+}-permeable nonselective cation channel. Proc Natl Acad Sci USA 98:1182–1187

Hanaoka K, Qian F, Boletta A et al 2000 Co-assembly of polycystin-1 and -2 produces unique cation-permeable currents. Nature 408:990–994

Hofmann T, Obukhov AG, Schaefer M, Harteneck C, Gudermann T, Schultz G 1999 Direct activation of human TRPC6 and TRPC3 channels by diacylglycerol. Nature 397:259–263

Hughes J, Ward CJ, Peral B et al 1995 The polycystic kidney disease 1 (*PKD1*) gene encodes a novel protein with multiple cell recognition domains. Nat Genet 10:151–160

Inoue R, Okada T, Onoue H et al 2001 The transient receptor potential protein homologue TRP6 is the essential component of vascular $\alpha(1)$-adrenoceptor-activated Ca^{2+}-permeable cation channel. Circ Res 88:325–332

Jungnickel MK, Marrero H, Birnbaumer L, Lemos JR, Florman HM 2001 Trp2 regulates entry of Ca^{2+} into mouse sperm triggered by egg ZP3. Nat Cell Biol 3:499–502

Kiselyov K, Mignery GA, Zhu MX, Muallem S 1999 The N-terminal domain of the IP3 receptor gates store-operated hTrp3 channels. Mol Cell 4:423–429

Kiselyov K, Shin DM, Wang YM, Pessah IN, Allen PD, Muallem S 2000 Gating of store-operated channels by conformational coupling to ryanodine receptors. Mol Cell 6:421–431

Kiselyov K, Shin DM, Muallem S 2003 Signalling specificity in CGPR-dependent Ca^{2+} signalling. Cell Signal 15:243–253

Lintschinger B, Balzer-Geldsetzer M, Baskaran T et al 2000 Coassembly of Trp1 and Trp3 proteins generates diacylglycerol- and Ca^{2+}-sensitive cation channels. J Biol Chem 275: 27799–27805

Liu XB, Wang WC, Singh BB et al 2000 Trp1, a candidate protein for the store-operated Ca^{2+} influx mechanism in salivary gland cells. J Biol Chem 275:3403–3411 [Erratum in J Biol Chem 275:9890–9891]

Lockwich TP, Liu XB, Singh BB, Jadlowiec J, Weiland S, Ambudkar IS 2000 Assembly of Trp1 in a signaling complex associated with caveolin-scaffolding lipid raft domains. J Biol Chem 275:11934–11942

Lockwich TP, Singh BB, Liu X, Ambudkar IS 2001 Stabilization of cortical actin induces internalization of transient receptor potential 3 (Trp3)-associated caveolar Ca^{2+} signaling complex and loss of Ca^{2+} influx without disruption of Trp3-inositol trisphosphate receptor association. J Biol Chem 276:42401–42408

Ma HT, Patterson RL, van Rossum DB, Birnbaumer L, Mikoshiba K, Gill DL 2000 Requirement of the inositol trisphosphate receptor for activation of store-operated Ca^{2+} channels. Science 287:1647–1651

Mengerink KJ, Moy GW, Vacquier VD 2002 SuREJ3, a polycystin-1 protein, is cleaved at the GPS domain and localizes to the acrosomal region of sea urchin sperm. J Biol Chem 277:943–948

Minke B, Cook B 2002 TRP channels proteins and signal transduction. Physiol Rev 82:429–472

Mochizuki T, Wu G, Hayashi T et al 1996 PKD2, a gene for polycystic kidney disease that encodes an integral membrane protein. Science 272:1339–1342

Montell C 2001 Physiology, phylogeny and functions of the TRP superfamily of cation channels. Science's STKE 90:RE1 *http://stke.sciencemag.org/cgi/content/full/sigtrans;2001/90/re1*

Moy GW, Mendoza LM, Schulz JR, Swanson WJ, Glabe CG, Vacquier VD 1996 The sea urchin sperm receptor for egg jelly is a modular protein with extensive homology to the human polycystic kidney disease protein, PKD1. J Cell Biol 133:809–817

Nauli SM, Alenghat FJ, Luo Y et al 2003 Polycystins 1 and 2 mediate mechanosensation in the primary cilium of kidney cells. Nat Genet 33:129–137

Parekh AB, Penner R 1997 Store depletion and calcium influx. Physiol Rev 77:901–930

Parnell SC, Magenheimer BS, Maser RL, Zien CA, Frischauf A-M, Calvet JP 2002 Polycystin-1 activation of c-Jun N-terminal kinase and AP-1 is mediated by heterotrimeric G proteins. J Biol Chem 277:19566–19572

Putney JW Jr 1990 Capacitative calcium entry revisited. Cell calcium 11:611–624

Putney JW Jr, Broad LM, Braun FJ, Lievremont JP, Bird, GSJ 2001 Mechanisms of capacitative calcium entry. J Cell Sci 114:2223–2229

Qian F, Germino FJ, Cai Y, Zhang X, Somlo S, Germino GG, 1997 PKD1 interacts with PKD2 through a probable coiled-coil domain. Nat Genet 16:179–183

Sandford R, Sgotto B, Aparicio S et al 1997 Comparative analysis of the polycystic kidney disease 1 (PKD1) gene reveals an integral membrane glycoprotein with multiple evolutionary conserved domains. Hum Mol Genet 6:1483–1489

Tang Y, Tang J, Chen ZG et al 2000 Association of mammalian Trp4 and phospholipase C isozymes with a PDZ domain-containing protein, NHERF. J Biol Chem 275:37559–37564

Tang J, Lin YK, Zang ZM, Tikunova S, Birnbaumer L, Zhu MX 2001 Identification of common binding sites for calmodulin and inositol 1,4,5-trisphosphate receptors on the carboxyl termini of trp channels. J Biol Chem 276:21303–21310

Tsiokas L, Kim E, Arnould T, Sukhatme VP, Walz G 1997 Homo- and heterodimeric interactions between the gene products of *PKD1* and *PKD2*. Proc Natl Acad Sci USA 94:6965–6970

Tsiokas L, Arnould T, Zhu C, Kim E, Walz G, Sukhatme VP 1999 Specific association of the gene product of PKD2 with the TRPC1 channel. Proc Natl Acad Sci USA 96:3934–3939

Vassilev PM, Guo L, Chen XZ et al 2001 Polycystin-2 is a novel cation channel implicated in defective intracellular Ca^{2+} homeostasis in polycystic kidney disease. Biochem Biophys Res Commun 282:341–350

Venkatachalam K, Ma HT, Ford DL, Gill DL 2001 Expression of functional receptor-coupled TRPC3 channels in DT40 triple receptor InsP3 knockout cells. J Biol Chem 276:33980–33985

Vennekens R, Voets T, Bindels RJM, Droogmans G, Nilius B 2002 Current understanding of mammalian TRP homologues. Cell Calcium 31:253–264

Walker RG, Willingham AT, Zuker CS 2000 A Drosophila mechanosensory transduction channel. Science 287:2229–2234

Wu G, Somlo S 2000 Molecular genetics and mechanism of autosomal dominant polycystic kidney disease. Mol Gen Metab 69:1–15

Xu SZ, Beech DJ 2001 TrpC1 is a membrane-spanning subunit of store-operated Ca^{2+} channels in native vascular smooth muscle cells. Circ Res 88:84–87

Zitt C, Zobel A, Obukhov AG et al 1996 Cloning and functional expression of a human Ca^{2+}-permeable cation channel activated by calcium store depletion. Neuron 16:1189–1196

DISCUSSION

Montell: If you remove the extracellular domain of polycystin 1, do you get constitutive activation?

Delmas: This is a good question. Actually, we have two constructs, one is a full-length *PKD1* cDNA and one is a N-terminally truncated form lacking most of the extracellular domain. These have been cloned in Jing Zhou's laboratory in Boston. We did see constitutive activity with both constructs, but the full-length clone was far less efficient.

Montell: Where does the antibody bind that you say mimics ligand binding?

Delmas: It is not clear what are the natural ligands of PKD1. So, we used the polyclonal anti-hPC1 antibody MR3 as a putative ligand of hPC1. This binds to the N-terminal extracellular domain of hPC1 near the receptor for egg jelly domain. Another antibody binds to the distal part of the protein. In our hands, binding close to the egg jelly domain gave the best results. Nevertheless, the reduced activity of antibodies acting on remote regions may also arise from improper folding, the extracellular domain being about 3000 amino acids long.

Montell: So antibodies that recognize distinct parts of the extracellular domain lead to constitutive activity, but removing the extracellular domain does not. With receptor tyrosine kinases you can get constitutive activation just by lopping off the extracellular domain. But with polycystin 1, it doesn't seem that the antibodies are mimicking ligand binding, unless they are coincidently recognizing different parts of the extracellular domain that come together in the 3D structure.

Scharenberg: A better analogy might be an antigen receptor, where you just have the tail sitting in there. If you lop off the extracellular domain, nothing happens. It doesn't have intrinsic enzymatic activity. Didn't someone take the C-terminus of PKD1, tack it onto a new transmembrane span, and show that if this is ligated then there is activation?

Delmas: I am not aware of such an experiment. But this is undoubtedly an interesting experiment to perform.

Muallem: You indicated that binding of your antibodies activated G_i and G_o. How does this happen?

Delmas: There is a G protein-binding site on the C-terminus of polycystin 1, but this domain is masked by its interaction with polycystin 2 (Delmas et al 2002). Binding of the antibody on the extracellular side of PKD1 leads to a conformational change of the protein complex, which unmasks the G protein-binding site. So, we have on the one hand, activation of polycystin 2 and on the other G protein activation. We looked at the activity of endogenous Ca^{2+} channels in order to sense the level of activated G proteins. These channels are known to be directly modulated by the $G\beta\gamma$ dimers of G proteins and hence provide a very good biosensor for free $G\beta\gamma$.

Montell: There is at least one paper suggesting that polycystin 1 may be a novel type of GPCR (Parnell et al 1998). No one knows exactly how many transmembrane domains it has: but they were suggesting that perhaps there were seven.

Muallem: So do you propose that PKD1 is a form of G protein-coupled receptor?

Delmas: Yes, I do, but perhaps an unorthodox one. There are two papers on this, our own (Delmas et al 2002) and another one by Calvet's group (Parnell et al 2002). We showed that overexpressing PKD1 activates $G_{i/o}$-type G proteins, while they showed activation of G_q-type G proteins in addition to the ones we described. Activation seems to be constitutive and was suppressed by deletion of the C-terminus of PKD1.

Muallem: Did trafficking to the membrane remain the same? Could it be a problem of trafficking? There is a paper by Koulen et al (2002) showing that PKD2 does stay in the ER.

Delmas: If you overexpress polycystin 2 in sympathetic neurons it does stick in the ER and doesn't go to the plasma membrane. It is also the case in most cell lines but you will find many conflicting data in the literature on native kidney epithelia. You get expression to the membrane when polycystin 2 is co-expressed with polycystin 1.

Freichel: Where do you isolate the neurons you work with from? What TRPs are expressed in them?

Delmas: I used sympathetic neurons isolated from rat superior cervical ganglia. By RT-PCR from the whole ganglion, we have detected TRPC1, 3 and 6 but I cannot guarantee they are all expressed in neurons. They could be in non-neuronal cells too.

Putney: Did you say that the embryonic kidney cells respond to MR3 without you having to transfect anything into them? Do you have a cell line that will respond like this?

Delmas: We used kidney epithelia of distal tubule origin (E15.5) and examined Ca^{2+} mobilization in these cells in response to MR3. These cells have been prepared in Jing Zhou's lab. They constitutively express polycystin 1 and polycystin 2 without us having to transfect anything in.

Putney: What do the single channels look like? Do they look like the PKD1/PKD2 channel or the one with TRP1 added?

Delmas: I did not chase the channel in kidney cells myself, and currently we don't know what the characteristics of the channels activated by MR3 are. The answer therefore should await further experiments, but they won't be trivial. In reconstitution experiments, we do know that MR3 activates a 100 pS cation channel that shares most of the properties of polycystin 2.

Putney: No, which of all of these does the native channel, which is activated by MR3, look like?

Delmas: I don't know that.

Zhu: What was the reason for expressing PC1/PC2 in the sympathetic neurons?

Delmas: Pragmatism! About five years ago I started to express many proteins in sympathetic neurons and in cell lines. Sympathetic neurons worked all the time. When we came to polycystin 1, which is a 14 kb cDNA, we couldn't get any expression at all in cell lines, so we decided to use sympathetic neurons.

Zhu: So you are basically using it as a carrier cell line. I'm assuming that there is no endogenous PC1 or PC2.

Delmas: That is correct. Sympathetic neurons are also very useful because they express endogenous ion channels, such as Ca^{2+} and KCNQ channels, and both can be used as biosensors for signalling pathways.

Westwick: Before you transfect in your TRPC6 and TRPC1, when you add bradykinin or the cholinergic agonist can you measure any channel activity?

Delmas: No. In most of the recordings we maintain the cells at −70 mV. There are very few active channels at this voltage. Some ion channels such as the M/KCNQ channels are modulated by bradykinin or oxotremorine M, but at −70 mV these channels are not activated. David Brown's lab has a great experience on sympathetic neurons, and high conductance cation channel has never been recorded in these cells.

Westwick: Do you know about the relative amounts of IP_3 generated by each of those agonists? Are they similar in terms of the time course and duration? I am looking for an alternative explanation for why you see a difference between bradykinin and the muscarinic receptor, when you can get similar effects by co-injecting IP_3 in.

Delmas: Well, I should point out that IP_3 only activates TRPC1, so it has differential effects on TRPC1 and TRPC6. About the efficiency of muscarinic and bradykinin receptors, the data we have come from population-cell analysis, and they seem to suggest that muscarinic and bradykinin receptors are as effective.

This is corroborated by our data showing that both receptors activate TRPC6 with the same speed at saturating concentrations. So, the most plausible way to explain the differential coupling to TRPC1 and TRPC6 is co-localization and clustering of signalling molecules. One important additional factor is that IP_3 receptors in sympathetic neurons seem to have a low affinity for IP_3. High concentrations of IP_3 are therefore required to activate these IP_3 receptors, and you can only get this if PLC is located close to the IP_3 receptor. When PLC is activated by remote receptors such as the muscarinic receptors, then there is a very weak activation.

Penner: This is the type 1 IP_3 receptor, which is the highest affinity receptor.

Delmas: This is not clear to me. Many subtypes can be expressed and multiple cofactors are known to alter the affinity of the IP_3 receptor.

Putney: Probably because the concentration of receptor is so high.

Muallem: There is a way to get to it. You indicated that TRPC3 is activated by both InsP3 and OAG. Do you get similar activation by bradykinin and muscarinic receptors?

Delmas: We do get good activation of TRPC3 by both receptors, but to assess their efficiency we need specific experiments, such as dose–response curves and we do not have these data yet.

Putney: Going back to John Westwick's point, one of the things about IP_3- and DAG-mediated responses is that when you activate PLC it takes much less PLC activity to maximally activate the DAG pathway than the IP_3 pathway. We find that the activation of release and store-operated channels is more or less proportional to receptor occupation. With the same receptor, much lower concentrations of agonists will produce maximal activation of the DAG pathway. Can that quantitative difference explain what you are interpreting as a spatial difference in these two receptors?

Muallem: In this case you shouldn't get the same amount of IP_3 or the same activation of PLC.

Putney: The agonist that gives TRP1 activation also gives TRP6 activation, but not the other way around.

Delmas: Correct. This may be an alternative; still we observed differential membrane expression and co-localization of the bradykinin receptor with the IP_3 receptor.

Putney: It seems to me that it can be explained simply by low PLC activation by the muscarinic agonists.

Penner: It is the other way round.

Delmas: Judging by the DAG-activated TRPC6, the activation of PLC is the same with muscarinic and bradykinin receptors. If the differential activation of TRPC1 has to do something with PLC, it should be downstream to it, i.e. diffusion, degradation and location of IP_3. Because the M1 receptor is not colocalized with the IP_3 receptor, we think that the lack of activation ofTRPC1 by the M1 receptor is due to mislocation.

Putney: I know that is your interpretation! I'm simply saying that our data show that if you use a very low concentration of bradykinin, it will look like acetylcholine: it ought to give you TRP6 activation and not TRP1 activation. If this were the case, it would suggest that an alternative explanation is just the degree of PLC activation.

Westwick: It is not just the degree of PLC activation because IP_3 can only come from PIP_2 while DAG can come from PIP_2, PIP and PI. You get a much higher concentration of DAG before you get a high concentration of IP_3.

Muallem: This wouldn't change the interpretation you are driving to. You are driving to the interpretation where you make very low DAG in one case, and you activate one of the TRPs and the other will not. However, eventually, when you use high enough agonist concentrations, the system is boosted as much as it can get. And if the two receptors are similarly coupled and can generate the same stimulus intensity, eventually you should be able to activate both channels.

Putney: It will be limited by how many receptors are there, for example.

Authi: Don't the experiments also suggest that there is a close association of the IP_3 receptor to the B2 receptor, but not to the muscarinic receptor? The PLC activities are similar, yet there is a difference with Ca^{2+} mobilization seen when the B2 receptor is stimulated but is not seen when the muscarinic receptor is activated.

Scharenberg: You have to be very careful with the IP_3 measurements. If you do IP_3 measurements using a receptor binding assay, you can see very rapid 15–20 s spikes. If you do it using lithium treatment and a binding resin, and let it accumulate over a long period (15–30 min), you can have stimuli which will give you the same result, yet if you measure the instantaneous IP_3 you can bind, it can be very different. This is a real problem in the DT40 system with the lyn-deficient DT40s. The Ca^{2+} signals are very different, but if you measure IP_3 accumulation over 30 min they look the same. If you measure IP_3 accumulation within a minute using a receptor binding assay then they are completely different.

Ambudkar: In your experiments you used oxotremerine M as an agonist. Are you using this because it is more M1 specific? It is a rather weak agonist compared with some other muscarinic agonists. Oxotremorine M is much weaker than carbachol as an agonist.

Delmas: Yes, this is right, but it is more specific than carbachol.

Ambudkar: You might want to try something like carbachol to stimulate those receptors?

Delmas: I don't think so. Carbachol is also a weak activator of nicotinic receptors, and they are highly expressed in sympathetic neurons. In our experiments we used the agonists at saturating concentrations.

Ambudkar: We have expressed TRPC1 using adenovirus *in vivo* in rat submandibular glands which express primarily the M3 subtype. TRPC1 coupled to the M3 muscarinic receptors in these cells and we got activation of TRPC1 through the M3 pathway. It also immunoprecipitated with the IP_3 receptor as well as $G\alpha_q$.

Penner: I don't know whether you have shown this or not. Most of the data you showed were in perforated patch. Do you lose all of the responses if you do a whole-cell experiment with BAPTA?

Delmas: The reason we used perforated patch is because sympathetic neurons in culture are quite sensitive to washout. For example, many transduction pathways including those described today are labile using the patch-ruptured configuration, so we used a protocol that avoided washout. We haven't tried using whole-cell recording with BAPTA.

Penner: The reason I am asking is because it could be that there may be some Ca^{2+} dependence. Channel activity may be a consequence of the Ca^{2+} signal.

Delmas: That's a possibility.

Nilius: Bradykinin also stimulated arachidonic acid signalling, so you will get an arachidonic acid signal also with bradykinin stimulation. It is very well known that in sympathetic ganglia there is also a high expression of TRPV4. This channel has a single channel conductance of 100 pS. How can you be sure that you don't have this arachidonic acid signalling in your channel?

Delmas: I am not aware of expression of TRPV4 in SCG neurons. About the stimulation of arachidonic signalling by bradykinin receptor, this is highly speculative. Arachidonic acid is a modulator of Ca^{2+} channels in these cells but bradykinin has never been shown to have any effect on Ca^{2+} channels. In mock transfected cells I don't see any channel with such a huge conductance.

Schilling: Does the bradykinin receptor co-immunoprecipitate with TRPC1?

Delmas: We tried to do this and failed. We are not sure that our antibody is very good. We tried to get TRPC1 from bradykinin receptor immunocomplex and could not get it.

Schilling: You showed a profile of MR3 activation of Ca^{2+} signalling. Is this in the presence of extracellular Ca^{2+}?

Delmas: Yes. In kidney epithelia the MR3 responses are lost when extracellular Ca^{2+} is removed. In sympathetic neurons, the story is more complex and we have in some instances detected the involvement of internal calcium stores in the MR3 responses, but I cannot say more about that right now.

Taylor: I am a little bit uncomfortable with the idea of very low affinity IP_3 receptors. If I remember your *Neuron* paper correctly, there was an effect of calmodulin inhibitors that broke down this preferential association. I am wondering whether, given that IP_3 kinase also seems to be specifically localized in these cells, there is a buffer barrier of IP_3 kinase that is providing the limitation.

Delmas: That is a possibility indeed. We did not check the possible role of IP_3 kinases in the differential activation of IP_3 receptors. However, we did apply a protein phosphatase inhibitor cocktail and did not see any significant change in the coupling or absence of coupling.

Authi: From the data you have shown, you are suggesting that there is a constitutive association of a TRPC protein with the IP_3 receptor. Also, I am getting this implication from the data that Indu Ambudkar has shown. When you get Ca^{2+} entry occurring, is there an increase in the association of the IP_3 receptor to the TRPC protein, or is it just the same?

Delmas: We haven't done these experiments, but this is a very good point.

Ambudkar: Work from Sage's laboratory shows an increase in this coupling in platelets (Rosado & Sage 2001). It has not been shown in other cell types.

Delmas: All I know is that in the sympathetic neuron the release of Ca^{2+} from stores is modest. If there is an effect in changing the affinity of the IP_3 receptor for the TRP channel, it is probably very sensitive to Ca^{2+}. Is there a cooperative effect by Ca^{2+} flowing through the TRPC channel? I just don't know.

Authi: One of the things you showed in your paper was that agents that disrupt the cytoskeleton appeared not to affect the TRPC–IP_3 receptor interaction. Is that correct?

Delmas: Yes, this is correct but in the paper we applied cytochalasin D for a short period of time. A short application at low concentrations disrupts the interaction between the bradykinin receptor and the IP_3 receptor but not between the IP_3 receptor and the bradykinin receptor. We know this is the case because after this treatment the microinjection of IP_3 can still activate TRPC1. Longer application, for example for 10 minutes, disrupts the residual interaction between the IP_3 receptor and TRPC1. Consequently, IP_3 becomes inefficient to activate TRPC1. So both the concentration and the timing of application of cytochalasin D and related compounds are important.

Putney: We did this experiment, looking at endogenous signalling without putting TRP into cells. We also got complete block of activation of entry through the agonist by cytochalasin D, but there was no effect on the activation by thapsigargin. It seemed that the cytoskeleton was involved in properly orienting or localizing the PLC close to the IP_3R, but whatever happened from that point on to the channel was not so dependent on those same structural relationships.

Delmas: We should be very cautious not to raise general statements from a particular type of cells. It is becoming clear that what we observe is very cell specific. For example, the coupling of TRPC to IP_3R requires a very close apposition of the plasma membrane and the endoplasmic membrane, but not all

cells possess such an architectural organization. Also, protein interacting with TRPC may recruit the channel in distinct signalling pathways, meaning a given TRPC may be activated by different means in a single cell.

Putney: Is that similar to what you found? After cytochalasin D treatment, could you still activate with IP_3?

Delmas: Yes, with a short application of cytochalasin D. If you apply the drug for more than 10 min, it tends to disrupt the interaction of the IP_3R and TRPC1.

Putney: Did you try thapsigargin?

Delmas: No, we did not try this.

Penner: If there is this interaction between TRP1 and the IP_3R when TRP1 is overexpressed, couldn't that interaction also happen in the intracellular stores? Under those conditions you might change the IP_3R function.

Penner: Does IP_3 sensitivity change?

Muallem: Not as far as we can see. It is incredible how resistant it is. We get a four or fivefold increase in the number of IP_3Rs in the cells. We permeabilize the cells and get beautiful increase in release, but if we go to the intact cells, release is completely normal. This raises some questions about the biochemistry. What we have started doing now is to try to get into surface labelling of proteins.

Montell: I might have missed something, but is there any evidence that TRPC1 and polycystin 2 interact *in vivo*?

Delmas: No, not yet.

Montell: Can you inhibit the endogenous current, which is activated by the MR3 antibodies, by introducing antibodies to TRPC1 into your patch pipette?

Delmas: This is a difficult experiment but we could try to do this. We plan to try to demonstrate that TRP is indeed part of the channel complex.

Gill: We have heard how the interaction between TRPs is quite conservative. This is the same for voltage-activated K^+ channels: they form complexes that are very conservative and only channels that are like each other interact. Considering how conservative these interactions are, doesn't it seem surprising that TRPC1 would interact with other dissimilar TRPC channels?

Delmas: Yes. This experiment doesn't show that it is happening *in vivo*, though. We have to take into account that we are overexpressing the channel and it remains to test this hypothesis in native cells.

Penner: If you simply overexpress any muscarinic receptor, you will get all sorts of signalling promiscuity that is not seen in the normal cell.

Putney: There is a stoichiometry issue here. To get the stoichiometry you see, do you have to put in a huge excess of TRPC1 plasmid?

Delmas: I typically used a 1:1 ratio of TRPC1 to polycystin 2.

Putney: Do you ever see any of the channels that you would have seen if you had not put TRPC1 in?

Delmas: So far I have seven positive patches, and these seven patches show me the same small conductance, which is far too small for homomeric polycystin 2. Naturally, we need more data to conclude.

Putney: Of course, you can't directly relate plasma concentration and protein concentration, but this suggests that the affinity is very strong. You are not going out of your way to push this association by overwhelming the channels with a lot of TRPC1.

Delmas: Perhaps, but there is another argument which suggests that the affinity is strong. In cells overexpressing the polycystin complex, it becomes difficult to activate TRPC1 by stimulating the bradykinin receptor. Actually, it is like TRPC1 is sequestered within the polycystin complex.

Putney: So you don't have any excess.

Delmas: No, it is as if all the TRPC1 has been sucked into the polycystin complex and none can interact any more with the IP_3 receptor and the bradykinin microcomplex.

Penner: Does bradykinin still release Ca^{2+}, though?

Delmas: Yes, it does.

Nilius: I am very surprised that in the sympathetic neurons there are not a lot of Ca^{2+}-activated channels such as SK channels.

Delmas: When we do cell attached patches on the cell body it is difficult to detect SK channels. I believe they are mainly in the dendrites. The BK channels are not activated by bradykinin receptor because this requires huge concentrations of Ca^{2+}.

Ambudkar: I have a question about the TRPC1 interaction with PKD1 and 2. Have you done any co-immunoprecipitations to see whether all the TRP1 in the plasma membrane is being pulled down with PKD1 or 2?

Delmas: I don't have any data on this, but we are trying to do this and also triple immunofluorescent labelling.

References

Delmas P, Nomura H, Li X et al 2002 Constitutive activation of G-proteins by polycystin-1 is antagonized by polycystin-2. J Biol Chem 277:11276–11283

Koulen P, Cai Y, Geng L et al 2002 Polycystin-2 is an intracellular calcium release channel. Nat Cell Biol 4:191–197

Parnell SC, Magenheimer BS, Maser RL et al 1998 The polycystic kidney disease-1 protein, polycystin-1, binds and activates heterotrimeric G-proteins in vitro. Biochem Biophys Res Commun 251:625–631

Parnell SC, Magenheimer BS, Maser RL, Zien CA, Frischauf A-M, Calvet JP 2002 Polycystin-1 activation of c-Jun N-terminal kinase and AP-1 is mediated by heterotrimeric G proteins. J Biol Chem 277:19566–19572

Rosado JA, Sage SO 2001 Activation of store-mediated calcium entry by secretion-like coupling between the inositol 1,4,5-trisphosphate receptor type II and human transient receptor potential (hTrp1) channels in human platelets. Biochem J 356:191–198

General discussion I

Montell: Going back to this issue of the complexity of the Ca^{2+} regulation of calmodulin with TRP and its effect on the inositol 1,4,5-trisphosphate receptor (IP_3R), there is another wrinkle to consider—the effects of PKC phosphorylation on the calmodulin interaction. In the case of TRPL, Warr & Kelly (1996) showed that the calmodulin–TRPL interaction was disrupted by PKA phosphorylation, which in turn was modulated by PKC phosphorylation. You can imagine scenarios whereby if Ca^{2+} levels go up, you can augment calmodulin binding to a TRPC protein and then, as the Ca^{2+} levels rise further, you get PKC phosphorylation and dissociation of calmodulin. Is there any possibility that this is occurring for any of the TRPCs?

Zhu: There are plenty of potential phosphorylation sites. In terms of PKC activation there is a great example from Flockerzi's group. TRPV6 (CaT1) is PKC phosphorylated. In terms of TRPL, the site we have identified is an additional site to the two calmodulin binding sites that are already known. I think we are going to have to deal with them individually: I don't think there is anything common at this point.

Nilius: So is this phosphorylation site in TRPV6 not conserved?

Zhu: It seems that the calmodulin binding sites we have isolated from the TRPCs are somewhat less conserved as compared with some of the other calmodulin binding sites. They are not that typical. Also, the calmodulin binding site that we have measured has a somewhat lower affinity than some of the classical calmodulin binding sites. I would think that the way they are modulated with respect to phosphorylation would be different among different TRPs. Especially for the ones that have multiple calmodulin binding sites, I would think that each might have some different function.

Penner: Anant has worked with relatively high Ca^{2+} concentrations in the cells. At least for I_{CRAC}, which is one store-operated channel, this doesn't seem to make a difference.

Parekh: This is a different protocol: activating CRAC simultaneously with IP_3 and high Ca^{2+}. If you do the other experiment, which is using high Ca^{2+} and then applying carbachol, there is a less effective activation of I_{CRAC}. Part of this is possibly because of Ca^{2+} inactivation of IP_3 receptors.

Putney: Those are problems with Ca^{2+} release. We clamp Ca^{2+} at 0.5 μM, and as long as we can get the Ca^{2+} released by using ionomycin, I_{CRAC} activation seems perfectly normal.

Parekh: We have done that, going up to 1 μM with EGTA to buffer Ca^{2+}. The thapsigargin-evoked CRAC is normal.

Nilius: That's another nice argument that CRAC is not CaT1.

Zhu: Zweifach & Lewis (1995) had an earlier experiment showing that depending on whether EGTA or BAPTA was used in the buffer there was a difference in the rate of Ca^{2+}-dependent inactivation.

Penner: That inactivation is channel limited.

Putney: It is thought to be a site with very low affinity close to the channel.

Zhu: This is clearly something that we haven't been able to resolve whether the so-called common IP_3R calmodulin-binding domain has anything to do with fast inactivation. We haven't been able to measure this kind of inactivation yet with a TRP.

Penner: Rich Lewis has looked at this inactivation mechanism.

Zhu: In his case it is still slow inactivation. What Louis Vaca reported was not a fast inactivation. With respect to voltage-gated Ca^{2+} channels, the calmodulin-binding domain is known to be involved in the fast inactivation. Also, with the CaT1, that was a fast inactivation. The holding potential that they had to use was 70 mV in order not to see fast inactivation ahead of time (Niemeyer et al 2001).

Nilius: There was a very detailed description of the speed of inactivation in TRPV5 and TRPV6 which is related to differences in the linker between transmembrane segments 2 and 3 (Nilius et al 2002).

Fleig: I have a comment with regard to the interaction of several TRPCs with IP_3R domains. It should be noted that at least part of these TRPC channels can be activated by different means not involving necessarily IP_3R conformational changes. With regard to the model where I_{CRAC} or store-operated channels are activated by conformational changes of the IP_3R which is then somehow sensed by the store-operated channel, isn't it more tempting to speculate that it is actually vice versa? That is, the store-operated channel undergoes a conformational change upon store-depletion which is then sensed by the IP_3R possibly modulating its function, like a feedback system.

Zhu: I don't want to give details about how IP_3 senses ER Ca^{2+}, but there is one study by Zorzato's group (Moccagatta et al 2002) in which they showed a protein that has a high-capacity Ca^{2+}-binding domain that is situated in the ER. This protein goes through the ER membrane and has a cytosolic N-terminal head that interacts with the IP_3R. He believes that this protein is serving as a mediator of Ca^{2+} sensing for the IP_3R. If this is true then the IP_3R could be undergoing some kind of conformational change whenever there is a depletion of Ca^{2+} from the ER. In turn, this conformational change could serve as a way of activating the TRP via conformational coupling.

Penner: That is a good point. If we assume that it is the SOC which actually passes Ca^{2+}—and we know that the IP_3Rs have a bell shaped dose–response curve for

Ca^{2+} — that should immediately lower the IP_3R activity down to zero, even though IP_3 would still be present.

Zhu: Yes. We don't really know what kind of conformation IP_3R has at that stage. Does it go back to its original state, or does it stay in a different conformation where the TRPC interaction site is still exposed? The other part of the story is that there is a possibility that a specific pool of IP_3Rs may be involved in activation of the store-operated channel but may not be so important for Ca^{2+} release. If this is true, then by opening these IP_3Rs this part of the cell doesn't really experience any big increase in Ca^{2+}. Therefore there won't be an effect on the calmodulin or anything else that is Ca^{2+} sensitive.

Putney: The problem with making it go the other way is where IP_3Rs are found in cells. In fact, the problem for conformational coupling is that it hasn't been possible to see any IP_3Rs that are particularly concentrated near the plasma membrane. This is not a refutation of the fact that the channels can be coupled with IP_3Rs; it just means that quantitatively most of the IP_3Rs in cells are not close to the channels.

Fleig: So why do they have an interaction domain?

Muallem: There is pretty good evidence that IP_3Rs can be close to the plasma membrane.

Putney: I don't deny that. I'm just suggesting that in terms of all of the IP_3Rs in the cell, only a small proportion are close to channels in the plasma membrane. Therefore, interaction from the channel to the receptor is unlikely to be a major determinant of release kinetics.

Penner: It will be important for the signalling properties of the particular IP_3R it interacts with. If we are assuming localized domains, then that IP_3R is as crucial as the SOC in performing the function that it serves.

Muallem: That is absolutely correct. I think that it is actually how it is happening. In some cells, such as salivary gland cells, we can see that a very tiny amount of the stores and therefore a small proportion of the IP_3Rs are going to be quite important.

Putney: If you are talking about Ca^{2+} coming in through the CRAC channels and regulating the IP_3Rs, and then a small pool of these regulate the CRAC channels, then not only do I agree with this, I have published it! I thought you were talking about a physical interaction such as that occurring between L-type channels and ryanodine receptors that would have an influence on Ca^{2+} release kinetics.

Penner: But if you close that channel you can no longer keep that store empty, because it will refill.

Putney: That is a different kind of interaction. It is an interaction through Ca^{2+}.

Westwick: I have a comment directed more at Craig Montell. Going back to the signalplex, is there actually a change in activation by light in terms of the members that are part of that signalplex, or is it just a phosphorylation change within it that causes a conformational change responsible for the channel activity?

Montell: The short answer is that we don't know if any of the interactions with INAD are light-dependent. There are at least three proteins that seem always to be bound to INAD: TRP, PLC and PKC. They are present in stoichiometric concentrations with INAD. If that interaction is disrupted they become mislocalized. Some of the other players, such as NINAC, may interact dynamically with INAD, though this is very hard to measure. When NINAC isn't bound to INAD its spatial distribution doesn't change: it isn't dependent on INAD for localization. We can see a readout for the NINAC/INAD interaction because there is an effect on termination, but it is hard to tell whether there is a small change in the localization of NINAC. We imagine that there are dynamic interactions with INAD, but we don't have any evidence for them.

Gudermann: What is the role of arrestin in your signalplex? I am confused. There is a lot of arrestin in the cell body which will then translocate to the microvilli. Usually, if you think in terms of receptor desensitization, binding of arrestin to the receptor is regarded as a fast event. If it has to move from the cell body into the microvilli it will take some time. Are timescales simply different in your model, so what is regarded fast for the β adrenergic receptor, is terribly slow for rhodopsin?

Montell: In the dark, about one third of the arrestin is in the rhabdomeres. It is not that there is no rhabdomeral arrestin. After a 5 minute exposure to light, about 80% of the arrestin is concentrated in the rhabdomeres. The arrestin movement doesn't seem to have a role in the rapid adaptation that occurs just after the lights are turned on. Rather, it seems to have a role in long-term adaptation. If flies have been kept in the dark for a long time and you expose the flies to a pulse of light, the termination of the photoresponse is slower than in flies that had been kept in the light. It is this longer-term adaptation that is affected by this arrestin movement, but not the short-term adaptation.

Hardie: To put it another way, even in the dark the amount of arrestin in the microvillus is probably sufficient to bind to the activated rhodopsin within say 100 ms which should be fast enough for its function under dim illumination. In the light adapted situation it is probably important that the rhodopsin is inactivated more rapidly, say within 20–30 ms. Then it may make sense to double or treble the arrestin concentration.

Montell: Light-dependent arrestin shuttling isn't some queer phenomenon that is going on just in fly photoreceptor cells. The same phenomenon is occurring in mammalian rods and cones. This was reported as long ago as 1985, although no one has described what its function is or how it is regulated. This sort of arrestin shuttling and long-term adaptation may also be occurring in hormonally stimulated cells. However, it is easier to observe arrestin translocation in rods, cones and fly photoreceptor cells because these cells are so polarized.

References

Moccagatta L, Treves S, Ronjat M, Mikoshiba K, Zhu X, Zorzato F 2002 Junctate interacts with the $InsP_3R$ and modulates Ca^{2+}entry. Biophys J 82:115A, Part 2

Niemeyer BA, Bergs C, Wissenbach U, Flockerzi V, Trost C 2001 Competitive regulation of CaT-like-mediated Ca^{2+} entry by protein kinase C and calmodulin. Proc Natl Acad Sci USA 98:3600–3605

Nilius B, Prenen J, Hoenderop JG et al 2002 Fast and slow inactivation kinetics of the Ca^{2+} channels ECaC1 and ECaC2 (TRPV5 and TRPV6). Role of the intracellular loop located between transmembrane segments 2 and 3. J Biol Chem 277:30852–30858

Warr CG, Kelly LE 1996 Identification and characterization of two distinct calmodulin-binding sites in the TRPL ion-channel protein of Drosophila melanogaster. Biochem J 314:497–503

Zweifach A, Lewis RS 1995 Rapid inactivation of depletion-activated calcium current (I_{CRAC}) due to local calcium feedback. J Gen Physiol 105:209–226

Activation, subunit composition and physiological relevance of DAG-sensitive TRPC proteins

Thomas Gudermann, Thomas Hofmann, Michael Mederos y Schnitzler and Alexander Dietrich

Institut für Pharmakologie und Toxikologie, Fachbereich Medizin, Philipps-Universität Marburg, Karl-von-Frisch-Str. 1, 35033 Marburg, Germany

Abstract. The classical transient receptor potential (TRP) protein family consists of seven members which share a common gating mechanism contingent on phospholipase C activation. While some family members are thought to be activated subsequent to emptying of intracellular calcium stores, others appear to be gated by as yet undefined lipid messengers. TRPC 3, 6 and 7 form a structural and functional TRPC subfamily characterized by their sensitivity towards diacylglycerols (DAGs). TRPC6 is a non-selective cation channel that is activated by DAG in a membrane-delimited fashion, independently of protein kinase C. Depletion of internal Ca^{2+} stores is not required for TRPC6 activity. TRPC6 mRNA and protein are abundantly expressed in smooth muscle cells and DAG-evoked Ca^{2+} transients can be observed in primary myocytes derived from lung and blood vessels. Thus, TRPC6 is a promising candidate for as yet unidentified non-selective cationic channels in smooth muscle cells potentially involved in vasoconstrictor-activated cation influx and myogenic tone of resistance arteries. Recent systematic studies revealed that TRPC proteins assemble into heteromultimers predominantly within the confines of distinct TRPC subfamilies. The known principles of channel complex formation will be instrumental in assessing the physiological role of distinct TRPC proteins in living cells.

2004 Mammalian TRP channels as molecular targets. Wiley, Chichester (Novartis Foundation Symposium 258) p 103–122

After binding to their cognate receptors on the cell membrane, many hormones, neurotransmitters and growth factors induce increases in $[Ca^{2+}]_i$ in response to phospholipase C (PLC) activation (Berridge et al 2000, Putney & Bird 1993). Apart from inositol 1,4,5-trisphosphate (IP_3)-mediated Ca^{2+} release from intracellular storage organelles, Ca^{2+} permeable plasma membrane ion channels are activated in a receptor- and PLC-dependent manner in most cells. Receptor-stimulated cation channels are gated in response to agonist-binding to a membrane receptor distinct from the channel protein itself. Thus, channel proteins function as integrating effectors receiving information from various classes of cell surface receptors such as G protein-coupled receptors and receptor tyrosine kinases.

Over the past couple of years a large family of mammalian homologues of the *Drosophila* transient receptor potential (TRP) visual transduction channel (Montell & Rubin 1989) have been identified (Clapham et al 2001, Hofmann et al 2000b, Montell 2001, Montell et al 2002a). Based on structural homology and on systematic glycosylation scanning analysis (Vannier et al 1998), TRP proteins are thought to be patterned according to the structural superfamily of six-transmembrane ion channels encompassing most voltage-gated K^+ channels, the cyclic nucleotide-gated channel family, and single transmembrane cassettes of voltage-activated Ca^{2+} and Na^+ channels. Both N- and C-termini of TRP proteins are thought to be located intracellularly, and a putative pore-forming region is bordered by transmembrane domains 5 and 6. Based on primary sequence homology the conventional TRP proteins can be assigned to three subfamilies, TRPC, TRPV and TRPM (Montell et al 2002b). The classical TRPs, the TRPC proteins, are receptor-operated cation influx channels and share the common feature of a gating mechanism contingent on PLC activation. Store-dependent and -independent activation mechanisms have been postulated for nearly each member of the TRPC family, and for several TRPC proteins this is still a moot issue (Hofmann et al 2000b).

The TRPC3/6/7 subfamily

The TRPC subfamily is composed of seven members which can be divided in four groups by means of sequence homology: TRPC1, TRPC4/5, TRPC3/6/7 and TRPC2. While TRPC4 and TRPC5 share approximately 65% identical amino acids, members of the TRPC 3/6/7 group are 70–80% identical. TRPC3 was originally cloned from human embryonic kidney cells (Zhu et al 1996). However, upon close examination of its mRNA expression profile, the gene was found to be predominantly expressed in brain. The full-length cDNA of mouse TRPC6 has been isolated from brain (Boulay et al 1997), while the human orthologue was cloned from placenta (Hofmann et al 1999). Murine TRPC6 is expressed as two splice variants, the shorter one devoid of a 54 amino acid sequence at the extreme N-terminus when compared with the predicted human protein. These findings conform to the isolation of three splice variants of rat TRPC6 differing by 52–68 amino acids at their N-termini (Zhang & Saffen 2001). As opposed to TRPC3, TRPC6 appears to be more widely expressed in extraneural tissues, for instance lung, ovary and spleen, and quite prominently in smooth muscle cells (for review see Hofmann et al 2000b). TRPC7 was identified as the last member of the TRPC3/6/7 subfamily and is found to be widely expressed predominantly in mouse heart, lung and eye with lower transcript levels in brain, spleen and testis (Okada et al 1999). By means of PCR screening, human TRPC7 was reported not to be endogenously present in HEK 293 cells (Riccio et al 2002).

Biophysical properties

TRPC3, TRPC6 and TRPC7 are non-selective cation channels displaying inward and outward rectification. TRPC3 and TRPC6 have a short mean open time of <0.1 ms and single channel conductances of about 66 and 35 pS, respectively (Hofmann et al 1999, Kiselyov et al 2000, Okada et al 1999, Zitt et al 1997). The relative ion permeability P_{Ca}/P_{Na} ranges from 3 to 6. Currents carried by either of the three ion channels are significantly suppressed upon addition of 100–200 μM La^{3+}. The sensitivity towards Gd^{3+}, however, might differ. While TRPC3-mediated cation entry is insensitive to 10 μM Gd^{3+}, an IC_{50} value of approximately 2 μM has been determined for the heterologously expressed recombinant protein as well as for the endogenous channel protein in portal vein myocytes (Inoue et al 2001). On the contrary, there was no significant effect of 100 μM Gd^{3+} on TRPC7-mediated Ca^{2+} influx in HEK293 cells (Okada et al 1999). However, the indicated Gd^{3+} concentrations have to be interpreted with great caution, because IC_{50} values for TRPC6 were determined electrophysiologically, while TRPC3 and TRPC7 were analysed by means of cation-sensitive fluorescent dyes. A systematic comparison of the sensitivity towards lanthanides applying one methodological approach for all family members is still lacking.

The non-specific cation channel blocker flufenamate may represent another pharmacological tool to differentiate between TRPC3, TRPC6 and TRPC7. In HEK293 cells expressing the recombinant mouse TRPC6 as well as in rabbit portal vein myocytes (Inoue et al 2001) and in A7r5 cells endogenously harbouring TRPC6 (Jung et al 2002), flufenamate has been reported to reversibly enhance currents mediated by TRPC6, whereas TRPC3 and TRPC7 were inhibited by the drug. The potentiating effect of flufenamate, however, could neither be reproduced in isolated smooth muscle cells from mouse brain arteries shown to express TRPC6 (M. Mederos y Schnitzler, U. Storch, T. Gudermann, unpublished results), nor in a HEK293 cell line permanently expressing TRPC6 (Basora et al 2003).

Members of the TRPC3/6/7 subfamily are subject to complex regulation by $[Ca^{2+}]_o$. For recombinant proteins, a potentiating effect of decreasing $[Ca^{2+}]_o$ on channel activity has been described for TRPC3 (Lintschinger et al 2000) and TRPC7 (Okada et al 1999) and also appears to apply to the endogenously expressed TRPC6 in A7r5 cells (Jung et al 2002), while ionic currents mediated by the recombinant mouse TRPC6 were increased in Ca^{2+}-containing as opposed to Ca^{2+}-free medium (Inoue et al 2001).

While the members of the TRPC3/6/7 subfamily share many common biophysical properties, they differ remarkably in their constitutive channel activity. When expressed in HEK293 cells, TRPC3 and TRPC7 display elevated basal channel activities which can be substantially potentiated by agonist challenge, but not by emptying of intracellular Ca^{2+} stores (summarized in: Trebak et al 2003b). On the contrary, TRPC6 impresses as a tightly receptor-regulated cation channel with negligible constitutive activity (Hofmann et al 1999). In light of approximately 80% identical amino acid residues within the TRPC3/6/7 subfamily, the molecular determinants for the disparate basal channel activities may additionally reside in post-translational modifications like *N*-linked glycosylation. While glycosylation at a single site in the first extracellular loop of TRPC3 has been demonstrated experimentally (Vannier et al 1998), a second potential *N*-linked glycosylation site can be discriminated in the predicted second extracellular loop of TRPC6 (Dietrich et al 2003). At present, the impact of post-translational modifications on the basal channel activity of TRPC3/6/7 family members is poorly understood, as is the cellular and physiological relevance of constitutive TRPC channel activity. Further cellular and *in vivo* studies with mutated TRPC genes will be enlightening in this regard.

Gating mechanisms

The first study on the expression of TRPC3 in HEK296 cells, entertained the notion that TRPC3 is a receptor-operated as well as a thapsigargin-activated cation channel (Zhu et al 1996), while TRPC3 appeared to be Ca^{2+}-activated, when CHO-K1 cells were chosen as a heterologous expression system (Zitt et al 1997). The issue as to whether TRPC3 should be regarded as a store- or a receptor-operated Ca^{2+} permeable channel has been controversial ever since (summarized in Trebak et al 2003b). In the majority of earlier studies, TRPC3 behaved as a receptor-operated cation channel, and there is experimental support for the hypothesis that the expression level of the channel protein may critically influence its functional characteristics: at a low expression level in DT40 chicken B lymphocytes, TRPC3 was found to be activated by depletion of Ca^{2+} stores, while at higher channel densities in the cell membrane TRPC3 activity was increased through receptor coupling to phospholipase C isoforms (Vazquez et al 2001, 2003).

The concept of a store-operated TRPC3 gained considerable support from the observation that when stably expressed at low levels in HEK293 cells, TRPC3 could be activated exclusively under conditions in which both IP_3 bound to its receptor and depleted Ca^{2+} store organelles were present. In excised membrane patches, TRPC3 currents activated under these conditions were abolished after extensive washing and could be recovered by addition of brain microsomal membranes (Kiselyov et al 1998). These findings lent credibility to the notion of 'conformational coupling' between TRPC3 and the IP_3 receptor. In the aftermath of this seminal series of experiments, interaction domains in the N-terminus of the IP_3 receptor as well as in the C-terminus of TRPC3 were identified (Boulay et al 1999, Zhang et al 2001). Subsequently, a region in the C-terminus of TRPC3 was defined and dubbed CIRB domain (calmodulin-IP_3 receptor-binding domain), because calmodulin and the interacting IP_3 receptor peptide compete for binding at this site. Collectively, these results suggest that calmodulin and the IP_3 receptor regulate TRPC3 channel activity in a competitive manner. However, this point of view is challenged, because in a DT40 cell line devoid of all three forms of IP_3 receptors, TRPC3 is activated by agonist to the same extent as in wild-type cells (Venkatachalam et al 2001) and in the same TRPC3-expressing HEK293 cell line previously used to develop the conformational coupling concept, TRPC3 activation was recently demonstrated to function independently of the IP_3 receptor (Trebak et al 2003a).

These recent findings notwithstanding, interaction of TRPC3 with IP_3 receptors may be involved in the assembly of cellular signalling complexes at the plasma membrane. In accord with this assumption, a systematic mutagenesis study recently revealed that the CIRB domain of TRPC3 is involved in channel targeting to the cell membrane without requiring functional interaction with either calmodulin or IP_3 receptors (Wedel et al 2003). Also, a remarkable body of evidence has accumulated to support the notion that TRPC3 like its close relative TRPC6 is activated by diacylglycerol (DAG) representing a *bona fide* second messenger (see below) (Hofmann et al 1999, Trebak et al 2003a). Thus, the issue of store- versus receptor-operated gating of TRPC3 still remains a highly contentious issue necessitating additional experimental avenues such as genetically modified mice to eventually come up with a solution for this scientific conundrum.

As opposed to the situation with TRPC3, store-operated activation of TRPC6 has never been much of an issue, although high-affinity IP_3 receptor peptide interaction with the TRPC6 C-terminus has been proven biochemically (Tang et al 2001). Functional analysis of mouse TRPC6 can be condensed to the statement that the latter protein functions as a receptor-activated, but not store-operated cation channel (Boulay et al 1997). Following transient transfection of the human channel in CHO-K1 cells, TRPC6 behaved as a receptor-activated non-selective cation channel insensitive to the depletion of internal stores by thapsigargin as well as to the addition of ionomycin or IP_3 (Hofmann et al 1999). Activation of G proteins by AlF_4^- infusion, however, resulted in TRPC6 activation which could by blocked by pretreatment with the PLC inhibitor

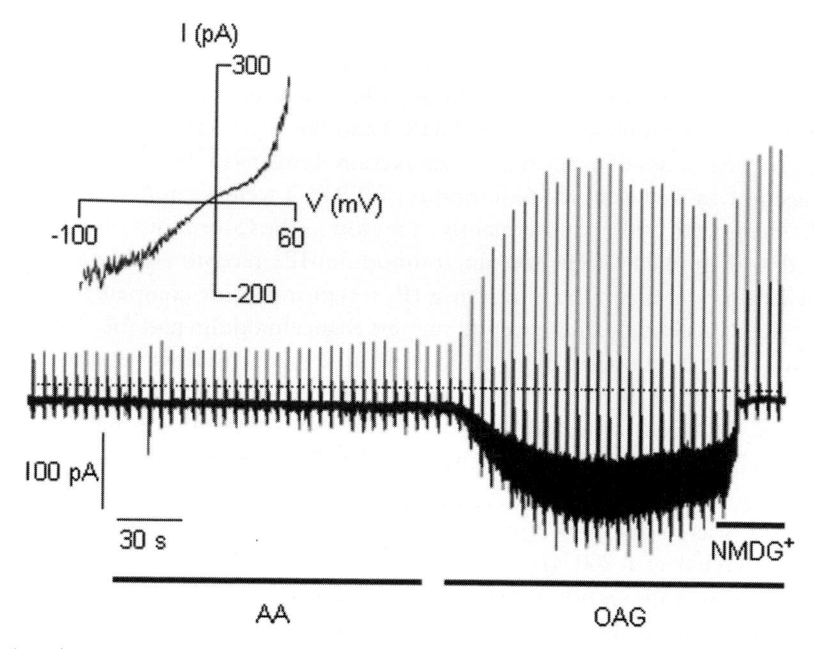

FIG. 1. Activation of TRPC6 by lipid mediators. HEK293 cells transiently expressing human TRPC6 were stimulated with 100 μM arachidonic acid (*AA*) and the 100 μM of the membrane-permeable diacylglycerol analogue 1-oleoyl-2-acetyl-sn-glycerol (*OAG*) for the times indicated. A representative whole cell current recorded at a holding potential of −60 mV is depicted. The regular spikes represent voltage ramps applied at 5 s intervals. A representative current trace obtained during a voltage ramp from −100 to +60 mV at maximal OAG stimulation is shown in the *inset*.

U73122. As TRPC6 activation depended on PLC activity, but could not be mimicked by IP_3 application or store depletion, the role of DAG as an additional second messenger produced by PLC as well as DAG metabolites, e.g. arachidonic acid released from DAG, were characterized further (Fig. 1). In isolated inside-out patches membrane-permeable (OAG) as well as naturally occurring DAGs (SAG, SLG) were able to activate TRPC6 in a membrane-delimited fashion (Hofmann et al 1999). The cellular relevance of the DAG effect was further substantiated by the observation that blockade of endogenous DAG metabolism by the DAG lipase inhibitor RHC80267 was sufficient to profoundly increase the TRPC6-dependent cation influx. Of note, DAG stimulation of TRPC6 is independent of PKC activity as deduced from the ineffectiveness of various PKC inhibitors or down-regulation of PKC by long-term phorbol ester pretreatment.

TRPC3 and TRPC7 are activated by DAG in a similar manner (Hofmann et al 1999, Okada et al 1999, Trebak et al 2003a), while other TRPC proteins are unresponsive to this lipid messenger. Whereas PKC activity is not required for

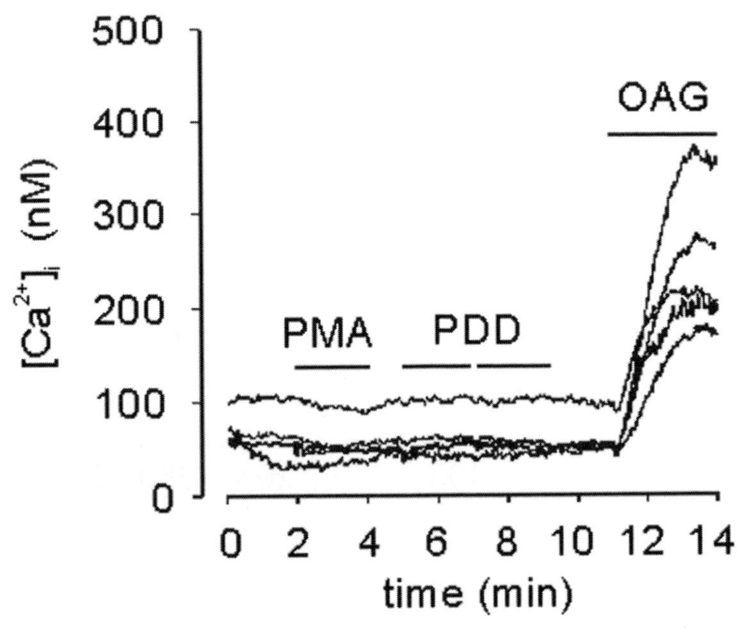

FIG. 2. Lack of phorbol esters to activate TRPC6. HEK293 cells transiently expressing human TRPC6 were loaded with the fluorescent dye fura-2 and stimulated with 10 μM phorbol-12-myristoyl-13-acetate (*PMA*), phorbol-12,13-didecanoate (*PDD*) and 100 μM 1-oleoyl-2-acetyl-sn-glycerol (*OAG*) for the times indicated. The free intracellular Ca^{2+} concentration $[Ca^{2+}]_i$ was monitored in single cells.

channel gating (Fig. 2), it appears to be involved in the negative regulation of TRPC3/6/7 family members. A 10–15 min preincubation with phorbol esters completely abrogates DAG-mediated TRPC3 (Okada et al 1999, Trebak et al 2003a), TRPC6 (Inoue et al 2001, Zhang & Saffen 2001) and TRPC7 activation (Okada et al 1999), suggesting that PKC negatively regulates these cation channels. As yet, serine and/or threonine phosphorylation sites relevant for the latter effect have not been identified in TRPC proteins.

In essence, TRPC3, TRPC6 and TRPC7 form a structural and functional subfamily of second-messenger activated cation channels coupling receptor/PLC signalling pathways to cation entry. Although the latter TRPC proteins are generally classified as DAG-responsive, it is still a matter of debate as to whether DAG is the *bona fide* physiological activator of native channel complexes. There is no doubt that all members of the TRPC3/6/7 family *can* be activated by DAG. When the basal turnover of cellular DAG is blocked by DAG lipase and DAG kinase inhibitors (Hofmann et al 1999, Okada et al 1999, Trebak et al 2003a), an increase in TRPC3/6/7 activity is invariably observed, demonstrating that endogenously generated DAG is sufficient for channel activation. Furthermore,

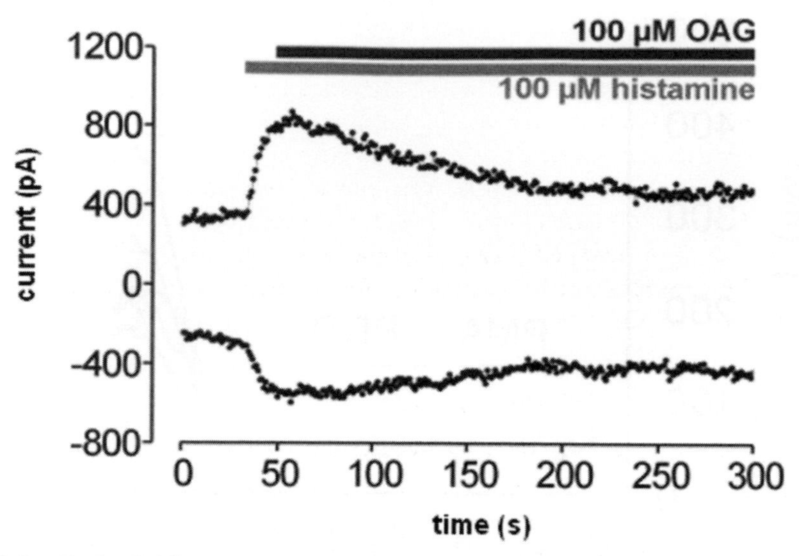

FIG. 3. Lack of additivity of agonist and OAG. HEK293 cells transiently expressing human TRPC6 in conjunction with the $G_{q/11}$-coupled H_1 histamine receptor were stimulated with 100 μM histamine and 100 μM 1-oleoyl-2-acetyl-sn-glycerol (*OAG*) for the times indicated by the horizontal bars over the current traces. Representative whole cell currents recorded at holding potentials of +60 (upper trace) and −60 mV (lower trace) are depicted.

low agonist concentrations insufficient to produce IP_3-induce Ca^{2+} release from intracellular stores elicit significant TRPC3 and TRPC7-mediated Ca^{2+} entry (Okada et al 1999, Trebak et al 2003a). Most notably, receptor agonists and OAG do not display additive effects on TRPC3 and TRPC6 current amplitudes (Trebak et al 2003a; Fig. 3) indicating that the same TRPC channels are activated by OAG and through phospholipase C-coupled receptors. Last but not least, the inhibitory impact of DAG metabolism blockers on the inactivation kinetics of agonist-dependent TRPC3 activity (Trebak et al 2003a) lends further credence to the notion that DAG generated in response to phospholipase C-coupling membrane receptors is indeed the activator of DAG-responsive TRPC channels.

At present, a direct interaction of DAG with TRPC3/6/7 proteins has not been demonstrated. A splice variant of rat TRPC6, TRPC6B, lacking 54 N-terminal amino acids, was reported not to respond to OAG challenge while the full-length TRPC6A did, implicating that the N-terminal region missing from the B isoform is necessary for DAG activation (Zhang & Saffen 2001). It should be noted, however, that none of the other DAG-responsive TRPC channels harbours an extended N-terminus like rat TRPC6A, and that the B isoform was recently shown to respond to OAG challenge (Jung et al 2003). In addition, N-terminal truncations of TRPC3 and TRPC6 which leave the ankyrin repeats intact yield correctly membrane targeted and fully functional ion channels, while larger N-terminal deletions result in intracellular retention of proteins (Hofmann et al 2002, Wedel et al 2003). Future detailed analyses will have to identify potential DAG/TRPC interaction sites. By analogy with the capsaicin interaction sites and TRPV1 (Jordt & Julius 2002), possible candidates in the TRPC3/6/7 family are supposedly located within the first intracellular loop and the neighbouring portions of transmembrane domains 2 and 3 which are oriented towards the cytoplasm. With a mapped DAG binding site still missing, TRPC3/6/7 activation by an additional DAG-binding, for instance C1 domain-containing protein cannot be formally excluded.

Subunit composition of TRP channels

In the study of receptor- or store-operated cation channels in native environments, it has turned out to be difficult to unequivocally ascribe cation channel properties measured in native settings to those of single, heterologously overexpressed TRPC channels. Moreover, some members of the TRPC family, such as TRPC2, seem to be poorly expressed in heterologous expression systems (Hofmann et al 2000a). Thus, heteromultimerization of subunits contributing to the pore properties of native TRPC channels represents an enticing possibility. Functional TRPC channels are thought to be composed of a tetrameric array of identical or different subunits. Biochemical and functional evidence has recently been provided in favour of a homo- and heterotetrameric architecture of TRP channels as worked out for TRPC6 (Hofmann et al 2002), TRPV1 (Kedei et al 2001) and TRPV5/6 (Hoenderop et al 2003).

The cell physiological relevance of heteromeric TRP channel multimers could clearly be demonstrated for certain combinations: coassembly of TRPL and TRPγ, members of the *Drosophila* TRPC family which are constitutively active ion channels when expressed alone, resulted in a tightly regulated, PLC-operated channel (Xu et al 2000). Coexpression of TRPC1 and TRPC3 was reported to give rise to a constitutively active cation conductance raising the possibility that these two TRPC channels might form heteromultimers with functional properties disparate from either channel alone (Lintschinger et al 2000). In neurons, TRPC1 and TRPC4 or TRPC5 are subunits of a receptor-operated heteromeric channel activated independently of store depletion. Coexpression of TRPC1/TRPC5, which have overlapping expression patterns in the hippocampus, gave rise to a novel non-selective cation channel with a voltage dependence similar to that of the NMDA receptor, but clearly set apart from any other reported TRPC channel (Strübing et al 2001). Thus, the heteromeric assembly of TRPC proteins greatly enhances the versatility, but also the complexity of TRPC channels in native environments.

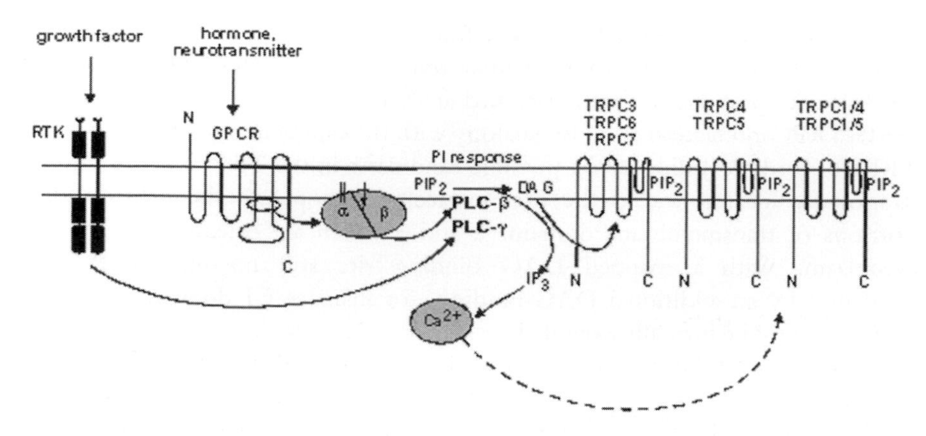

FIG. 4. TRPC channel complexes as integrators of phospholipase C-mediated signals. Cellular signals emanating from receptor tyrosine kinases (RTKs) and from heptahelical G-protein-coupled receptors converge upon distinct permissive combinations of TRPC channels as indicated. Phospholipase Cβ (*PLCβ*) and γ (*PLCγ*) isoforms engaged in the classical phosphatidyl inositol response generate second messengers like diacylglycerol (*DAG*) and inositol trisphosphate (IP_3) which are proposed to activate the channels in a membrane-delimited fashion or by emptying of intracellular Ca^{2+} stores. Note that not all possible TRPC permutations are permissive in living cells.

Recently, the basic principles of TRPC channel homo- and hetero-multimerization in living cells were defined by means of a combination of different experimental approaches: cellular cotrafficking of TRPC subunits, differential functional suppression by dominant-negative subunits, fluorescence resonance energy transfer (FRET) between labelled TRPC subunits, and co-immunoprecipitation (Hofmann et al 2002). The outcome of these experiments was the realization that TRPC2 does not interact with any other known TRPC protein and that TRPC1 has the propensity to form complexes together with TRPC4 and TRPC5, commensurate with the aforementioned electrophysiological findings (Strübing et al 2001). All other TRPCs assemble into multimers only within the narrow confines of TRPC subunits, i.e. TRPC4/5 or TRPC3/6/7 (Fig. 4). By way of a systematic co-immunoprecipitation strategy, the combinatorial rules of TRPC assembly were confirmed in isolated rat brain synaptosomal preparations (Goel et al 2002). However, there might be certain exceptions to this general paradigm: in embryonic, but not in adult rat brain previously unrecognized channel heteromers consisting of TRPC1 pus TRPC4/5 plus TRPC3/6 were identified. This novel combination of TRPC subunits could be reconstituted in HEK293 cells, and was sensitive to TRPC5 dominant negative suppression exclusively when TRPC1 was present (Strübing et al 2003). These findings are at odds with previous observations (Goel et al 2002, Hofmann et al

2002), but may be explained by assuming additional embryonic cell-specific factors which favour and stabilize certain TRPC combinations which are non-permissive in adult tissues. The definition of combinatorial rules governing TRPC complex assembly in cells and native tissues will be invaluable to decipher the puzzling complexity of phospholipase C-dependent cation conductances, thus aiding in the conclusive assessment of the physiological roles of distinct TRPC proteins *in vivo*.

Physiological roles

At present, the available information on the physiological role of DAG-activated TRPC channels is still fairly scant. However, there are a few remarkable leads: TRPC3 is primarily expressed in the mammalian brain in a narrow time window around the time of birth and may participate in activity-dependent changes that occur at this time in development. A close correlation between the spatial and temporal expression pattern of TRPC3 and the receptor tyrosine kinase TrkB activated by brain-derived neurotrophic factor (BDNF) indicated that TRPC3 may be part of a BDNF-induced signalling cascade centrally involving PLCγ and generation of IP_3 (Li et al 1999). In line with this hypothesis, TRPC3 protein could be detected in TrkB immunoprecipitates. In pontine neurons, BDNF activates a Ca^{2+}-dependent, non-selective cation current reminiscent of TRPC3. However, with regard to single channel conductance, mean open time and sensitivity towards IP_3 the channel characterized in pontine neurons clearly differs from heterologously expressed TRPC3. Therefore, at present it is not clear whether TRPC3 in close association with TrkB is the sole molecular correlate mediating BDNF-dependent cation currents or whether other channel subunits or accessory proteins may impart the distinct biophysical properties observed in pontine neurons (Li et al 1999).

DAG- and α_1-adrenoceptor-activated Ca^{2+}-permeable cation channels were described in human prostate cancer epithelial cells (Sydorenko et al 2003, Thebault et al 2003). Based on RT-PCR analysis, TRPC3 is the most likely candidate mediating this response (Sydorenko et al 2003). Future studies will have to address the issue as to whether TRPC-mediated Ca^{2+} influx impacts on prostate cancer cell proliferation, thereby highlighting a novel therapeutic target.

TRPC6 is highly expressed in smooth muscle cells. Together with TRPC3 and TRPC1 it makes up the major complement of cation channels which may underlie the well characterized receptor-operated Ca^{2+}-permeable non-selective cation channels in vascular and airway smooth muscle cells (Inoue et al 2001, Jung et al 2002, Welsh et al 2002, Xu & Beech 2001). More than 20 years have elapsed since the suggestion was made that receptor activation could lead to calcium entry into smooth muscle cells by mechanisms independent of membrane depolarization, and the concept of receptor-operated cation channels was put forward (Large 2002). Receptor-stimulated cation channels are gated in response to agonist binding to a membrane receptor distinct from the channel protein itself. In airway smooth muscle cells, voltage-independent Ca^{2+}-permeable cation channels conduct the lion's share of the Ca^{2+} required for agonist-induced bronchoconstriction. This physiological situation is reflected by the lack of effectiveness of blockers of voltage-gated Ca^{2+} channels in pathophysiological states of increased airway smooth muscle tone such as asthma. In vascular smooth muscle cells, Ca^{2+} influx through non-selective cation channels represents only a minor portion of the overall vasoconstrictor-induced Ca^{2+} entry, but the agonist-induced cation influx is thought to be required for cell membrane depolarization resulting in the activation of voltage-gated Ca^{2+} channels (Large 2002).

A case has recently been made for TRPC6 being the molecularly identified correlate of the vasoconstrictor-activated Ca^{2+}-permeable cation channels (Inoue et al 2001). As characterized in native vascular smooth muscle cells, the latter channels are activated by vasoconstrictors acting at G protein-coupled receptors linked to PLC and by DAG independent of protein kinase C (Large 2002). The biophysical and pharmacological properties of α_1-adrenoceptor-activated non-selective cation channels in rabbit portal vein myocytes have been found to conform to those of TRPC6 expressed in HEK293 cells (Inoue et al 2001). Down-regulation of endogenously expressed TRPC6 in primary portal vein myocytes by way of pretreatment with antisense oligonucleotides markedly inhibited TRPC6-like currents, thus further substantiating the concept of TRPC6 mediating agonist-induced, store-depletion-independent Ca^{2+} entry in vascular myocytes. This notion is further supported by the recent biophysical characterization of vasopressin-activated cation channels in the rat aortic smooth muscle cell line A7r5 (Jung et al 2002).

In addition, TRPC6 has been posited to play a central role in the intravascular pressure-induced depolarization and constriction of brain small arteries and arterioles (Welsh et al 2002). The biophysical characterization of smooth muscle cells after hypo-osmotic swelling as well as the measurement of myogenic tone in isolated resistance arteries pretreated with antisense oligonucleotides directed at TRPC6 strongly supports the assumption that TRPC6 plays an essential role in the regulation of myogenic tone.

Endogenous DAG-activated cation channels in general and TRPC6 specifically have in the meantime been detected in other tissues and cells, in particular in thrombocytes, neutrophils, and lymphocytes. TRPC6 protein is abundantly present in the plasma membrane of blood platelets which display DAG-stimulated Ca^{2+} entry independently of PKC (Hassock et al 2002). Thus, TRPC6 represents the first identified non-store-operated cation channel in platelets. In Jurkat cells as well as in human peripheral blood T lymphocytes, DAG elicits the influx of extracellular cations independently of store depletion. TRPC6 protein was demonstrated in purified plasma membrane fractions of T lymphocytes (Hassock et al 2002), and by RT-PCR TRPC6 mRNA could be detected in both Jurkat cells, peripheral lymphocytes and in neutrophils. At present, cellular functions of blood cells specifically relying on TRPC6 activation have not been examined in great detail.

Future perspectives

TRPC3, TRPC6, and TRPC7 constitute a unique structural and functional TRPC subfamily (see Fig. 4). Since the initial reports on the cloning of the genes, the question as to whether these channels represent store- or receptor-operated cation channels has been surrounded by much controversy, and a comprehensive integrating concept to settle the issue is still elusive. Also, the activation of all three channels by DAG is still not unanimously accepted as the physiological gating mechanism. A major caveat hampering the search for a definite answer to these questions is the fact that at a molecular level channel activation by DAG is not understood.

The controversial issue of channel gating notwithstanding, all members of the TRPC3/6/7 family share many biophysical properties. Therefore, the question arises whether the three gene products are functionally redundant or whether they serve unique and indispensable cellular roles. The genetic inactivation of TRPC3, TRPC6 and TRPC7 in mice will certainly provide invaluable tools to address these issues *in vivo*.

A major drawback for all attempts to define the physiological role of TRPC channels is the complete lack of specific channel blockers. Due to the fact that a number of TRP proteins emerged as attractive novel drug targets, the advent of specific channel blockers can be awaited with optimism.

As outlined above, TRPC6 may be an important novel target for new drug therapies aimed at reducing vascular smooth muscle tone to treat human diseases associated with exuberant vasoconstriction, such as hypertension and vasospasm. Furthermore, in the lung receptor-operated Ca^{2+} entry plays a pivotal role in many cell types like airway smooth muscle cells, neutrophils and lymphocytes which entertain the pathophysiology of asthma and chronic obstructive pulmonary disease. Considering that TRPC6 is functionally active in all these cells, the channel may turn out to be a highly attractive drug target for the treatment of some of the most common chronic diseases.

Acknowledgements

The author's own work reported herein was funded by the Deutsche Forschungsgemeinschaft. Whenever possible, review articles rather than original reports were listed. The author apologizes to all researchers whose work could not be cited due to space limitations.

References

Basora N, Boulay G, Bilodeau L, Rousseau E, Payet MD 2003 20-hydroxyeicosatetraenoic acid (20-HETE) activates mouse TRPC6 channels expressed in HEK293 cells. J Biol Chem 278:31709–31716

Berridge MJ, Lipp P, Bootman MD 2000 The versatility and universality of calcium signalling. Nat Rev Mol Cell Biol 1:11–21

Boulay G, Brown DM, Qin N et al 1999 Modulation of Ca^{2+} entry by polypeptides of the inositol 1,4, 5-trisphosphate receptor (IP3R) that bind transient receptor potential (TRP): evidence for roles of TRP and IP3R in store depletion-activated Ca^{2+} entry. Proc Natl Acad Sci USA 96:14955–14960

Boulay G, Zhu X, Peyton M et al 1997 Cloning and expression of a novel mammalian homolog of Drosophila transient receptor potential (Trp) involved in calcium entry secondary to activation of receptors coupled by the Gq class of G protein. J Biol Chem 272: 29672–29680

Clapham DE, Runnels LW, Strübing C 2001 The TRP ion channel family. Nat Rev Neurosci 2:387–396

Dietrich A, Mederos y Schnitzler M, Emmel A, Kalwa H, Hofmann T, Gudermann T 2003 N-linked protein glycosylation is a major determinant for basal TRPC3 and TRPC6 channel activity. J Biol Chem 278:47842–47852

Goel M, Sinkins WG, Schilling WP 2002 Selective association of TRPC channel subunits in rat brain synaptosomes. J Biol Chem 277:48303–48310

Hassock SR, Zhu MX, Trost C, Flockerzi V, Authi KS 2002 Expression and role of TRPC proteins in human platelets: evidence that TRPC6 forms the store-independent calcium entry channel. Blood 100:2801–2811

Hoenderop JG, Voets T, Hoefs S et al 2003 Homo- and heterotetrameric architecture of the epithelial Ca^{2+} channels TRPV5 and TRPV6. EMBO J 22:776–785

Hofmann T, Obukhov AG, Schaefer M, Harteneck C, Gudermann T, Schultz G 1999 Direct activation of human TRPC6 and TRPC3 channels by diacylglycerol. Nature 397:259–263

Hofmann T, Schaefer M, Schultz G, Gudermann T 2000a Cloning, expression and subcellular localization of two novel splice variants of mouse transient receptor potential channel 2. Biochem J 351:115–122

Hofmann T, Schaefer M, Schultz G, Gudermann T 2000b Transient receptor potential channels as molecular substrates of receptor-mediated cation entry. J Mol Med 78:14–25

Hofmann T, Schaefer M, Schultz G, Gudermann T 2002 Subunit composition of mammalian transient receptor potential channels in living cells. Proc Natl Acad Sci USA 99:7461–7466

Inoue R, Okada T, Onoue H et al 2001 The transient receptor potential protein homologue TRP6 is the essential component of vascular alpha(1)-adrenoceptor-activated Ca^{2+}-permeable cation channel. Circ Res 88:325–332

Jordt SE, Julius D 2002 Molecular basis for species-specific sensitivity to "hot" chili peppers. Cell 108:421–430

Jung S, Muhle A, Schaefer M, Strotmann R, Schultz G, Plant TD 2003 Lanthanides potentiate TRPC5 currents by an action at extracellular sites close to the pore mouth. J Biol Chem 278:3562–3571

Jung S, Strotmann R, Schultz G, Plant TD 2002 TRPC6 is a candidate channel involved in receptor-stimulated cation currents in A7r5 smooth muscle cells. Am J Physiol Cell Physiol 282:C347–C359

Kedei N, Szabo T, Lile JD et al 2001 Analysis of the native quaternary structure of vanilloid receptor 1. J Biol Chem 276:28613–28619

Kiselyov K, Xu X, Mozhayeva G et al 1998 Functional interaction between InsP3 receptors and store-operated Htrp3 channels. Nature 396:478–482

Kiselyov KI, Shin DM, Wang Y, Pessah IN, Allen PD, Muallem S 2000 Gating of store-operated channels by conformational coupling to ryanodine receptors. Mol Cell 6:421–431

Large WA 2002 Receptor-operated Ca^{2+}-permeable nonselective cation channels in vascular smooth muscle: a physiologic perspective. J Cardiovasc Electrophysiol 13:493–501

Li HS, Xu XZ, Montell C 1999 Activation of a TRPC3-dependent cation current through the neurotrophin BDNF. Neuron 24:261–273

Lintschinger B, Balzer-Geldsetzer M, Baskaran T et al 2000 Coassembly of Trp1 and Trp3 proteins generates diacylglycerol- and Ca2+-sensitive cation channels. J Biol Chem 275:27799–27805

Montell C 2001 Physiology, phylogeny, and functions of the TRP superfamily of cation channels. Sci STKE 2001:RE1

Montell C, Rubin GM 1989 Molecular characterization of the Drosophila trp locus: a putative integral membrane protein required for phototransduction. Neuron 2:1313–1323

Montell C, Birnbaumer L, Flockerzi V 2002a The TRP channels, a remarkably functional family. Cell 108:595–598

Montell C, Birnbaumer L, Flockerzi V et al 2002b A unified nomenclature for the superfamily of TRP cation channels. Mol Cell 9:229–231

Okada T, Inoue R, Yamazaki K et al 1999 Molecular and functional characterization of a novel mouse transient receptor potential protein homologue TRP7. Ca^{2+}-permeable cation channel that is constitutively activated and enhanced by stimulation of G protein-coupled receptor. J Biol Chem 274:27359–27370

Putney JW Jr, Bird GS 1993 The inositol phosphate-calcium signaling system in nonexcitable cells. Endocr Rev 14:610–631

Riccio A, Mattei C, Kelsell RE et al 2002 Cloning and functional expression of human short TRP7, a candidate protein for store-operated Ca2+ influx. J Biol Chem 277:12302–12309

Strübing C, Krapivinsky G, Krapivinsky L, Clapham DE 2001 TRPC1 and TRPC5 form a novel cation channel in mammalian brain. Neuron 29:645–55

Strübing C, Krapivinsky G, Krapivinsky L, Clapham DE 2003 Formation of novel TRPC channels by complex subunit interactions in embryonic brain. J Biol Chem 278:39014–39019

Sydorenko V, Shuba Y, Thebault S et al 2003 Receptor-coupled, DAG-gated Ca2+-permeable cationic channels in LNCaP human prostate cancer epithelial cells. J Physiol 548:823–836

Tang J, Lin Y, Zhang Z, Tikunova S, Birnbaumer L, Zhu MX 2001 Identification of common binding sites for calmodulin and inositol 1,4,5-trisphosphate receptors on the carboxyl termini of trp channels. J Biol Chem 276:21303–21310

Thebault S, Roudbaraki M, Sydorenko V et al 2003 Alpha1-adrenergic receptors activate Ca^{2+}-permeable cationic channels in prostate cancer epithelial cells. J Clin Invest 111:1691–1701

Trebak M, St JBG, McKay RR, Birnbaumer L, Putney JW, Jr 2003a Signaling mechanism for receptor-activated canonical transient receptor potential 3 (TRPC3) channels. J Biol Chem 278:16244–16252

Trebak M, Vazquez G, Bird GS, Putney JW 2003b The TRPC3/6/7 subfamily of cation channels. Cell Calcium 33:451–461

Vannier B, Zhu X, Brown D, Birnbaumer L 1998 The membrane topology of human transient receptor potential 3 as inferred from glycosylation-scanning mutagenesis and epitope immunocytochemistry. J Biol Chem 273:8675–8679

Vazquez G, Lievremont JP, St JBG, Putney JW, Jr 2001 Human Trp3 forms both inositol trisphosphate receptor-dependent and receptor-independent store-operated cation channels in DT40 avian B lymphocytes. Proc Natl Acad Sci U S A 98:11777–11782

Vazquez G, Wedel BJ, Trebak M, St John Bird G, Putney JW Jr 2003 Expression level of the canonical transient receptor potential 3 (TRPC3) channel determines its mechanism of activation. J Biol Chem 278:21649–21654

Venkatachalam K, Ma HT, Ford DL, Gill DL 2001 Expression of functional receptor-coupled TRPC3 channels in DT40 triple receptor InsP3 knockout cells. J Biol Chem 276:33980–33985
Wedel BJ, Vazquez G, McKay RR, St JBG, Putney JW Jr 2003 A calmodulin/inositol 1,4,5-trisphosphate (IP3) receptor-binding region targets TRPC3 to the plasma membrane in a calmodulin/IP3 receptor-independent process. J Biol Chem 278:25758–25765
Welsh DG, Morielli AD, Nelson MT, Brayden JE 2002 Transient receptor potential channels regulate myogenic tone of resistance arteries. Circ Res 90:248–250
Xu SZ, Beech DJ 2001 TrpC1 is a membrane-spanning subunit of store-operated Ca^{2+} channels in native vascular smooth muscle cells. Circ Res 88:84–87
Xu XZ, Chien F, Butler A, Salkoff L, Montell C 2000 TRPgamma, a drosophila TRP-related subunit, forms a regulated cation channel with TRPL. Neuron 26:647–257
Zhang L, Saffen D 2001 Muscarinic acetylcholine receptor regulation of TRP6 Ca^{2+} channel isoforms. Molecular structures and functional characterization. J Biol Chem 276:13331–13339
Zhang Z, Tang J, Tikunova S et al 2001 Activation of Trp3 by inositol 1,4,5-trisphosphate receptors through displacement of inhibitory calmodulin from a common binding domain. Proc Natl Acad Sci USA 98:3168–3173
Zhu X, Jiang M, Peyton M et al 1996 TRP, a novel mammalian gene family essential for agonist-activated capacitative Ca^{2+} entry. Cell 85:661–671
Zitt C, Obukhov AG, Strübing C et al 1997 Expression of TRPC3 in Chinese hamster ovary cells results in calcium-activated cation currents not related to store depletion. J Cell Biol 138:1333–1341

DISCUSSION

Montell: Even though the *in vitro* data that TRPC3 and TRPC6 are activated by DAG look very good, TRPC3 may not be activated by DAG *in vivo*. Several years ago we looked at an endogenous TRPC3 conductance in pontine neurons. This conductance, although activated by BDNF and through a PLCγ pathway was not at all activated by OAG or DAG. It also wasn't activated in the whole-cell mode by passive Ca^{2+} release using thapsigargin. IP_3 did activate the endogenous TRPC3 current but it wasn't as long-lasting as the BNDF induced current. I am interested to hear from you that TRPC3 and 6 interact. Are the TRPC3/6 heteromultimers activated by DAG or OAG? The obvious resolution of this conundrum is that TRPC3 is activated by DAG in your experiments because you are analysing TRPC3 homomultimers and we were looking at heteromultimers consisting of TRPC3 and some other channel, such as TRPC6. Now that there is a TRPC6 knockout mouse in your lab, have you looked for an endogenous wild-type conductance that is not in the knockout mouse that you would presume to be TRPC6. Is this current activated by DAG?

Gudermann: We have only done some quick and dirty experiments trying to co-express TRPC3 and TRPC6, and then comparing activation mechanisms. We found no major differences. But if we really want to know whether there is an effect, we have to generate concatamers in order to build our channel complex. We are in the process of doing this. We have the constructs, but we haven't analysed them yet. I know that there are discrepancies between what you found in pontine neurons and what we see in heterologous cell systems, but we feel quite confident about our results. If you look at Dr Mori's data on smooth muscle cells, for instance, he sees a very high degree of similarity between what he describes as the α_1-adrenoceptor-activated non-selective cation channel and TRPC6. This is exactly what we see. Then, as far as knockout mice are concerned, we have looked in isolated smooth muscle cells at whether we can find channels there that are activated by OAG. We do find them there. According to our interpretation, this is TRPC3 that we see there.

Montell: You showed that TRPC3 RNA is up-regulated. Is the protein up-regulated?

Gudermann: The protein is there. I am not sure that we can say that there is more protein there compared with the wild-type on the basis of our Western blots. We are working on this. But in terms of function we see increased basal activity in smooth muscle cells from these knockout mice, and we have an ion channel very similar to TRPC3 that can be activated by OAG.

Montell: I would argue that in the knockout mice which lack TRPC6, that if indeed TRPC3 is up-regulated and if TRPC3 normally heteromultimerizes with TRPC6, that now you are just looking at a TRPC3 homomultimer which perhaps doesn't normally exist. This could be an artificial situation that is being generated *in vivo*.

Gudermann: In wild-type smooth muscle cells a number of labs have found OAG-induced cation currents.

Nilius: Is there increased basal Ca^{2+} in the knockouts?

Gudermann: There maybe is slightly increased basal Ca^{2+}. When we look at the basal influx of cations, however, there is a dramatic difference in that smooth muscle cells from knockout mice show a much higher basal cation influx compared to wild-type animals.

Gill: Craig Montell, do you think the TRPC3/6 heteromultimer is more sensitive to OAG, and when you homomultimerize with TRPC3 it is not so sensitive?

Montell: I was just asking: I don't actually know. I was proposing that perhaps the 3/6 heteromultimer was not activated by OAG. But it sounds like something is being activated by OAG *in vivo*.

Gill: So there is definitely an endogenous OAG activation, but not a big difference in the knockouts versus the wild-type.

Hardie: You showed that TRPC6 was not activated by arachidonic acid. Have you checked whether TRPC3 and 7 are also specifically activated by DAG and not by arachidonic acid?

Gudermann: No, we haven't done this. We only did one experiment on TRPC6.

Nilius: We tried TRPC3 and it is not activated by arachidonic acid (B. Nilius, M. Kamouchi, unpublished results).

Ambudkar: You showed a trace from a TRPC6 expression experiment in which you stimulated with OAG followed by agonist and in the reverse order. It looked as if when OAG was followed by the histamine, the OAG signal was going up and then it started coming down when histamine was added. In the reverse experiment it stayed up. Does this mean anything? Is it a consistent result? It almost looked as if the agonist was bringing down the OAG response.

Gudermann: These types of experiments haven't been done extensively. Our main focus has been whether or not we see additional activation. One can hypothesize that activation then contributes to the off rate, but we have not explored that.

Hardie: Can it be Ca^{2+}-dependent inactivation?

Gudermann: We haven't looked in detail.

Ambudkar: The basal activity is increased in the knockout mice. I don't understand the basis of this increase, i.e. is it mediated via spontaneously activated TRPC3 or TRPC6 channels? During α_1-stimulation are you sure the effect is not on the basal component. Do you see high basal activity due to TRPC3 or C6 even in the wild-type.

Gudermann: We never get high basal activity with TRPC6.

Gill: In the knockout, when you have this overexpression of TRPC3, did you have more basal activity?

Gudermann: Yes.

Westwick: What is the obvious phenotype for the TRPC6 knockout?

Gudermann: I showed you the phenotype that we are really sure of.

Westwick: Do they have normal leukocyte counts and blood pressure, for example?

Gudermann: Whatever else we looked at is not dramatically different from the wild-type. We started to look at the cardiovascular phenotype and there also appears to be an increased sensitivity to vasoconstrictors. But it is far too early to tell you anything in detail.

Nilius: Do you have any idea about a possible binding site for DAG?

Gudermann: No. We took the obvious approach of trying to construct chimeric channels, changing amino acids. We constructed a lot of these chimeras and they are retained intracellularly. Many of these didn't make it to the cell membrane.

Nilius: Are your channels modulated by Ca^{2+}?

Gudermann: We haven't studied Ca^{2+}-dependent inactivation systematically. In most of our electrophysiological experiments Ca^{2+} is buffered to a low concentration.

Schilling: I have a question about the mechanism of DAG activation. There is another hypothesis: that these channels are activated not by generation of IP_3 or DAG, but actually by hydrolysis of phosphatidyl inositol 4,5-bisphosphate (PIP_2). The idea is that PIP_2 binding favours the closed state of the channels and that PIP_2

hydrolysis favours activation. I have two specific questions. First, what is the effect of inhibiting PLC activity with U73122 on OAG activation of TRPC6? Is OAG activating a PLC that could then be depleting the cell membrane of PIP_2?

Gudermann: If we add OAG afterwards we see manganese influx.

Schilling: For *Drosophila* TRPL expressed in insect Sf9 cells, we saw dramatic channel activation by receptor stimulation. We also observed dramatic activation by SAG or linoleic acid. Both the increase in TRPL currents and Ca^{2+} signal observed with SAG and LLA, were partially and specifically blocked by U73122 and not by U73343. Apparently the effect of SAG and LLA on TRPL occurred at least in part through activation of PLC. This led to the hypothesis that perhaps it is the hydrolysis of PIP_2 that is responsible for activation of these channels, rather than the generation of IP_3 or DAG. In fact, application of PIP_2 to TRPL single channels in inside-out patches caused inhibition of channel activity. The question is whether the same mechanism would apply to the mammalian TRPCs: that is, are they inhibited by interaction with PIP_2 and subsequently activated by hydrolysis of PIP_2 via a specific PLC in close association with the channel? I'm not suggesting that DAG doesn't play a role. It could remain bound to the channel, but the hydrolysis of channel associated PIP_2 by PLC is actually the initial event that activates the channel. My second question is what is the effect of DAG lipase inhibitor on PIP_2 concentrations in the membrane?

Gudermann: We did not measure PIP_2 concentrations in the membrane. The only experiment we did was to add PIP_2 in inside-out patches. With this approach we didn't see a big effect, but we can't rule out the hypothesis you put forward.

Montell: In photoreceptor cells that express a derivative of PLC that doesn't bind to INAD, and is therefore not tethered close to the TRP channels, you get normal activation. This would argue that it is not just hydrolysis of a localized PIP_2 which activates TRP.

Hardie: The experiment referred to by Bill Schilling has actually been reported for TRPC3. Namely, U73122 had no effect on OAG activation of the TRPC3 channels in DT40 cells (Venkatachalam et al 2001).

Li: You said that the TRPC6 knockout mice had an increased vasoconstrictor response. In terms of the TRPC profile in vascular smooth muscle cells do you still see the enhanced expression of the TRPC3 that you saw in the airway smooth muscle cells?

Gudermann: At the mRNA level, yes.

Li: What about TRPC7?

Gudermann: There is not much change.

Zhu: Do you know the source of endogenous DAG? In your experiment using the DAG lipase inhibitor you saw increased activity. Do you think there is a basal turnover of PLC? Do you know whether that is specifically phosphatidylinositol-PLC, or could it be PLC that breaks down PC (phosphatidylcholine), for example?

Gudermann: Most probably the latter. There is a basal production of DAG.

Zhu: Did you test it at the other glycosylation site? You said it is only the first site glycosylated. What if you have only the second site: do you also get baseline increase?

Gudermann: Yes. Removal of any glycosylation site will increase basal activity (Dietrich et al 2003).

Authi: Have you looked to see whether DAG binds to any component of TRPC3, 6 or 7 protein? Are there any defined binding sites?

Gudermann: There are no canonical known binding sites for DAG on these proteins. Even if one believes in DAG, it is still an open question whether DAG binds to the channels or binds to a protein which then interacts with the channels. We try to look at complexes that are formed to see whether any of the other so far unknown players is responsible for the DAG effect.

Gill: There is a site that seems to come up whenever we look through the sequence: this is a DAG kinase-like sequence near the N-terminus.

Gudermann: There is a hydrophobic stretch at the very N-terminus. When we deleted this out we had a channel that was nicely trapped intercellularly. This was then our tool to do the co-trafficking experiments.

References

Dietrich A, Mederos y Schnitzler M, Emmel A, Kalwa H, Hofmann T, Guderman T 2003 N-linked protein glycosylation is a major determinant for basal TRPC3 and TRPC6, channel activity. J Biol Chem 278:47842–47852

Venkatachalam K, Ma H-T, Ford DL, Gill DL 2001 Expression of functional receptor-coupled TRPC3 channels in DT40 triple receptor InsP3 knockout cells. J Biol Chem 276: 33980–33985

Signalling mechanisms for TRPC3 channels

James W. Putney Jr, Mohamed Trebak, Guillermo Vazquez, Barbara Wedel and Gary St.J. Bird

Calcium Regulation Section, Laboratory of Signal Transduction, National Institute of Environmental Health Sciences, National Institutes of Health, Department of Health and Human Services, PO Box 12233, Research Triangle Park, NC 27709, USA

Abstract. The putative ion channel subunits TRPC3, TRPC6 and TRPC7 comprise a structurally related subgroup of the family of mammalian TRPC channels. As is the case for the founding member of the TRPC family, *Drosophila* TRP, the ion channels formed by these proteins appear to be activated in some manner downstream of phospholipase C (PLC). Earlier studies indicating that TRPC3 could be activated by depletion of intracellular stores (i.e. that it is a store-operated channel, SOC) were subsequently shown to be attributable to constitutive activity of the channels. Studies on the mechanism of activation of TRPC6 and TRPC7 indicated that PLC-dependent activation involved diacylglycerol and was independent of G proteins or inositol 1,4,5-trisphosphate (IP_3). Although TRPC3 can also be activated by diacylglycerols, there is evidence suggesting that these channels can be activated by IP_3 and the IP_3 receptor through a conformational coupling mechanism. We have re-examined the activation mechanism for TRPC3 in mammalian cells by using HEK293 cell lines stably expressing human TRPC3. Our data indicate that, like TRPC6 and TRPC7, TRPC3 is activated by PLC-generated diacylglycerol and is independent of G proteins or IP_3. However, in an avian pre-B cell line, TRPC3 can function either as a diacylglycerol-activated channel, or as a SOC. The mechanism of regulation of TRPC3 in this cell line appears to be related to the level of expression of the protein.

2004 Mammalian TRP channels as molecular targets. Wiley, Chichester (Novartis Foundation Symposium 258) p 123–139

Mammalian homologues of *Drosophila* TRP

TRP is a *Drosophila* photoreceptor mutant incapable of maintaining a sustained receptor potential in response to photo stimulation (Cosens & Manning 1969). Because insect photoreceptors utilize a phospholipase C (PLC) signalling system, this phenotype suggested to Hardie & Minke (1993) that TRP might function as a component of a PLC-dependent Ca^{2+} entry pathway. When the sequence of TRP was determined (Montell & Rubin 1989), homology to the sequence of mammalian voltage-dependent Ca^{2+} channels was noted (Phillips et al 1992). Subsequently, a variety of strategies led to the identification and cloning of seven mammalian homologues of *Drosophila* TRP, now termed TRPC1–TRPC7. Within the TRPC family, four subfamilies can be identified based on structural and functional similarity: TRPC1, TRPC2, TRPC3/6/7 and TRPC4/5. In addition, the TRPC family is a component of a larger superfamily of more distantly related TRP-like genes that encode for ion channels with a variety of cellular functions (Birnbaumer et al 1996, Harteneck et al 2000, Montell et al 2002, Minke & Cook 2002, Clapham et al 2001, Vennekens et al 2002, Zitt et al 2002).

The predominant mechanism by which PLC-linked receptors activate Ca^{2+} entry across the plasma membrane is a process known as *capacitative Ca^{2+} entry* (CCE) or *store-operated Ca^{2+} entry* (SOCE) (Putney 1997). CCE occurs when inositol 1,4,5-trisphosphate (IP_3) or some other signal discharges Ca^{2+} from intracellular endoplasmic reticulum (ER) stores; the subsequent fall in ER Ca^{2+} concentration then signals to the plasma membrane activating capacitative Ca^{2+} entry, or store-operated channels (SOCs). Neither the mechanism for this retrograde signalling, nor the molecular identity of the SOCs is known with certainty (see Putney & McKay 1999, Barritt 1999, Putney et al 2001, Venkatachalam et al 2002). There are two general theories for the mechanism of signalling CCE. One theory suggests the formation and release of a diffusible messenger from ER which would activate SOCs (Randriamampita & Tsien 1993). The second theory is known as conformational coupling (Irvine 1990, Berridge 1995); specialized IP_3 receptors lying just below the plasma membrane are suggested to make direct contact with Ca^{2+} channels in the plasma membrane. Depletion of Ca^{2+} from the underlying ER would cause a change in the conformation of the IP_3 receptor, which would interact with SOCs, resulting in their activation.

Initial studies suggested that TRP channels might be involved in CCE. However, in most instances, subsequent studies have demonstrated that the apparent SOCE results from constitutive entry (Zhu et al 1998, McKay et al 2000, Trebak et al 2002). However, TRPC3 was clearly shown to behave as a SOC in one study (Vazquez et al 2001), although in most instances it is clearly not store-operated. The mechanisms by which TRPC3 is regulated under different experimental, and possibly biological conditions is the primary focus of this review.

Regulation of TRPC3 by phospholipase C

TRPC3 and TRPC1 were the first of the mammalian TRPs shown to be activated by an agonist (Zhu et al 1996). Transient expression of TRPC3 substantially augmented Ca^{2+} entry in response to activation of PLC-linked receptors.

Thapsigargin, an inhibitor of the ER Ca^{2+} pump, activates CCE, and this response appeared to be increased as well (Zhu et al 1996). However, in a subsequent study (Zhu et al 1998), it was shown that the apparent increase in entry in response to thapsigargin could be attributed to an enhanced entry through constitutively active channels. In fact, the apparent increased entry of Ca^{2+} in thapsigargin-treated TRPC3 cells likely results from a combination of constitutive entry and diminished buffering of Ca^{2+} by the poisoned ER pumps (Trebak et al 2002). Currently, the majority of published reports on TRPC3 activation conclude that the channel is not regulated by depletion of Ca^{2+} stores, at least in mammalian cells (Zitt et al 1997, Zhu et al 1998, Li et al 1999, Ma et al 2000, McKay et al 2000, Venkatachalam et al 2001).

A number of reports have concluded that gating of TRPC3 channels involves IP_3 and the IP_3 receptor. Kiselyov et al (1998) studied TRPC3 channels stably expressed in HEK293 cells, and demonstrated that single TRPC3 channels could be activated by a cholinergic agonist. Upon excision of the patch, channel activity diminished. This activity could be restored by addition of IP_3 to the bath. With more prolonged washing of the excised patch, IP_3 was not sufficient to activate the channels, rather both IP_3 and the IP_3 receptor were required. In a subsequent study (Kiselyov et al 1999), this same group demonstrated that the N-terminal, IP_3-binding domain of the IP_3 receptor was sufficient for activation of the channels. These authors proposed that TRPC3 is regulated by a mechanism similar to the conformational coupling mechanism proposed for SOCs. In the studies of Kiselyov et al, it would seem that IP_3 plays a more significant and direct role in gating the channels, and a role for depletion of Ca^{2+} stores is not well established.

Despite the uncertainty regarding the relationship of TRPC3 gating to mechanisms of CCE, from the work of Kiselyov et al (Kiselyov et al 1998, 1999), the role of IP_3 and the IP_3 receptor would seem solidly established. Following on this idea, Boulay et al (1999) used a GST-pulldown strategy to identify an IP_3 receptor binding sequence in the C-terminal region of TRPC3, and two TRPC3 binding sequences in the IP_3 receptor. Transient expression of the peptide sequence from TRPC3 inhibited both agonist- and thapsigargin-induced entry. One of the IP_3 receptor-binding peptides inhibited entry, while the other potentiated the response. The results are consistent with the conclusions of Kiselyov et al (1998, 1999): direct, protein–protein interaction between TRPC3 and underlying IP_3 receptors is involved in the TRPC3 activation mechanism. Subsequent work demonstrated that a calmodulin-binding domain substantially overlaps with the IP_3 binding domain, suggesting a regulatory mechanism through which IP_3 receptor and calmodulin compete for a common calmodulin–IP_3 receptor binding (CIRB) site (Zhang et al 2001).

The other product of PLC-mediated breakdown of phosphatidylinositol 4,5-bisphosphate (PIP_2) is diacylglycerol (DAG). All three members of the

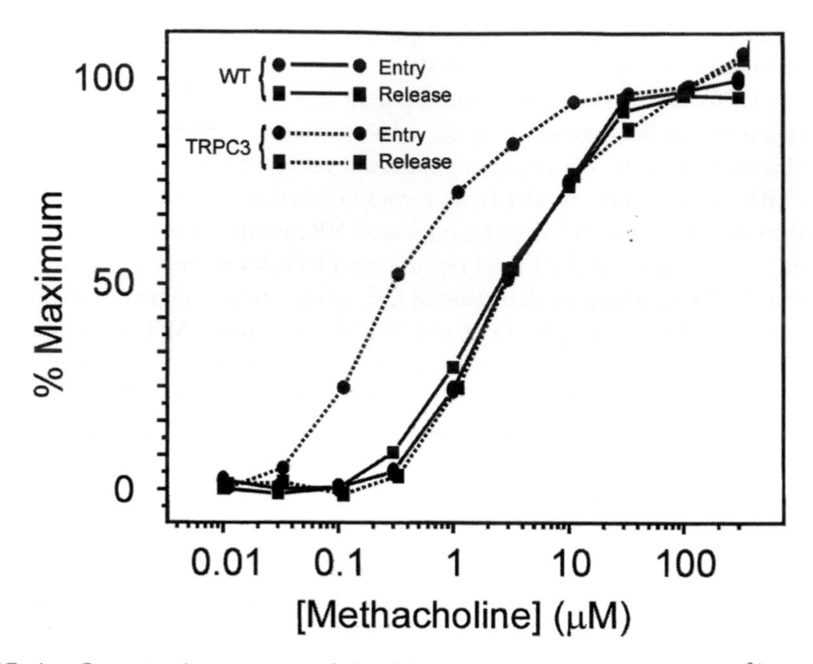

FIG. 1. Concentration-response relationships for methacholine-stimulated Ca^{2+} release and entry through endogenous CCE channels in wild-type (WT) HEK-293 cells, and release and entry through TRPC3 channels in TRPC3-expressing HEK-293 cells. For the WT cells, the curves for release and entry are superimposable, consistent with the view that for CCE, it is the extent of Ca^{2+} store depletion that is responsible for entry. For the TRPC3 cells, the curve for entry lies one order of magnitude to the left of that for release, suggesting that signals other than IP_3-mediated release may be involved in the activation of TRPC3 channels. Redrawn from data originally presented in Trebak et al (2003).

TRPC3/6/7 subfamily are activated by membrane permeant DAG, such as 1-oleoyl-2-acetyl-*sn*-glycerol (OAG) (Hofmann et al 1999, Okada et al 1999). In experiments utilizing whole-cell patch-clamp, no activation of TRPC6 or TRPC7 by IP_3 was observed (Hofmann et al 1999, Okada et al 1999, Inoue et al 2001). This distinct difference from TRPC3 is surprising given the close structural homology between TRPC3, TRPC6 and TRPC7.

The dose–response relationships for muscarinic receptor agonist activation of both CCE and TRPC3 are shown in Fig. 1. For non-transfected cells, the dose–response relationship for activation of release and entry coincide, consistent with the idea that the entry occurs through CCE and is causally related to the extent of depletion of intracellular Ca^{2+} stores. For TRPC3 cells, entry is considerably more sensitive to low concentrations of agonist than is release. This implies that entry can be activated by phospholipase C activation that is insufficient to produce enough IP_3 to activate IP_3 receptors. Thus, we sought to determine whether signals other than IP_3 might be responsible for entry in TRPC3-transfected HEK293 cells.

We measured TRPC3 activity by examining fluorescent signals from intracellular Ca^{2+} indicators, or by measuring membrane current with the patch-clamp technique in the whole-cell configuration. We introduced IP_3 into TRPC3-expressing HEK293 cells by including it in the patch pipette in concentrations ranging from 2 μM to 1 mM. The non-leak subtracted current in TRPC3-expressing cells was significantly greater than in wild-type cells (control: 65 ± 7 pA; TRPC3: 178 ± 14 pA). The presence of IP_3 in the pipette did not further increase this current (161 ± 17 pA), nor did it increase Ca^{2+} entry (Trebak et al 2003). However, subsequent addition of a phospholipase C-coupled agonist caused a dramatic increase in membrane current and in $[Ca^{2+}]_i$ (Fig. 2; Trebak et al 2003). Inclusion of the IP_3 receptor antagonist, heparin, in the pipette completely blocked the release of Ca^{2+} stores by a muscarinic agonist, as well as Ca^{2+} entry in wild-type cells. However, the activation of Ca^{2+} entry in TRPC3 cells was unaffected.

These results indicate that the mechanism by which agonists for phospholipase C-linked receptors activate TRPC3 in HEK293 cells does not involve IP_3. One possibility is that a component of the heterotrimeric G protein which couples receptors to PLC is also capable of activating TRPC3. However, we found that epidermal growth factor, which activates PLC through tyrosine phosphorylation, was also an efficient activator of TRPC3. Thus, given the observation that DAG is an efficient activator of all three members of the TRPC3/6/7 subfamily, DAG produced from PLC cleavage of PIP_2 would seem to be a likely candidate for the signal linking PLC to TRPC3 channels in HEK293 cells. In support of this conclusion, inhibitors of DAG metabolism caused activation of TRPC3, and slowed the off-rate of agonist-activated TRPC3 when the PLC-linked receptors were blocked (Trebak et al 2003).

TRPC3 can be store-operated or DAG-regulated, depending on expression level

To further investigate the mechanism(s) by which TRPC3 channels are regulated, we expressed TRPC3 in the DT40 avian B lymphocyte cell line, as well as in a variant of DT40 lacking IP_3 receptors (DT40-IP_3RKO) (Sugawara et al 1997). In DT40, CCE does not require IP_3 receptors (Sugawara et al 1997). Surprisingly, and in marked contrast to the results in HEK293 cells, in DT40 transient expression of TRPC3 resulted in the appearance of a store-operated Ba^{2+} entry which was not observed in control or mock-transfected cells (Vazquez et al 2001) (Fig. 3). This entry was reduced by about 50% in DT40-IP_3RKO cells, leading us to conclude that TRPC3 appears to form SOCs, some of which are dependent on IP_3R, and

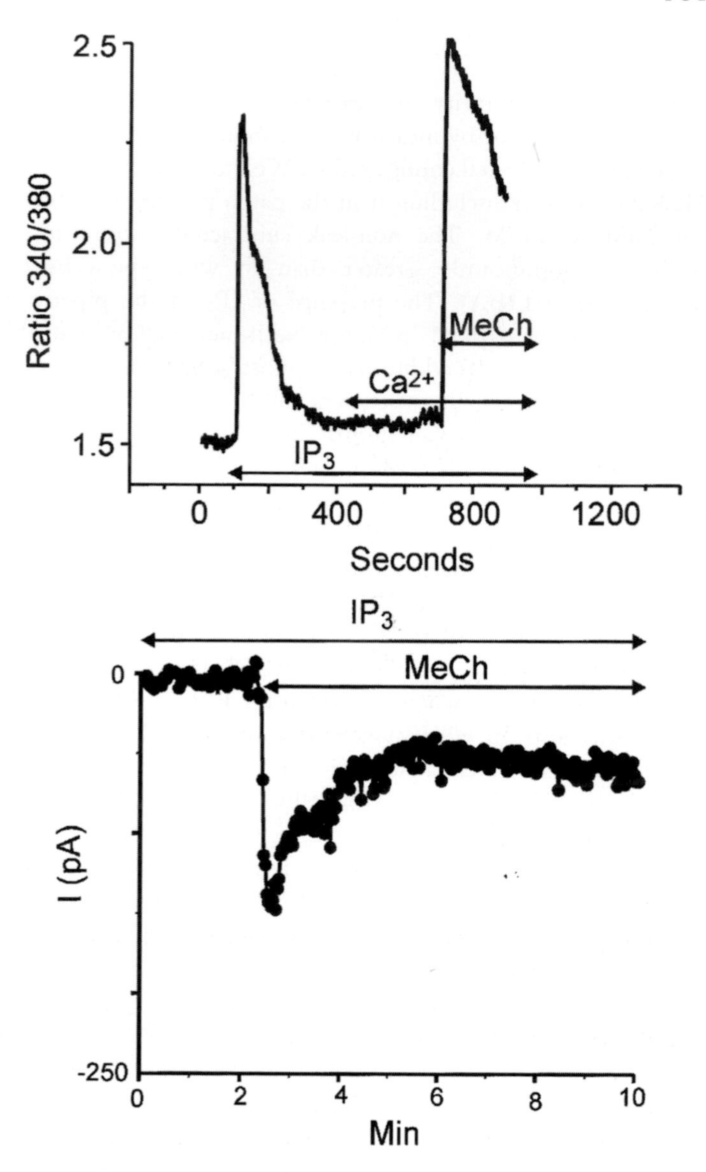

FIG. 2. IP_3 does not activate TRPC3 channels in HEK293 cells. (*Top*) IP_3 (400 μM) was introduced via a patch pipette into HEK293 cells stably expressing TRPC3, in the absence of extracellular Ca^{2+} and in the presence of 5 μM Gd^{3+} to block endogenous, SOCs. IP_3 caused rapid release of Ca^{2+} stores, but subsequent addition of Ca^{2+} failed to reveal any IP_3-dependent Ca^{2+} entry. Subsequent addition of methacholine (MeCh, 100 μM) resulted in rapid activation of Ca^{2+} entry. (*Bottom*) Methacholine (100 μM) increases inward current in TRPC3-expressing HEK293 cells in the presence of large quantities (400 μM) of IP_3. Redrawn from data originally presented in (Trebak et al 2003).

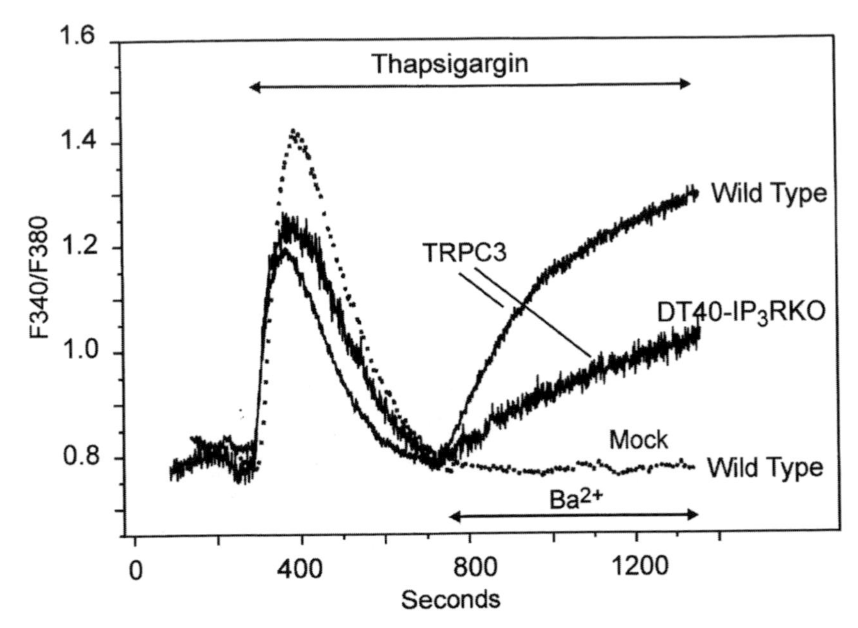

FIG. 3. Store-operated TRPC3 in DT40 B lymphocytes. DT40 B lymphocytes were transiently transfected with TRPC3. Mock (pcDNA3) transfected cells showed no store-operated Ba^{2+} entry in response to store depletion with thapsigargin. TRPC3-transfected cells exhibited thapsigargin-activated Ba^{2+} entry in both wild type and DT40-IP_3RKO cells. In the latter instance, store-operated Ba^{2+} was approximately half of that in wild-type cells. Redrawn from data originally presented in Vazquez et al (2001).

some of which, like the endogenous channels, are not (Vazquez et al 2001). However, utilizing a similar protocol, Venkatachalam et al (2001) transiently expressed TRPC3 in DT40 cells, but did not observe store-operated entry. Rather, these investigators reported the appearance of a Ba^{2+}-permeable receptor-operated entry. Significantly, receptor-activated TRPC3-dependent entry was also observed in DT40-IP_3RKO cells. These investigators concluded that TRPC3 forms receptor-operated channels in DT40 and the activation of these channels is likely signalled by DAG (Venkatachalam et al 2001).

Therefore, our results indicating a store-operated TRPC3 in DT40 conflict not only with our own findings in HEK293 cells, but with those reported by Venkatachalam et al with DT40 cells. In studies examining the trafficking of GFP-fusion TRPC3 channels in HEK293 and DT40 cells, we found that the level of expression of TRPC3 in DT40 was considerably less than that for HEK293 cells. Thus, we attempted to increase the expression level of TRPC3 in DT40 by two strategies. In the experiments in which TRPC3 behaved as a SOC, we used a construct of TRPC3 cloned in pcDNA3, under control of a CMV promoter.

To increase expression, we subcloned the TRPC3 cDNA into a different vector under control of an avian β actin promoter (Vazquez et al 2002), a more efficient promoter for avian cells. A second, simpler strategy to increase TRPC3 expression involved utilization of a 10-fold greater amount of the plasmid. Both strategies resulted in a loss of store-operated behaviour of TRPC3, and appearance of receptor-activated activity (Vasquez et al 2003). Significantly, this receptor-regulated TRPC3 activity was similar in DT40 and DT40-IP_3RKO cells.

The conclusion is that TRPC3 can behave as a SOC or as a channel regulated by DAG. The determining factor for the nature of the TRPC3 channel appears to be the expression level. The ability of TRPC3 to form non-store-operated, receptor activated channels in DT40-IP_3RKO cells is consistent with the conclusion from the studies in HEK293 cells that DAG is the likely signal for TRPC3 activation, and demonstrates that neither IP_3 nor the IP_3 receptor is necessary. What then is the significance of the CIRB domain? To address this question, we fused TRPC3 to a green fluorescent protein and constructed a series of C-terminal deletion mutants. Deletion of residues downstream of C-terminal to the CIRB domain had no effect on TRPC3-dependent Ca^{2+} entry. However, deletion of a major portion of the CIRB domain resulted in a channel that showed no TRPC3-mediated Ca^{2+} entry, and when these cells were examined by confocal microscopy, it was apparent that the CIRB-deleted channels did not traffic to the plasma membrane properly. This might indicate that interaction with the IP_3 receptor plays a role in proper targeting of TRPC3. However, as noted above, in either its store-operated or DAG-regulated mode, TRPC3 is indeed expressed at the plasma membrane in DT40-IP_3RKO cells. In addition, as was the case for HEK293 cells, we found that the CIRB-deletion TRPC3 did not traffic properly to the plasma membrane in DT40 and in DT40-IP_3RKO cells (Wedel et al 2003).

Conclusions

We have examined the regulation of TRPC3 as a representative of the TRPC3/6/7 subgroup of TRPC channels. By examining the behaviour of the channel protein expressed in two different cellular environments, and under conditions of different levels of expression, we conclude that TRPC3 can function as either a SOC or as a DAG-regulated channel. We cannot yet say with certainty which, or even if either of these two behaviours reflects the physiological function of TRPC3/6/7. All TRPCs appear to be non-selective cation channels, and thus it is unlikely that they would be involved in formation of the highly Ca^{2+}-selective channels that underlie the extensively studied store-operated current, I_{crac} (Hoth & Penner 1993). However, there are clear examples of endogenous store-operated non-selective cation channels (Trepakova et al 2001). Likewise, there are examples of endogenous DAG-activated non-selective cation channels (Tesfai et al 2001, Jung et al 2002). It would be very intriguing if, under physiological conditions, TRPC3 (and/or TRPC6 and 7) channels could act in two completely different modes of regulation, depending on the expression environment.

References

Barritt GJ 1999 Receptor-activated Ca^{2+} inflow in animal cells: a variety of pathways tailored to meet different intracellular Ca^{2+} signalling requirements. Biochem J 337:153–169

Berridge MJ 1995 Capacitative calcium entry. Biochem J 312:1–11

Birnbaumer L, Zhu X, Jiang M et al 1996 On the molecular basis and regulation of cellular capacitative calcium entry: roles for Trp proteins. Proc Natl Acad Sci USA 93:15195–15202

Boulay G, Brown DM, Qin N et al 1999 Modulation of Ca^{2+} entry by TRP-binding polypeptides of the inositol 1,4,5-trisphosphate receptor (IP3R). Evidence for roles of TRP and the IP3R in store depletion-activated Ca^{2+} entry. Proc Natl Acad Sci USA 96:14955–14960

Clapham DE, Runnels LW, Strübing C 2001 The TRP ion channel family. Nat Rev Neurosci 2:387–396

Cosens DJ, Manning A 1969 Abnormal electroretinogram from a *Drosophila* mutant. Nature 224:285–287

Hardie RC, Minke B 1993 Novel Ca^{2+} channels underlying transduction in *Drosophila* photoreceptors: implications for phosphoinositide-mediated Ca^{2+} mobilization. Trends Neurosci 16:371–376

Harteneck C, Plant TD, Schultz G 2000 From worm to man: three subfamilies of TRP channels. Trends Neurosci 23:159–166

Hofmann T, Obukhov AG, Schaefer M, Harteneck C, Gudermann T, Schultz G 1999 Direct activation of human TRPC6 and TRPC3 channels by diacylglycerol. Nature 397:259–263

Hoth M, Penner R 1993 Calcium release-activated calcium current in rat mast cells. J Physiol 465:359–386

Inoue R, Okada T, Onoue H et al 2001 The transient receptor potential protein homologue TRP6 is the essential component of vascular α1-adrenoceptor-activated Ca^{2+}-permeable cation channel. Circ Res 88:325–332

Irvine RF 1990 'Quantal' Ca^{2+} release and the control of Ca^{2+} entry by inositol phosphates — a possible mechanism. FEBS Lett 263:5–9

Jung S, Strotmann R, Schultz G, Plant TD 2002 TRPC6 is a candidate channel involved in receptor-stimulated cation currents in A7r5 smooth muscle cells. Am J Physiol 282:C347–C359

Kiselyov K, Xu X, Mozhayeva G et al 1998 Functional interaction between $InsP_3$ receptors and store-operated Htrp3 channels. Nature 396:478–482

Kiselyov K, Mignery GA, Zhu MX, Muallem S 1999 The N-terminal domain of the IP_3 receptor gates store-operated hTrp3 channels. Mol Cell 4:423–429

Li H-S, Xu X-ZS, Montell C 1999 Activation of a TRPC3-dependent cation channel through the neurotrophin BDNF. Neuron 24:261–273

Ma H-T, Patterson RL, van Rossum DB, Birnbaumer L, Mikoshiba K, Gill DL 2000 Requirement of the inositol trisphosphate receptor for activation of store-operated Ca^{2+} channels. Science 287:1647–1651

McKay RR, Szymeczek-Seay CL, Lièvremont J-P et al 2000 Cloning and expression of the human transient receptor potential 4 (TRP4) gene: localization and functional expression of human TRP4 and TRP3. Biochem J 351:735–746

Minke B, Cook B 2002 TRP channel proteins and signal transduction. Physiol Rev 82:429–472

Montell C, Rubin GM 1989 Molecular characterization of the *Drosophila trp* locus: a putative integral membrane protein required for phototransduction. Neuron 2:1313–1323

Montell C, Birnbaumer L, Flockerzi V 2002 The TRP channels, a remarkably functional family. Cell 108:595–598

Okada T, Inoue R, Yamazaki K et al 1999 Molecular and functional characterization of a novel mouse transient receptor potential protein homologue TRP7. Ca^{2+}-permeable cation channel that is constitutively activated and enhanced by stimulation of G protein-coupled receptor. J Biol Chem 274:27359–27370

Phillips AM, Bull A, Kelly LE 1992 Identification of a *Drosophila* gene encoding a calmodulin-binding protein with homology to the *trp* phototransduction gene. Neuron 8:631–642

Putney JW Jr 1997 Type 3 inositol 1,4,5-trisphosphate receptor and capacitative calcium entry. Cell Calcium 21:257–261

Putney JW Jr, McKay RR 1999 Capacitative calcium entry channels. Bioessays 21:38–46

Putney JW Jr, Broad LM, Braun F-J, Lièvremont J-P, Bird G StJ 2001 Mechanisms of capacitative calcium entry. J Cell Sci 114:2223–2229

Randriamampita C, Tsien RY 1993 Emptying of intracellular Ca^{2+} stores releases a novel small messenger that stimulates Ca^{2+} influx. Nature 364:809–814

Sugawara H, Kurosaki M, Takata M, Kurosaki T 1997 Genetic evidence for involvement of type 1, type 2 and type 3 inositol 1,4,5-trisphosphate receptors in signal transduction through the B-cell antigen receptor. EMBO J 16:3078–3088

Tesfai Y, Brereton HM, Barritt GJ 2001 A diacylglycerol-activated Ca^{2+} channel in PC12 cells (an adrenal chromaffin cell line) correlates with expression of the TRP-6 (transient receptor potential) protein. Biochem J 358:717–726

Trebak M, Bird GStJ, McKay RR, Putney JW Jr 2002 Comparison of human TRPC3 channels in receptor-activated and store-operated modes. Differential sensitivity to channel blockers suggests fundamental differences in channel composition. J Biol Chem 277:21617–21623

Trebak M, Bird GStJ, McKay RR, Birnbaumer L, Putney JW Jr 2003 Signaling mechanism for receptor-activated TRPC3 channels. J Biol Chem 278:16244–16252

Trepakova ES, Gericke M, Hirakawa Y, Weisbrod RM, Cohen RA, Bolotina VM 2001 Properties of a native cation channel activated by Ca^{2+} store depletion in vascular smooth muscle cells. J Biol Chem 276:7782–7790

Vazquez G, Lièvremont J-P, Bird GStJ, Putney JW Jr 2001 Human Trp3 forms both inositol trisphosphate receptor-dependent and independent store-operated cation channels in DT40 avian B-lymphocytes. Proc Natl Acad Sci USA 98:11777–11782

Vazquez G, Wedel BJ, Bird GStJ, Joseph SK, Putney JW Jr 2002 An inositol 1,4,5-trisphosphate receptor-dependent cation entry pathway in DT40 B lymphocytes. EMBO J 21:4531–4538

Vasquez G, Wedel BJ, Trebak M, St John Bird G, Putney JW Jr 2003 Expression level of the canonical transient receptor potential 3 (TRPC3) channel determines its mechanism of activation. J Biol Chem 278:21649–21654

Venkatachalam K, Ma H-T, Ford DL, Gill DL 2001 Expression of functional receptor-coupled TRPC3 channels in DT40 triple receptor InsP3 knockout cells. J Biol Chem 276: 33980–33985

Venkatachalam K, van Rossum DB, Patterson RL, Ma H-T, Gill DL 2002 The cellular and molecular basis of store-operated calcium entry. Nat Cell Biol 4:E263–E272

Vennekens R, Voets T, Bindels RJ, Droogmans G, Nilius B 2002 Current understanding of mammalian TRP homologues. Cell Calcium 31:253–264

Zhang Z, Tang J, Tikunova S et al 2001 Activation of Trp3 by inositol 1,4,5-trisphosphate receptors through displacement of inhibitory calmodulin from a common binding domain. Proc Natl Acad Sci USA 98:3168–3173

Wedel BJ, Vasquez G, McKay RR, St J Bird G, Putney JW Jr 2003 A calmodulin/inositol 1,4,5 triphosphate (IP3) receptor-binding region targets TRPC3 to the plasma membrane in a calmodulin/IP3 receptor-independent process. J Biol Chem 278:25758–25765

Zhu X, Jiang M, Peyton M et al L 1996 *trp*, a novel mammalian gene family essential for agonist-activated capacitative Ca^{2+} entry. Cell 85:661–671

Zhu X, Jiang M, Birnbaumer L 1998 Receptor-activated Ca^{2+} influx via human Trp3 stably expressed in human embryonic kidney (HEK)293 cells. Evidence for a non-capacitative calcium entry. J Biol Chem 273:133–142

Zitt C, Obukhov AG, Strübing C et al 1997 Expression of TRPC3 in Chinese hamster ovary cells results in calcium-activated cation currents not related to store depletion. J Cell Biol 138:1333–1341

Zitt C, Halaszovich CR, Lückhoff A 2002 The TRP family of cation channels: probing and advancing the concepts on receptor-activated calcium entry. Prog Neurobiol 66:243–264

DISCUSSION

Gill: One thing I think important is that the TRP channel is either receptor or store operated under these conditions, but at the lower concentrations it is not receptor operated, only store operated. Is that correct?

Putney: To accumulate these data we are using an imaging system, looking at multiple cells that have been co-transfected with marker. We can also look at cells that are positive for the marker. We are accumulating about 50–100 different cells. No '10 μg/ml' cells were ever receptor activated. The only clear cut experiment to see that in is in the IP_3 receptor knockout cells. We never got a positive from the 10 μg/ml group. Then, using 2-aminoethoxydiphenyl borane (2APB) block to isolate store-operated entry with 100 μg/ml we never saw a cell that showed store-operated behaviour. We only did those two concentrations. We have done two other manipulations to increase expression of TRPC3. One is to drive it by a better DT40 promoter, a β actin promoter. In this case we only need 10 μg and we get the high-expressing phenotype. The other is that we get a lot better protein expression if we let the protein synthesis occur after expression at lower temperatures. This also gives us the high-expressing phenotype by only using 10 μg.

Gill: Both phenotypes are sensitive to OAG. So even when it is expressed in the store-operated mode, OAG can get to it. But when it is expressed in this mode, the DAG from receptor activation of PLC does not get to it.

Putney: I can imagine lots of reasons for that. The easiest is that there is not enough in the right place.

Muallem: I want to underscore what you said about the behaviour of TRPC3. In our hands it does behave differently. Craig Montell has already mentioned his results on TRPC3 and Patrick Delmas indicated that they also saw that TRPC3 channels behave as SOCs when they expressed them in neurons. Last week we repeated some of our experiments and got the same behaviour, and we remember Mike Zhu's data about the specific sites for binding IP_3 receptors to the TRPC3 channels. Last week in a meeting in Halle, Marcus Hoth presented data which he is allowing me to cite. He isolated Jurkat cell mutants that don't have little store-operated Ca^{2+} influx and I_{CRAC}. Then he analysed the TRPC channels and found that TRPC3 is missing. Only when he expressed TRPC3 back did he completely restore Ca^{2+} influx and also I_{CRAC}. These TRPC3 channels therefore have the ability to behave in different ways according to the environment they are put in. I would also remind all of you that HEK cells are not homogeneous. These clones vary from lab to lab. You need to be able to keep the same lines going in order to reproduce experiments. There is no question that TRPC3 is activated by DAG and its activation is not that simple, but it seems to be able to interact not only with itself, but perhaps with other TRPC channels to give very different behaviour than TRPC3 alone. When it comes to regulation by IP_3 receptors my major point is whether or not it can be regulated by interactions with IP_3 receptors, and we believe that it can, of all TRPC channels TRPC3 shows the most extensive interaction with IP_3 receptors.

Putney: You haven't said anything that I can disagree with. One thing that would be helpful is if people really believe that TRPC3 can be activated through an IP_3/IP_3 receptor mechanism, it would be useful to compare all three members of the subgroup under those conditions. One of the more bothersome issues is why TRPC3 should be so different from 6 and 7. This may not be the case: it may be that the lab that sees the IP_3 effect studies 3, and another lab that sees mostly DAG studies 6. This may be the difference. But it would help clarify things if we knew whether we are talking about something that has to do with different proteins or merely different ways of doing the experiments, or different environments.

Muallem: I agree that these things need to be clarified. For us, looking at the sequence over the last week or so it seems that the hormone binding motif is quite different in both groups of channels. Whether this is the answer or not, I don't know.

Putney: We tried to look at the calmodulin–IP_3 receptor binding (CIRB) domain in TRP3. We ran into the same brick wall we always do with these: as soon as we take that out it doesn't traffic properly. For any mutants that we made, all point mutations were of no consequence or showed improper trafficking.

Westwick: If you go way back into the late 70s and early 80s there were quite a lot of papers showing that phorbol esters and DAG could block receptor-activated IP_3 formation and manganese influx. Your work may provide some explanation for that: it was definitely a PKC-dependent effect. You could block it with a variety of more selective PKC inhibitors than the staurosporine compound. In your system, where you are adding OAG, if we accept that OAG is activating by whatever means TRP3,6 or 7, it is also going to activate PKC. You have demonstrated with your phorbol ester experiments that you are getting phosphorylation at these sites on TRPC3. If you did your experiments with the agonists in the presence of a PKC inhibitor, would you get enhanced Ca^{2+} influx? You have two opposing effects.

Putney: We are doing those now. A frustration has been for us to demonstrate that this phorbol ester regulation of these channels has a counterpart in physiology.

Westwick: If you add DAG in the presence of a PKC inhibitor, what happens to TRP channel activity?

Penner: Those experiments have been done. Activation of PKC appears to inhibit store-operated Ca^{2+} entry in RBL cells, but it does not do so in Jurkat cells.

Westwick: It is the exact opposite in Jurkat cells: it actually enhances.

Putney: Are we wandering from the issue here, which is TRPC3 regulation?

Penner: No, but your TRPC3 regulation is the store-operated mode, so may you use the PKC regulation to address the physiological context.

Putney: TRPC3 is completely blocked by phorbol ester, whether it is in HEK or in the store operated mode in DT40. But the endogenous SOCs in DT40 and in HEK are completely insensitive. This appears to result from a direct phosphorylation of that ion channel subunit by PKC. This to me is the strongest evidence that in HEK and DT40 cells TRPC3 is not a component of their SOCs. Going back to your question about the physiological regulation, one thing that is interesting about this phorbol ester effect is that it has the right kind of pharmacology in terms of which phorbol esters do it and so on, but the dose–response curve is a bit high: it takes a lot of phorbol ester to completely block.

Westwick: That depends on the cell type.

Putney: It takes hundreds of nanomolar concentrations to get complete inactivation. This is why I think OAG works. It is not producing that much C kinase activation and yet it is much better at activating TRPC3. The only evidence that we have for a physiological role of C kinase is the following experiment. The mutant, which cannot be phosphorylated by protein kinase C, produces a more sustained response to agonist than does the wild-type. The wild-type runs downhill very slowly and the mutant stays up. This is the cleanest experiment because any other experiment with pharmacology has the complicating factor that the phospholipase C pathway itself is much more sensitive to up- and down-regulation by protein kinase C than is TRPC3 itself.

Westwick: I agree, but also IP_3 metabolism is modified by PKC. PKC activates IP_3 phosphatases.

Putney: The cleanest experiment is surgically just to take out the regulation of TRP3, which is achieved by mutating the one site. When this is done, the kinetics are more sustained and not slowly inactivating, suggesting that there is a physiological role for this inhibition by PKC.

Penner: I would like to add a little bit to the situation where we need to distinguish where these channels really are. Part of the controversy could be resolved if we assume that these ion channels could be in a store-operated compartment. If you would activate them in there you could get a store-operated mechanism simply by introducing a Ca^{2+}-permeable channel that releases Ca^{2+} from that particular store that then triggers the mechanism of SOCE. Your channels may be in different compartments — in the plasma membrane and in stores — and thereby manifest different types of behaviours.

Putney: The evidence that this is not the case in this instance is that in the store-operated mode the tool we are using is thapsigargin. If you treat cells with thapsigargin, wait a while in the absence of Ca^{2+}, and then add huge amounts of ionomycin, there is not a ripple in the baseline. We cannot find any evidence for retained Ca^{2+} in the wild-type cells in response to thapsigargin. Yet no barium-permeable channels are activated by thapsigargin.

Penner: This is because you haven't transfected them.

Putney: We haven't measured the current, but as far as we can tell the pharmacological and biophysical properties of the channels are different from the wild-type channels.

Penner: That is not the point. You can make another store-operated trigger if you include a new release channel.

Putney: In your simplest model you are asking whether we can explain the whole thing by having the same store-operated channels activated to a greater degree. They are not the same channels at all.

Penner: My question was, can you explain your data simply by insertion of TRPC3 into the store-dependent compartment?

Putney: No, I cannot.

Gill: You have raised the possibilities that they might be in different compartments or in different configurations. Perhaps both are important.

Ambudkar: You suggest that there is a low level of protein expression with less cDNA and higher levels of protein when you use higher concentrations of cDNA. How much protein is actually made under these two conditions, and how much protein do you think is required to form receptor-operated channels versus the SOCs?

Putney: I don't have numbers but I can show pictures. We have done this with a green fluorescent protein fusion protein. The images are consistent with all of the techniques that we use which we assume give us a higher level of expression. By comparison, the low expressors we can barely see, but it does appear to be there, and in the right place. There is a lot more of it with 100 μg than 10 μg in DT40, and much, much more of it in HEK cells.

Ambudkar: It was shown earlier that in stably transfected HEK-293 cells, where the expression is fairly low, TRPC3s can form store-operated channels.

Putney: We can't reproduce that.

Muallem: This is a critical point, and one that we raised in the original paper. Expression levels are a critical problem. When we expressed TRPC3 at high levels we were seeing too much spontaneous activity. The only way we could reduce spontaneous activity was by overexpression of the IP_3 receptor.

Putney: You are using cells from Lutz Birnbaumer's laboratory. This is a clone that has much higher expression levels than our cells.

Zhu: I believe that there is a small component of store-operated channels even in that cell line (Zhu et al 1998). This has to be analysed very carefully.

Muallem: My point is that the currents that we get from those cells are comparable to the currents which we get when we do transient transfections with low levels of TRPC3. We usually get a maximum current of 100–250 pA.

Putney: I can't argue that it couldn't be possible to go to HEK cells, get the correct expression level and get store-operated behaviour. But we have never seen it. We have seen the phenomenon that Mike Zhu mentioned, and this is a Ca^{2+} buffering artefact: if you use barium you won't see this.

Zhu: It may not be correct to assume that every channel that contains TRPC3 is Ba^{2+} permeable. If the store-operated channel formed by the heterologous expression of TRPC3 has to be a heteromultimer, then its biophysical properties could be very different from those of the homomultimeric TRPC3. Therefore neither Ba^{2+} permeability nor insensitivity to low concentrations of Gd^{3+} should be considered a common feature for channels that contain TRPC3. Thus I am not convinced that phenomenon discussed here is a Ca^{2+} buffering artefact.

Groschner: Could it be that at very high levels of TRPC3 expression you have both components, but you also have spontaneous activity of additional homomultimers, resulting in complete depolarization of the cells that would shut down other Ca^{2+} entry pathways? Do you have complete depolarization?

Putney: Even though we can't readily measure the endogenous pathway by patch-clamp, the thapsigargin experiments show that—with all the usual caveats—the steady-state Ca^{2+} entry through CCE looks to be almost exactly the same in the TRPC3-expressing cells. Even though you might expect them to be somewhat depolarized by whatever compensatory mechanism, they are certainly not completely depolarized. That response is completely abrogated by high K^+ if you do artificially depolarize.

Westwick: I have a question about the PKC regulatory site. I thought you said that when you get rid of that site, there is a longer Ca^{2+} influx or a slower turning off of the response. This was with the activation of the muscarinic receptor. What happens if you activate the B cell receptor in those cells?

Putney: We haven't put that variant into the B cells.

Authi: A quick point regarding physiology and protein kinase C. In platelets we know that there is TRPC6 in the plasma membrane. We now know that there is some TRPC7 but we haven't quite defined its location. In platelets if we put in an inhibitor of PKC we enhance an agonist-mediated barium entry, which we know is not mediated through the SOC pathway. So PKC inhibition does enhance barium entry.

Nilius: Your data on the modest overexpression of TRPC3 reminds me of the results in David Clapham's *Nature* paper (Yue et al 2001), but see also the challenge in Voets et al (2001). It is more or less the same protocol. It is intriguing that you can rescue the store operation by assuming that you have heteromultimers. What is your evidence for this?

Putney: The only evidence for this is that we form a channel that is blocked by gadolinium, whereas with high expression we get a channel that isn't blocked by gadolinium. My speculation is that this indicates that the actual composition of the channel, which depends on TRPC3 transfection in some way, is different from just a homomultimer of TRPC3.

Penner: I think it is dangerous to extrapolate from a Ca^{2+} signal whether or not your pharmacology of the channel has really changed.

Putney: I agree, it is not the best way to do the experiment.

Nilius: We are all doing these kinds of overexpression experiments. What we are really doing is pushing our protein in the secretory pathway. In the Golgi, for example, there is no SERCA. They have thapsigargin-insensitive Ca^{2+} pumps. The situation could be completely different just by accentuating the secretory pathway. It's a confusing situation. It could be that due to overexpression that channels may be mainly localized in the Golgi and may function as release channels, but they are then in a thapsigargin-insensitive compartment.

Putney: There are other possibilities. One that is interesting is that ordinarily, when one thinks of getting different phenotypes as you change the expression level, you would think that low expression is the physiological situation and high expression is where you start getting artefacts: this is where muscarinic receptors start activating adenylyl cyclase, for example. Yet there is more evidence for the receptor-operated mode actually being the physiological function of the TRPC3/6/7 than there is for the store-operated mode. These latest results are one exception. We entertain the possibility that a single gene product—a channel subunit—could participate in the formation physiologically of different kinds of channels. What we are doing now is protein chemistry via molecular biology. We are studying them in both instances under artificial conditions. The next step is to see whether either of these behaviours happen in the real word.

Montell: Everyone in this room, to varying degrees, uses overexpression systems. We all agree that they are valuable. I think your talk illustrates something we are all thinking but no one is saying: that despite the fact that we think they are valuable we need to say explicitly that they also have obvious limitations. Among the people who are studying TRP channels, we are seeing few attempts to find endogenous conductances either by suppressing currents from primary cells, or at least suppressing endogenous currents from cell lines. It seems like people should be putting more effort into this area.

Muallem: It is being done to some extent. There are data for TRPC1 and TRPC3.

Montell: Let's see by the end of this meeting whether there will be a single talk where someone is describing an endogenous conductance.

Nilius: Wait 20 minutes!

References

Voets T, Prenen J, Fleig A et al 2001 CaT1 and the calcium release-activated calcium channel manifest distinct pore properties. J Biol Chem 276:47767–477670

Yue L, Peng JB, Hediger MA, Clapham DE 2001 CaT1 manifests the pore properties of the calcium-release-activated calcium channel. Nature 410:705–709

Zhu X, Jiang M, Birnbaumer L 1998 Receptor-activated Ca^{2+} influx via human Trp3 stably expressed in human embryonic kidney (HEK)293 cells. Evidence for a non-capacitative Ca^{2+} entry. J Biol Chem 273:133–142

Diversity of TRP channel activation

Bernd Nilius and Thomas Voets

KU Leuven, Department of Physiology, Campus Gasthuisberg, Herestraat 49, B-3000 Leuven, Belgium

Abstract. Calcium entry controls a plethora of short- and long-term cell functions. With respect to the activation mechanisms of Ca^{2+} entry channels they are subdivided in the thoroughly characterized voltage-gated Ca^{2+} channels and the non-voltage gated Ca^{2+} channels. The latter group includes cation channels of the 'transient receptor potential' (TRP) superfamily, which consists of three subfamilies (TRPC, TRPV, TRPM). Activation of TRP channels is not yet completely understood, but examples of activation mechanisms of TRP channels from all three subfamilies will be discussed. The main focus is on the members of the TRPV subfamily, among which the TRPV4 channel shows a surprising gating promiscuity. It can be activated by cell swelling, heat and phorbol esters. Endogenous activators of the channel have not yet been described. It will be shown that lipid messengers related to arachidonic acid are endogenous TRPV4 activators. For TRPV5 and 6, the only highly Ca^{2+} selective channels within the TRP super-family, a voltage-dependent gating mechanism will be discussed, which includes an open pore block by Mg^{2+} and a highly Ca^{2+}-sensitive mechanism of inactivation. Regulation of channel availability by interaction with a protein bound to a site at the C-terminus of both channels will be demonstrated. Functional consequences of these different mechanisms of gating will be discussed.

2004 Mammalian TRP channels as molecular targets. Wiley, Chichester (Novartis Foundation Symposium 258) p 140–154

Many cell functions depend on influx of extracellular Ca^{2+}, which is triggered by a variety of mechanical and chemical signals. The superfamily of cation channels related to the 'transient receptor potential' channels (TRP) in *Drosophila*, TRP channels, provide pathways for Ca^{2+} entry or might be involved in the regulation of Ca^{2+} entry (Clapham et al 2001). The mechanisms that gate TRP channels are highly heterogeneous and still controversial. Most of the functional studies on TRP channels were probed in heterologous expression systems. Only very few studies on native cells are available. Among the mechanisms of TRP channel activation, receptor-activated gating via activation of a G protein-coupled receptor (GPCR)–phospholipase C (PLC) plays an important role, which is best described for the TRPC subfamily. Another intriguing mechanism of gating involves the depletion of intracellular Ca^{2+} stores. TRP channels, including TRPC1–7 and TRPV6 have been suggested to form such store-operated Ca^{2+}

channels (SOCs) or Ca^{2+} release-activated Ca^{2+} channels (CRACs) (see for an extensive review Venkatachalam et al 2002). However, the molecular identification of such channels is still elusive. Other TRP channels, such as TRPV1 and TRPV4, are gated by binding of agonists such as capsaicin (TRPV1) or phorbol esters (TRPV4) (Gunthorpe et al 2002, Watanabe et al 2002a), by radicals (TRPM2) (Hara et al 2002), nucleotides (TRPM2,7) (Perraud et al 2001, Runnels et al 2001) or an elevation of intracellular Ca^{2+} (TRPM4) (Launay et al 2002). Additionally, physical signals can also cause activation of TRP channels, such as mechanical forces for TRPV4 or heat for TRPV1–4 and cold for TRPM8. In addition, some channels in the TRP family (TRPV5,6) are characterized by a high-affinity open pore block by Mg^{2+} or Ca^{2+} and require removal of these cations by applying an increased driving force to remove the divalent cations from the binding site in the pore. This mechanism is therefore highly voltage dependent. Some novel gating mechanisms will be discussed in more detail.

Are TRPCs SOCs or CRACs? The TRPC4 knockout approach

One of the most fascinating mechanisms of activation of Ca^{2+} entry, which mainly occurs in non-excitable cells, is the gating by depletion of intracellular Ca^{2+} stores (SOCs). So far, all TRPCs (TRPC1–7) have been described as (potential) SOCs. A novel approach to study the functional role of TRPC channels in endothelial cells has been made possible by the development of the first TRPC channel transgenic mouse lacking a *Trpc4* (Freichel et al 2001). The *Trpc4* gene in mice may encode a membrane protein that previously has been described as a store-depletion activated, highly Ca^{2+}-selective cation channel (Philipp et al 2000), but controversially also as a non-selective cation channel that is not activated by Ca^{2+} store depletion but by activation of the PLC pathway (Schaefer et al 2000). In mice lacking TRPC4, agonist-induced Ca^{2+} entry through both store-operated channels and by PLC mediated channel stimulation is diminished dramatically in aortic endothelial cells (Freichel et al 2001). As a consequence, vasorelaxation is markedly reduced, showing that TRPC4 is an important component of store-operated channels in native endothelial cells and that these channels directly provide a Ca^{2+}-entry pathway essentially contributing to the regulation of blood vessel tone. In a recent paper that expands the description of an endothelial phenotype of the *Trpc4* knockout mouse, a role for TRPC4 in controlling the endothelial barrier function is demonstrated (Tiruppathi et al 2002). Studies were made in lung vascular endothelial cells from wild-type and $Trpc4^{-/-}$ mice. Thrombin increased endothelial permeability by activating a G protein-coupled proteinase-activated receptor 1 (PAR1), which resulted in a prolonged Ca^{2+} transient secondary to release of Ca^{2+} from intracellular stores. This store-dependent Ca^{2+} influx was drastically reduced in the *Trpc4* knockout mice, which was associated with a lack of thrombin-induced actin stress fibre formation and a reduced endothelial cell retraction response. Likewise, the increase in microvessel permeability upon stimulation of perfused mouse lungs with the PAR1 agonist peptide was drastically reduced in *Trpc4*$^{-/-}$ mice. Thus, TRPC4-dependent Ca^{2+} entry is a key determinant of microvascular permeability and likely a central player in endothelial cell function.

Both studies on the *Trpc4*$^{-/-}$ mouse present compelling evidence that TRPC4 is part of the signalling pathway that is involved in store-operated Ca^{2+} entry. However, it has to be mentioned that these results do not prove that TRPC4 is a CRAC or SOC channel itself. Because pore properties for Ca^{2+} selective TRP channels differ (Nilius et al 2001, Voets et al 2001), TRPC4 might be a component of the channel complex rather than forming part of the pore.

TRPV6 (CaT1) is not CRAC

TRPV6 (previously known as CaT1 and ECaC2) and the homologous TRPV5 (ECaC1) are probably the only highly Ca^{2+} selective ($P_{Ca}/P_{Na} > 100$) members of the TRP superfamily of cation channels related to the vanilloid receptor VR1 (TRPV1) (Hoenderop et al 2000, Vennekens et al 2000). Like all other Ca^{2+}-selective channels, they become permeable to monovalent cations when all extracellular divalent cations are removed (Nilius et al 2000, Yue et al 2001). A single aspartate residue in the pore region of TRPV5 and TRPV6 between transmembrane domains 5 and 6 determines open pore block by Mg^{2+}, Ca^{2+} selectivity and rectification (Nilius et al 2001). Monovalent TRPV6 (and TRPV5) currents are characterized by voltage-dependent opening at negative potentials and extremely strong inward rectification (Voets et al 2001). We have shown that the voltage-dependent gating of TRPV6 depends on intracellular Mg^{2+}. Both TRPV5 and TRPV6 show Mg^{2+}-dependent gating and Mg^{2+} contributes to strong inward rectification of TRPV6 (Voets et al 2001). Yue et al (2001) proposed that TRPV6 (CaT1) manifests the pore properties of CRAC in RBL cells and might be CRAC or at least part of the CRAC pore. Several features of TRPV6 are indeed identical with CRAC. However, in contrast to TRPV6 the activity of CRAC channels is affected by intracellular Mg^{2+} (Voets et al 2001, Kozak et al 2002). In tetra-concatamers, mutation of only one aspartate in the pore of one subunit renders the channels Mg^{2+} insensitive, diminishes rectification and dramatically reduce Ca^{2+} permeation (Hoenderop et al 2003, Voets et al 2003). Therefore, this site might be even considered dominant negative concerning the pore properties. It seems very difficult to reconcile these data with the hypothesis that TRPV6 or heteromers of TRPV6 are CRAC (see Fig. 1). TRPV5/6 are further distinguished from CRAC in the following features: sensitivity to store-depleting agents; inward rectification in the absence of divalent cations; relative permeability to Na^{+} and Cs^{+}; effect of

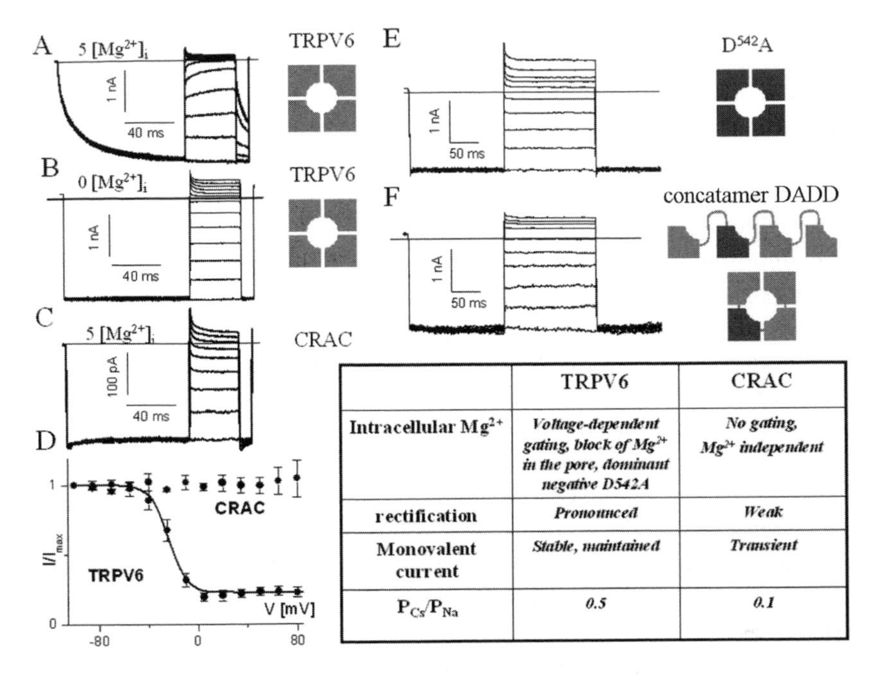

	TRPV6	CRAC
Intracellular Mg^{2+}	*Voltage-dependent gating, block of Mg^{2+} in the pore, dominant negative D542A*	*No gating, Mg^{2+} independent*
rectification	*Pronounced*	*Weak*
Monovalent current	*Stable, maintained*	*Transient*
P_{Cs}/P_{Na}	*0.5*	*0.1*

FIG. 1. (A–C) Current traces for TRPV6 and CRAC in divalent free solution (DVF). Holding potential is +20 mV. A voltage step to −100 mV is applied to completely activate the channels. Test steps were applied from, −100 to +100 mV. Note the slow gating for TRPV6 in the presence of intracellular Mg^{2+} (A), the voltage dependence after the test step and the pronounced rectification. In the absence of Mg^{2+} all these three features disappear (B). CRAC, activated by ionomycin in RBL cells under the same conditions as shown in A, B, also lacks these three typical TRPV5,6 features, even in the presence of intracellular Mg^{2+} (C). Plot of the initial current after the test step against the test potentials results in a Boltzmann-type inactivation curve for TRPV6 indicating the voltage dependence of removal of the pore-blocking Mg^{2+}. This pore-dependent mechanism is absent in CRAC (D). The same lack of all three pore features of TRPV5,6 were obtained by expression of the pore mutant D542A (E). The same feature occurs in concatamers in which only the second subunit carries the D542A pore mutation (F) indicating that D542A exerts a negative dominant effect on TRPV5,6 gating and permeation. The table summarizes the most striking differences between TRPV6 and CRAC (see also Voets et al 2001, Hoenderop et al 2003, Voets et al 2003).

2-aminoethoxydiphenyl borate (2APB). These results imply that the pores of TRPV6 and CRAC are not identical and indicate that TRPV6 is a Mg^{2+}-gated channel not directly related to CRAC. The molecular nature of SOC/CRAC is still elusive (Voets et al 2001).

Promiscuous gating of TRPV4

TRPV4 shares some similarities with the capsaicin receptor TRPV1, which is activated by several ligands, protons and heat, and is modulated by PKC. Thus, it is not too surprising that TRPV4 is also ligand-activated. The first identified TRPV4 ligand was the phorbol-derivative 4α-phorbol 12,13-didecanoate (4αPDD). This TRPV4 specific ligand, which does not affect PKC, activates outwardly rectifying cation currents and Ca^{2+} transients in native endothelial cells (Watanabe et al 2002a). Importantly, TRPV4 is also activated by arachidonic acid and the endocannabinoids anandamide (arachidonoylethanolamide, AEA) and 2-arachidonyl glycerol (2AG) (Watanabe et al 2003). AEA is likely to play an important role in the control of the vascular tone (Zygmunt et al 1999) and potentially in shock (Wagner et al 1998). The most likely activator of TRPV4 is the P450 product epoxyeicosatrienoic acid (5′,6′-EET) (Watanabe et al 2003). This complex signalling cascade certainly contributes to the functional significance of TRPV4 as a putative Ca^{2+} influx channel modulated by the integrated input of multiple physical and chemical stimuli. It can be anticipated that activation of TRPV4 by AEA, AA and EETs plays an important functional role in endothelium, e.g. it might be involved in the Ca^{2+} induced release of NO or even regulation of endothelium-derived hyperpolarization. Interestingly, not all pharmacological effects of anandamide on endothelium can be explained by activation of CB1 or CB2 receptors. Therefore, TRPV4 might be the missing target for some cardiovascular anadamide actions.

Our findings may also provide a possible mechanism for the activation of TRPV4 by cell swelling. It has been reported that cell swelling induces activation of PLA_2 and production of AA, and that metabolites of AA are released upon cell swelling (e.g. Thoroed et al 1997). AA and/or its metabolites might therefore be the stimulus for swelling-induced activation of TRPV4. Indeed, we have recently shown that inhibition of PLA_2 prevents activation of TRPV4 by cell swelling, whereas activation by other stimuli (4αPDD, heat, AA) is unaffected (Vriens et al 2004).

TRPV1–4 are able to sense temperatures over a range from $\sim$20 to $>$50 C (for a review see Gunthorpe et al 2002). TRPV4 is activated by warm temperatures, with a surprisingly low activation threshold of around 27 °C (Güler et al 2002, Watanabe et al 2002b). Likely, TRPV4 channels are therefore permanently open in endothelial cells at body temperature, and may thus contribute to modulation of $[Ca^{2+}]_i$. An elevated $[Ca^{2+}]_i$ in endothelial cells at body temperature has indeed been described (Watanabe et al 2002b) and could functionally be important for regulating the production of NO. Cooling of peripheral blood vessels could therefore induce vasoconstriction, and vice versa warming up could induce vasodilatation (Minson et al 2001). It is therefore not unlikely that TRPV4 acts as a cold and a warm endothelial cell receptor. Furthermore, a role of TRPV4 may be suggested in mediating the inflammatory response, e.g. by changing barrier properties.

Until now, TRPV4 is the only TRP channel proposed to act as a mechanotransducer (Liedtke et al 2000). Although data are so far not really supporting this function (see Strotmann et al 2000), it remains an interesting possibility that TRPV4 might be involved in mechano-sensing by endothelial cells, potentially via a mechano-stimulation of PLA_2 and subsequent activation by arachidonic acid.

Finally, the sensitivity of TRPV4 to endocannabinoids may provide a new molecular target for cannabinoids causing modulation of synaptic function. In this respect it is interesting to recall that anandamides act as retrograde messengers in the brain (Wilson & Nicoll 2002) and that the gene locus for the human TRPV4 channel is associated with bipolar affective disorder (Delany et al 2001).

TRPs good for anything? An endothelial approach

To this point, many functional properties of TRP channels are only known from studies in heterologous overexpression systems. Direct assessment of the role of TRP channels through gene-targeting in mice has revealed a role for TRPC2 in pheromone-dependent behaviour (Stowers et al 2002), for TRPC4 in endothelial-dependent vasorelaxation and endothelial cell (EC) barrier function (Freichel et al 2001, Tiruppathi et al 2002) and for TRPV1 in pain sensation (Caterina et al 2000, Gunthorpe et al 2002). For other channels, a function has been inferred from their functional properties and/or tissue distribution, which for example led to the view that TRPV5 and TRPV6 are responsible for Ca^{2+} reabsorption in kidney and intestine (Hoenderop et al 2000), that TRPM6 is involved in Mg^{2+} reabsorption (Schlingmann et al 2002) and TRPM5 in taste reception (Perez et al 2002).

Because our own interest is mainly focused on the function of endothelial cells, below we shortly review the possible functional significance of TRP channels in the vascular bed.

TRPs are involved in NO function. Many groups (see for a review Nilius & Droogmans 2001) have convincingly demonstrated that Ca^{2+} influx is necessary for the release and elevation of NO. The release of NO is not only involved in the control of relaxation of vascular smooth muscle cells, but also inhibits platelet aggregation, proliferation of smooth muscle cells (an anti-atherosclerotic function), the migration of monocytes and leucocytes, and the expression of adhesion molecules. NO produced by eNOS is a fundamental determinant of cardiovascular homeostasis and regulates systemic blood pressure, vascular remodelling and angiogenesis. NO release is activated via the Ca^{2+}/calmodulin-dependent pathway, and according to the above-described data on the TRPC4 knockout the sustained release of NO requires TRPC4 channel activation. This is so far the strongest evidence for the involvement of a TRP channel in endothelial function. The TRPC4 knockout model also hinted at an important role for this channel in controlling the EC barrier function. An increase in EC paracellular permeability, e.g. a decrease in EC barrier function and an increase in the hydraulic permeability of the EC layer, essentially depends on an increase in $[Ca^{2+}]_i$. Agonists like thrombin or histamine, which transiently increase $[Ca^{2+}]_i$ via Ca^{2+} release and Ca^{2+} influx, induce a transient increase in vascular permeability. Cytokines and growth factors induce a more sustained change. This increase in permeability is due to changes in the cytoskeleton, cellular contraction and cell–cell coupling. A decrease in EC barrier function is mediated by EC contraction, which is controlled by an endothelial Ca^{2+}/calmodulin-dependent non-muscle myosin light chain kinase (MLCK) (Verin et al 1998) that is activated by Ca^{2+} influx through a Ca^{2+} channel. From the knockout data, TRPC4 appears to be a major constituent of this influx channel (Tiruppathi et al 2002).

Endothelial cells sense mechanical forces, such as *biaxial tensile stress* (stretch, pressure) and *shear stress*. Many endothelial responses are modulated by changes in blood flow and blood pressure, such as secretion of prostacyclin (PGI_2), NO, expression of tissue plasminogen activator (tPA), plasminogen-activator inhibitor (PAI-1), several adhesion molecules (VCAM-1, the intercellular adhesion protein 1), growth inhibitors and growth factors (platelet-derived growth factor, PDGF), endothelin, monocyte-chemoattractant protein (MCP-1), eNOS, and activation of early response genes and small G proteins. Some of these effects, including the regulation of gene expression, are due to an increase in $[Ca^{2+}]_i$. Again, a connection to TRP channels is feasible. As discussed above, TRPV4, which is highly expressed in EC and functions as a Ca^{2+} entry channel in native cells, may link its activation to changes in cell shape (cell swelling), which, as discussed above, may involve the PLA_2-dependent production of AA.

Finally, TRPs are sensors for oxidant stress. TRPM2 is a G protein, adenosine 5′-diphosphoribose (ADPR) and nicotinamide adenine dinucleotide (NAD)-activated non-selective cation channel (Sano et al 2001, Hara et al 2002). Activation of TRPM2 is suppressed by intracellular ATP (Perraud et al 2001). Moreover, TRPM2 can be activated by oxidant stress and also by arachidonic acid, which depends on a *ISXXTKE* arachidonate recognition sequence (ARS) (Hara et al 2002). It is well known that this channel functions in blood cells and might be important for cell–cell interaction between EC and neutrophils (Sano et al 2001). Whether TRPM2 functions as an oxidant stress sensing and/or arachidonic acid-sensing Ca^{2+} channel in EC is still unclear, but it remains an intriguing further direction for TRP function exploration in the cardiovascular system. It has been recently shown that activation of Ca^{2+} influx by oxidative stress via a Ca^{2+}-permeant, non-selective cation channel requires the platelet-endothelial cell adhesion molecule (PECAM)-1/CD-31, which is known as a critical modulator of neutrophil–EC transmigration. The PECAM-dependent channel activation also required Src family tyrosine kinase activity (Ji et al 2002). TRP-channels have been proposed as possible candidates for this novel mechanism of regulating neutrophil–EC interaction, as well as for oxidant-mediated endothelial responses and injury.

Conclusion

TRP channels integrate various physical and chemical stimuli. Several TRP members might be potential candidates for store-operated Ca^{2+} entry. However, the identification of the best-described SOC, the Ca^{2+} release-activated Ca^{2+} channel (CRAC), is still elusive. Our results indicate that TRPV6, which has been put forward as a possible candidate, is unlikely to underlie CRAC. TRPV4 is an intriguing example of a channel that can function as an integrator of diverse physical and chemical stimuli. Our data provide evidence that endogenous lipid activators of TRPV4 channels are at least in part responsible for these unique properties, and point out multiple opportunities for TRPV4 to participate in different cell functions. Unfortunately, only few studies explicitly show the role of TRP channel activation in native cells. Clearly, we are still at the very beginning of the challenging adventure to unravel the significance and variability of TRP channel functions.

Acknowledgements

T.V. is a postdoctoral Fellow of the Fund for Scientific Research — Flanders (Belgium) (F.W.O.-Vlaanderen). We thank H. Watanabe, J. Vriens, R. Wondergem, and G. Droogmans (KU Leuven) as well as R. Bindels, J. Hoenderop (KU Nijmegen) and V. Flockerzi (Hamburg) for many helpful discussions and an exciting collaboration. This work was supported in part by the Belgian Federal Government, the Flemish and the Onderzoeksraad KU Leuven (GOA 99/07, F.W.O. G.0237.95, F.W.O. G.0214.99, F.W.O. G. 0136.00).

References

Caterina MJ, Leffler A, Malmberg AB et al 2000 Impaired nociception and pain sensation in mice lacking the capsaicin receptor. Science 288:306–313

Clapham DE, Runnels LW, Strubing C 2001 The trp ion channel family. Nat Rev Neurosci 2:387–396

Delany NS, Hurle M, Facer P et al 2001 Identification and characterization of a novel human vanilloid receptor-like protein VRL-2. Physiol Genomics 4:165–174

Freichel M, Suh SH, Pfeifer A et al 2001 Lack of an endothelial store-operated Ca^{2+} current impairs agonist-dependent vasorelaxation in TRP4-/- mice. Nat Cell Biol 3:121–127

Güler A, Lee H, Shimizu I, Caterina MJ 2002 Heat-evoked activation of TRPV4 (VR-OAC). J Neurosci 22:6408–6414

Gunthorpe MJ, Benham CD, Randall A, Davis JB 2002 The diversity in the vanilloid (TRPV) receptor family of ion channels. Trends Pharmacol Sci 23:183–191

Hara Y, Wakamori M, Ishii M et al 2002 LTRPC2 Ca^{2+}-permeable channel activated by changes in redox status confers susceptibility to cell death. Mol Cell 9:163–173

Hoenderop JGJ, Muller D, Suzuki M, van Os CH, Bindels RJM 2000 Epithelial calcium channel: gate-keeper of active calcium reabsorption. Curr Opin Nephrol Hypertens 9:335–340

Hoenderop JGJ, Voets T, Hoefs S et al 2003 Homo- and heterotetrameric architecture of the epithelial Ca^{2+} channels TRPV5 and TRPV6. EMBO J 22:776–785

Ji G, O'Brien CD, Feldman M et al 2002 PECAM-1 (CD31) regulates a hydrogen peroxide-activated nonselective cation channel in endothelial cells. J Cell Biol 157:173–184

Kozak JA, Kerschbaum HH, Cahalan MD 2002 Distinct properties of CRAC and MIC channels in RBL cells. J Gen Physiol 120:221–235

Launay P, Fleig A, Perraud AL, Scharenberg AM, Penner R, Kinet JP 2002 TRPM4 Is a Ca^{2+}-activated nonselective cation channel mediating cell membrane depolarization. Cell 109:397–407

Liedtke W, Choe Y, Marti-Renom MA et al 2000 Vanilloid receptor-related osmotically activated channel (VR-OAC) a candidate vertebrate osmoreceptor.Cell 103:525–535

Minson CT, Berry LT, Joyner MJ 2001 Nitric oxide and neurally mediated regulation of skin blood flow during local heating. J Appl Physiol 91:1619–1626

Nilius B, Droogmans G 2001 Ion channels and their functional role in vascular endothelium. Physiol Rev 81:1415–1459

Nilius B, Vennekens R, Prenen J, Hoenderop JG, Bindels RJ, Droogmans G 2000 Whole-cell and single channel monovalent cation currents through the novel rabbit epithelial Ca^{2+} channel ECaC. J Physiol 527:239–248

Nilius B, Vennekens R, Prenen J, Hoenderop JG, Droogmans G, Bindels RJ 2001 The single pore residue Asp542 determines Ca^{2+} permeation and Mg^{2+} block of the epithelial Ca^{2+} channel. J Biol Chem 276:1020–1025

Perez CA, Huang L, Rong M et al 2002 A transient receptor potential channel expressed in taste receptor cells. Nat Neorosci 5:1169–1176

Perraud AL, Fleig A, Dunn CA et al 2001 ADP-ribose gating of the calcium-permeable LTRPC2 channel revealed by Nudix motif homology. Nature 411:595–599

Philipp S, Trost C, Warnat J et al 2000 TRP4 (CCE1) protein is part of native calcium release-activated Ca^{2+}-like channels in adrenal cells. J Biol Chem 275:23965–23972

Runnels LW, Yue L, Clapham DE 2001 TRP-PLIK a bifunctional protein with kinase and ion channel activities. Science 291:1043–1047

Sano Y, Inamura K, Miyake A et al 2001 Immunocyte Ca^{2+} influx system mediated by LTRPC2. Science 293:1327–1330

Schaefer M, Plant TD, Obukhov AG, Hofmann T, Gudermann T, Schultz G 2000 Receptor-mediated regulation of the nonselective cation channels TRPC4 and TRPC5. J Biol Chem 275:17517–17526

Schlingmann KP, Weber S, Peters M et al 2002 Hypomagnesemia with secondary hypocalcemia is caused by mutations in TRPM6 a new member of the TRPM gene family. Nat Genet 31:166–170

Stowers L, Holy TE, Meister M, Dulac C, Koentges G 2002 Loss of sex discrimination and male-male aggression in mice deficient for TRP2. Science 295:1493–1500

Strotmann R, Harteneck C, Nunnenmacher K, Schultz G, Plant TD 2000 OTRPC4 a nonselective cation channel that confers sensitivity to extracellular osmolarity. Nat Cell Biol 2:695–702

Thoroed SM, Lauritzen L, Lambert IH, Hansen HS, Hoffmann EK 1997 Cell swelling activates phospholipase A(2) in Ehrlich ascites tumor cells. J Membr Biol 160:47–58

Tiruppathi C, Freichel M, Vogel SM et al 2002 Impairment of store-operated Ca^{2+} entry in TRPC4(−/−) mice interferes with increase in lung microvascular permeability. Circ Res 91:70–76

Venkatachalam K, van Rossum DB, Patterson RL, Ma HT, Gill DL 2002 The cellular and molecular basis of store-operated calcium entry. Nat Cell Biol 4:E263–E272
Vennekens R, Hoenderop JG, Prenen J et al 2000 Permeation and gating properties of the novel epithelial Ca^{2+} channel. J Biol Chem 275:3963–3969
Verin AD, Gilbert McClain LI, Patterson CE, Garcia JGN 1998 Biochemical regulation of the nonmuscle myosin light chain kinase isoform in bovine endothelium. Am J Resp Cell Mol Biol 19:767–776
Voets T, Prenen J, Fleig A et al 2001 CaT1 and the calcium release-activated calcium channel manifest distinct pore properties. J Biol Chem 276:47767–47770
Voets T, Janssens A, Prenen J, Droogmans G, Nilius B 2003 Mg^{2+}-dependent gating and strong inward rectification of the cation channel TRPV6. J Gen Physiol 121:245–260
Vriens J, Watanabe H, Janssens A, Droogmans G, Voets T, Nilius B 2004 Cell swelling, heat and chemical agonists use distinct pathways for the activation of the cation channel TRPV4. Proc Natl Acad Sci USA, in press
Wagner JA, Varga K, Kunos G 1998 Cardiovascular actions of cannabinoids and their generation during shock. J Mol Med 76:824–836
Watanabe H, Davis JB, Smart D et al 2002a Activation of TRPV4 channels (hVRL-2/mTRP12) by phorbol derivatives. J Biol Chem 277:13569–13577
Watanabe H, Vriens J, Suh SH, Benham CD, Droogmans G, Nilius B 2002b Heat-evoked activation of TRPV4 channels in an HEK293 cell expression system and in native mouse aorta endothelial cells. J Biol Chem 277:47044–47051
Watanabe H, Vriens J, Prenen J, Droogmans G, Voets T, Nilius B 2003 Anandamide and arachidonic acid use epoxyeicosatrienoic acids to activate TRPV4 channels. Nature 424: 434–438
Wilson RI, Nicoll RA 2002 Endocannabinoid signaling in the brain. Science 296:678–682
Yue L, Peng JB, Hediger MA, Clapham DE 2001 CaT1 manifests the pore properties of the calcium-release-activated calcium channel. Nature 410:705–709
Zygmunt PM, Petersson J, Andersson DA et al 1999 Vanilloid receptors on sensory nerves mediate the vasodilator action of anandamide. Nature 400:452–457

DISCUSSION

Muallem: I was very interested in the temperature-mediated channel activation. When you use EET can you still activate the channel with temperature?

Nilius: It is difficult for me to answer that. It seems that all the effects I have shown you are potentiated with higher temperature.

Hardie: Is the effect of temperature blocked by the enzyme inhibitors you mentioned?

Nilius: We haven't done that yet.

Fleig: With the endogenous channel, can heat alone activate TRPV?

Nilius: Yes.

Fleig: Is there an additive effect if you add EET at that point?

Nilius: We haven't done that experiment yet. There is an additive effect if we add 4αPDD.

Fleig: Is it a dose–response curve effect?

Nilius: It could be. We aren't sure what the activation mechanism by heat is yet. I don't want to say too much about the heat mechanism. We think it could be a ligand.

Fleig: It isn't in the inside–out patch in the overexpression system. I have a follow-up comment about TRPM4. Voltage clearly is very important with this channel, but Ca^{2+} is still the gating mechanism and voltage is acting as a modulator. You can't activate the channel by voltage.

Nilius: That is correct. It is completely different for Ca^{2+} activated K^+ channels, BKCa, which can be activated just by high voltage.

Fleig: With regards to voltage sensors, this immediately brings to mind skeletal muscle EC coupling. When you say 'voltage sensor' do you think that the gating currents are sufficient to activate something?

Nilius: That is our idea.

Schilling: Under physiological conditions, in the endothelial cell is TRPV4 active at 37 °C?

Nilius: Yes.

Schilling: So it would tend to depolarize the cells.

Nilius: Endothelial cells are depolarized cells, unless you activate the Ca^{2+} and K^+ channels, or you have a nice inward rectifier. There are many different endothelial cells and the spectrum of K^+ channels they have varies widely.

Schilling: How widespread is the distribution of TRPV4 in endothelial cells from different sites or different species?

Nilius: We don't know precisely. We have expression data only from mouse aorta cells (Wissenbach et al 2000).

Kunze: Is this always constitutively active? In endothelial cells, the membrane potential is generally sitting near –60 mV because of a strong inwardly rectifying K^+ channel, but at times the membrane potential of a confluent monolayer moves to –20 mV with a concomitant increase in a cation current. The mechanism behind this has always been of interest to us. I wonder whether the TRP channels may be responsible for the shift to the depolarized state. And if so, what turns them on?

Nilius: All the cultured endothelial cells have this huge inward rectifier. I believe this is not present in the native cells.

Kunze: In our hands it is present in the native cells, but even in the presence of this rectifier there is, sometimes, a large current that drives membrane potential to around –20 mV.

Nilius: You have also seen that this channel desensitizes through this one mechanism. The other mechanism we have clearly shown is that it is quite nicely regulated by intracellular and extracellular Ca^{2+}. It goes down when Ca^{2+} levels rise. There is therefore an intracellular Ca^{2+}-dependent inhibition of TRPV4, which is much less sensitive than TRPV5 and TRPV6. So we even know a site for this Ca^{2+}-dependent inactivation. There seems also to be a Ca^{2+}-dependent short activation. There are many regulators.

Westwick: Does 5′,6′-EET activate all these channels, including TRPV1?

Nilius: We are very happy because we have seen it in TRPV4, so we never tried it on TRPV1. I suspect that it doesn't activate TRPV1. Arachidonic acid doesn't activate TRPV1. For the activation of TRPV1 a special structure is required. We haven't tried TRPV3 and TRPV2 either.

Benham: Hwang et al (2000) found that TRPV1 was activated by the HPTE and HETE group of eicosanoids. They don't say whether they actually tried the EETs. We haven't tried them. Trying to work out what these channels are doing at 37 °C is a different situation from the room temperature at which we mostly do our *invitro* experiments. Activation is a complex picture for the TRPVs because of their temperature sensitivity.

Muallem: The temperature issue is not clear to me yet. You said that the desensitization means that it is inactivating at a higher temperature.

Nilius: It is gating. It is quite a good Ca^{2+} entry channel. This desensitization might be at least partially a Ca^{2+}-dependent inactivation process. We have a mutant at the C-terminus, and with this mutant we delay it.

Muallem: Is it ever active at 37 °C?

Nilius: Yes, it will be active. This basal activity will be regulated and is dependent on the level of intracellular Ca^{2+}.

Benham: In the Caterina paper (Guler et al 2002) they answered the question about what is actually happening if you keep cells at 37 °C for a while and then try going up and down 2 °C as might be seen physiologically. There they do see an elevation in Ca^{2+} when they go from 37–39 °C, and a decrease in Ca^{2+} going the other way. Whatever the desensitization equilibrium state is, it is still capable of responding to small changes in temperature.

Muallem: Is there adaptation to different temperatures?

Nilius: We go from 20–45 °C. There is no doubt that the heat response desensitizes TRPV4 (Watanabe et al 2002). In endothelial cells, at 37 °C we almost always see that Ca^{2+} goes up. For me there is no doubt that TRPV4 in endothelial cells plays a role in Ca^{2+} signalling.

Putney: The question was whether there was a difference between keeping cells at 37 °C and just acutely raising the temperature to 37 °C.

Nilius: There would be a difference.

Putney: Are there data? Would you expect that if they are kept at 37 °C then, they are not so active as when you acutely put them at 37 °C?

Benham: Yes, one implication you might get from Bernd Nilius' experiment going from 20–37 °C is that the channel is then maximally active at 37 °C. This doesn't make sense because you would have an enormous resting Ca^{2+} influx. Perhaps the reality is that you settle down to a steady-state equilibrium where the channels are largely desensitized in a temperature-adapted state, but they can nevertheless respond to small changes. The temperature effect is therefore still important, but not so that it gives a maximal probability of opening continuously. Certainly, in Caterina's paper (Guler et al 2002) their focus was on the preoptic hypothalamus where temperature-sensing neurons are present. They have shown that TRPV4 is expressed there, so their hypothesis is that it is responsible for the temperature-dependent depolarization seen in those neurons.

Groscher: You mentioned that TRPV4 could be for mechano-sensing in the endothelium, so it is responsive to cell swelling. Is it also responsive to shear stress, which is a more physiological stimulus?

Nilius: Yes. The response to shear stress is potentiated by temperature. It is also potentiated by 4αPDD.

Zhu: Is TRPM4 the non-selective cation channel?

Nilius: Yes.

Zhu: That is activated by depolarization. On the other hand we were talking about TRPV6, which is a hyperpolarizing activated channel.

Nilius: That is correct.

Zhu: Related to this, when you mutate that aspartic acid into something else (alanine, asparagine, or methionine) you only have to change one to mess up the Ca^{2+} selectivity. The suggestion is that there you have a symmetric pore. This is different from the concept of the voltage-gated channel, which in order for it to be Ca^{2+} selective, has to be asymmetric.

Nilius: For the L-type channel the pore motif is EEEE, which is symmetric. For the T-type Ca^{2+} channel it is EEDD.

Montell: Did you completely eliminate the Ca^{2+} selectivity?

Nilius: Yes. This is surprising, and for me it represents clear-cut evidence that CRAC cannot be a heteromultimer between TRPV6 and another subunit. If in the tetramer one aspartate D541 is mutated, Ca^{2+} selectivity is lost. You have to make a heteromultimer in which you can substitute this unique pore residue.

Zhu: So in order for them to be Ca^{2+} selective they are most likely symmetric pores.

Putney: What this is clear evidence for is that there is only one pore, if that is the way the pore is formed, as opposed to four pores that form when a tetramer is assembled. I suppose this was almost a settled issue anyway, but it wasn't for a long time.

Nilius: For the TRPV channels we have quite extensive data now about the tetramerization of TRPV5 and TRPV6 (Hoenderop et al 2003). There is no doubt.

Ambudkar: When you increase the temperature, is the result you see reflecting a true temperature effect on the channel or is it due to an effect on a regulatory mechanism or channel trafficking?

Nilius: I don't know. We haven't done this experiment. Our main focus is not the temperature sensing mechanism. Most cells have a fairly constant ambient temperature. We have a nice mechanism. If you go outside and it is cold, the peripheral circulation closes down to a degree and your hands go white. This is because TRPV4 is closed.

Barritt: I have a question of general cell physiology. If you think of cell volume control and stretch-activated channels, and consider all the TRP proteins, is TRPV4 the most likely candidate?

Nilius: So far, in the whole TRP family, the only candidate that has been proposed for this mechanism is TRPV4. In our hands, activation of TRPV4 by cell swelling is the least reliable method. All the other methods are much more robust.

Barritt: That doesn't argue that cell swelling is not a physiological pathway.

Nilius: I have been working in this field for around 10 years, and I was always very unhappy to imagine that my cells would be exposed to a 50% hypotonic solution. One's cells will never be challenged by the experimental stimulus of a 50% decrease in osmolarity!

Muallem: In looking at regulation by volume, one has to be very careful. Once you swell cells by exposing them to 150 milliosmolar solution, I don't know many channels that you would not activate. Most channels will display some activity.

Montell: In the case of OSM-9, it clearly allows the worms to sense changes in osmolarity in addition to being a mechanosensor.

Westwick: What cells are you doing your TRPV4 work with?

Nilius: HEK cells.

Westwick: So these effects are not dependent on the presence of other endogenous TRPVs.

Nilius: That is correct.

Fleig: I wanted to comment on cell swelling. To some extent it is a matter of definition. If you say 'cell swelling' this implies cell diameter increase without vesicle fusion. That is, cell swelling does not increase capacitance.

Zhu: Is this how swelling is defined in your experiments: by change of capacitance?

Nilius: When we swell cells we don't see an increase in capacitance. In our hands most swelling of cells is kind of an unfolding. Even then we see activation of so-called 'swelling-regulated anion channels' under iso-volumic conditions. Such iso-volumic activation comprises a reduction in ionic strength and perfusion with GTPγS. Cell volume doesn't change under these conditions but we see activation of the channel. In our hands the most critical part is that when the cells swell water enters and the intracellular ionic strength is reduced. This is the signal for activation of volume regulated anion channels.

Penner: Does this work in the whole-cell configuration? Is the ionic strength reduced?

Nilius: Yes.

Penner: You would have to do this very quickly.

Nilius: There is a nice paper addressing this (Voets et al 1999). We have calculated the change in ionic strength using a set of differential equations. We have a system so that we can measure the volume and current at the same time. When we have a patched cell, we see the volume going up fast due to osmotic water influx. Due to the reduced osmolarity in the cell, osmolytes from the pipette will equilibrate with the cell interior and again there is water influx, therefore, the cell continues to swell. The critical part for this exchange is your access resistance. The current follows the ionic strength but not volume.

References

Guler AD, Lee HS, Iida T, Shimizu I, Tominaga M, Caterina M 2002 Heat-evoked activation of the ion channel, TRPV4. J Neurosci 22:6408–6414

Hoenderop JG, Voets T, Hoefs S et al 2003 Homo- and heterotetrameric architecture of the epithelial Ca^{2+} channels TRPV5 and TRPV6. EMBO J 22:776–785

Hwang SW, Cho H, Kwak J et al 2000 Direct activation of capsaicin receptors by products of lipoxygenases: endogenous capsaicin-like substances. Proc Natl Acad Sci USA 97:6155–6160

Voets T, Droogmans G, Raskin G, Eggermont J, Nilius B 1999 Reduced intracellular ionic strength as the initial trigger for activation of endothelial volume-regulated anion channels. Proc Natl Acad Sci USA 96:5298–5303

Watanabe H, Vriens J, Suh SH, Benham CD, Droogmans G, Nilius B 2002 Heat-evoked activation of TRPV4 channels in a HEK293 cell expression system and in native mouse aorta endothelial cells. J Biol Chem 277:47044–47051

Wissenbach U, Bodding M, Freichel M, Flockerzi V 2000 Trp12, a novel Trp related protein from kidney. FEBS Lett 485:127–134

General discussion II

Montell: Thomas Gudermann, it might be interesting to take advantage of the TRPC6 knockout mice to consider whether TRPC6 has a role as a molecular anchor. If TRPC6 has such a role and is in a complex with a scaffold protein coupled to phospholipase C (PLC), one might imagine that the stability of PLC levels is dependent on TRPC6. Have you looked at PLC levels in the TRPC6 knockout mice?

Gudermann: No. It's an interesting hypothesis, but I don't think there are any data yet that support this.

Montell: Some TRPs do not get to the plasma membrane very effectively. In any of these cases, if you add an additional glycosylation site, does this facilitate plasma membrane localization?

Gudermann: In the TRPC3/6 mutants that we analysed we didn't find any differences between wild-type and mutants in terms of P current densities. In fact, TRPC3 and 6 are the only members of the TRPC family that are glycosylated. With a number of channels, such as the splice variant that is expressed in the VNO of TRPC2, we have a hard time getting it to go to the cell membrane in HEK cells.

Putney: Is it possible that the degree of constitutive activity has to do with whether or not channels are coupled to the signalling machinery? So a channel that is coupled to the signalling machinery which can be activated through PLC and diacylglycerol (DAG) may not be constitutively active, but one which is out on its own may have more constitutive activity. If this were the case, your peak current, which depends only on those coupled receptors, would look the same, but in the case of TRPC3 you might have more uncoupled channels.

Gudermann: That is a possibility. It would reconcile many of the data we have seen. At low expression there is interaction between TRPC3 and proteins such as calmodulin and inositol 1,4,5-trisphosphate receptor (IP_3R) which silences the cell, and yet with TRPC6 this isn't seen so clearly because it is tightly regulated: only if you overexpress it will you see an increase in constitutive activity because all these protein contacts are not possible any more.

Penner: I noticed that in one of your experiments, when you applied thapsigargin and measured manganese quenching, you levelled off at a plateau. Then when you came in and stimulated TRPC6, you saw an additional manganese quench. How is that possible? I would assume that if the store-operated Ca^{2+} entry pathway were present, at some point all the fura in the cytosol would have been saturated.

Doesn't this argue that you are opening pathways into intracellular compartments which also contain fura? Fura goes into every compartment. But the store-operated mechanism only gives you the manganese quenching of the cytosol. What you are left with in terms of plateauing off to the fluorescence level at the bottom is really not the fluorescence in the cytosol. Is that a possibility?

Gudermann: The flattening occurred at 380 nm. If you look at manganese quenching, this goes on and on.

Penner: So you can actually bring it all the way down.

Gudermann: Yes. If you then add histamine to the TRPC6 expressing cells, you see a massive influx from the outside which completely quenches your signal.

Putney: Your 380 nm signal changed almost immediately when you added thapsigargin, but your 360 nm signal only increased after a delay. This was true in both the wild-type and the TRP-transfected cells. Could it be that the store-operated channels (SOCs) are not particularly manganese permeant, and you are secondarily activating a Ca^{2+}-activated channel?

Muallem: An alternative is that the stores that regulate SOCs are not as leaky as we think they are, and when thapsigargin is added it takes some time to deplete them.

Hardie: We have all heard that when TRPC3, 6 and 7 are heterologously expressed they have constitutive basal activities. Is there any clear indication that endogenous TRPs are constitutively active? You showed that TRPC3 became constitutively active in the TRPC6 knockout, but is there any constitutive activity at all in the wild-type?

Gudermann: It is difficult to say. We can clearly distinguish between TRPC6 and TRPC3. The kind of constitutive activity that you see is much higher in TRPC3 compared with TRPC6.

Hardie: You said you saw an increase in constitutive activity in the TRPC6 knockout.

Gudermann: That was in terms of measuring the barium influx in isolated cells.

Hardie: You need to have some way of seeing endogenous current, of which TRPC6 is the best characterized. Does anyone know whether these are ever constitutively active under physiological conditions, or is it only when you express them?

Gill: At least in DT40 cells where there are two very different signalling pathways — the G protein-coupled receptor which couples to phospholipase C (PLC)β and the B cell receptor complex — both of these very different receptor complexes activate the TRPs almost identically. The overexpressed TRPs are not likely to be in a highly controlled environment. It seems more likely that they are there and they are activated by DAG. It doesn't seem that they have to be within one single complex.

Montell: When we were looking at TRPC3 a few years ago, we found that the endogenous TRPC3 in the pontine neurons is in a complex with the brain-derived neurotrophic factor (BDNF) receptor TrkB. We never followed this up, but it shows that the first protein involved in this signalling cascade and the last protein, TRPC3, are in the same complex. Perhaps it would be better to consider whether PLCγ rather than the β subunit is part of this complex.

Gill: The biggest phenotype difference in those neurons is that they were not sensitive to DAG. Perhaps this indicates that they are in a much more regulated, controlled environment, where manipulating DAG is not the primary way of activating these things.

Putney: What is the rigorous evidence that the current measured in these cells is TRPC3?

Montell: It is the antibody experiment.

Putney: But it was not affected by 1-oleoyl-2-acetyl-*sn*-glycerol (OAG). You could take those as conflicting pieces of evidence and cast your vote I guess.

Penner: Was your diacylglycerol (DAG) good?

Montell: We expressed TRPC3 in HEK cells and we could activate it with OAG or DAG.

Gill: There is a huge amount of TRPC3 in those cells. If by chance that happens to bind antibody really well, it is an unusual situation. Could it just be the fact that you have antibody bound to a complex, and it is just activating another channel?

Montell: No. We added BDNF in the whole cell configuration, and then prepared inside–out patches and then took Ca^{2+} away. We completely lost the channel activity. Two things are normally needed for activation: BDNF on the outside and Ca^{2+} on the inside. If you add the TRPC3 antibodies the identical conductance returns, except the mean open time increases significantly.

Gill: I agree with your earlier point, that it is really important to look at the physiological activation of the TRPs. This is one of the best examples of seeing TRPC3 activated *in vivo*.

Zhu: That channel has strikingly different single channel properties from the over-expressed channels.

Freichel: I have a question about the RT-PCR study looking at TRP channel expression levels. As far as I remember, it was a comparison with the housekeeping gene. Does that mean that TRPC4/5 are not expressed at all, or that they are expressed at a lower level than the housekeeping gene.

Gudermann: It was all normalized to the expression of the housekeeping gene. There was very low expression of TRPC4 but we couldn't detect TRPC5 at all.

Benham: With the TRPC6 knockout you kind of focused on the hypothesis that it is the up-regulation of TRPC3 in the absence of TRPC6 that gives the different phenotype. I wonder whether you could just argue that assuming that a TRPC6/C3 heteromer is the native state, then if you lose the partner you will end up with

TRPC3 homomers and this rather than expression level causes the different phenotype? Has anyone looked at the over-expression of TRPC3/6 concatamers to see whether this produces an overexpression phenotype or a more native phenotype?

Gudermann: That would be a logical explanation. We are making these concatamers now.

Putney: The question is, does a mixture look like 6 or 3?

Ambudkar: We have been talking about TRPC3s interacting with TRPC6 and 7, but there are a couple of reports indicating an interaction between C1 and C3 (Lintschinger et al 2000, Xu et al 1997). We did some immunoprecipitations (IPs) in our lab using endogenous TRP1 antibody and TRPC3-expressing 293 cells. We can pull down TRPC3 with TRPC1 from these cells. Is it possible that under some conditions TRPC3 could form complexes with TRPC1? It is something to think about.

Putney: Of the various experimental approaches, co-IP is probably the least rigorous proof. Lots of things can come down in co-IP for reasons other than direct interaction.

Ambudkar: I agree with you. But we do have these previous data that need to be resolved.

Montell: If we co-express TRPC1 and TRPC3 in the same cells they can be co-IPed, but we never compared that interaction with TRPC3 and TRPC6. It is possible that C1 and C3 interact, but based on the FRET (fluorescence energy resonance transfer) experiments by Hofmann and colleagues there seems to be a preference for interactions among members within the same subfamily.

Ambudkar: I am not questioning the FRET data. I am trying to think more in terms of explaining the different function of TRPC3 in different sets of experiments, i.e. whether it is determined by the type of TRPC channel it interacts with.

Montell: It is possible that heteromultimeric interactions occur even among TRP proteins, which do not display the strongest affinities.

Zhu: I still want to challenge the idea of DAG stimulation of TRPC3 and C6. Don Gill has a publication (Ma et al 2000) showing that DAG-activated TRPC3 was not blocked by 2-aminoethoxydiphenyl borate (2APB), but the carbachol-activated TRPC3 was. Jim Putney used a slightly lower concentration of 2APB, but this did not block TRPC3 at all in DT40 cells.

Putney: You are right that there is a conundrum there. What we find in HEK cells is slightly different from Don Gill. 30 μM will block about 50–60% of the TRPC3 activity, but higher concentrations won't block any more. It has no effect at all on OAG. There is a disconnect. In DT40 cells 2APB doesn't block the store operated TRPC3 channels, but inhibits it by just 15%. I don't know what 2APB is doing but my guess is that it has to do with the spatial relationship between the PLC and the diacylglycerol that is produced by it, and the TRP channels. This is not an issue with OAG. There is a precedent for that. We have shown that in the SOCs, cytochalasin D will completely block the activation of PLC by agonists but not the activation of SOC by IP_3 or thapsigargin. It blocked the release of Ca^{2+} due to activation of PLC receptors without blocking PLC activation. The interpretation is that if we change the spatial relationship between the receptor and the signal and its effector, we lose the connectivity.

Zhu: One of the papers published on the TRPC6 splice variant indicated that it was activated by agonist stimulation but not by OAG (Zhang & Saffen 2001).

Putney: This work has been re-done but not replicated.

Gudermann: It was changed in the N-terminus. When we remove the N-terminus from TRPC6 it doesn't make it to the membrane.

Putney: I'd add that the sequence that they thought was conferring OAG sensitivity is only present in TRPC6 and not the other members of that group.

Montell: One of the interesting things about the point mutation that changed the Ca^{2+} selectivity of TRPV6 is that it showed that TRP really is an ion channel. Pore-loop mutations in TRPC1 have been generated previously in Indu Ambdukar's lab and also shown to alter the selectivity. These experiments are important since no one has put a TRP into a lipid bilayer and formally demonstrated that it is an ion channel.

References

Lintschinger B, Balzer-Geldsetzer M, Baskaran T et al 2000 Coassembly of Trp1 and Trp3 proteins generates diacylglycerol- and Ca^{2+}-sensitive cation channels. J Biol Chem 275:27799–27805

Ma HT, Patterson RL, van Rossum DB, Birnbaumer L, Mikoshiba K, Gill DL 2000 Requirement of the inositol trisphosphate receptor for activation of store-operated Ca^{2+} channels. Science 287:1647–1651

Xu XZ, Li HS, Guggino WB, Montell C 1997 Coassembly of TRP and TRPL produces a distinct store-operated conductance. Cell 89:1155–1164

Zhang L, Saffen D 2001 Muscarinic acetylcholine receptor regulation of TRP6 Ca^{2+} channel isoforms. Molecular structures and functional characterization. J Biol Chem 276:13331–13339

Regulation of *Drosophila* TRP channels by lipid messengers

Roger C. Hardie

Cambridge University Department of Anatomy, Downing Street, Cambridge CB2 3DY, UK

Abstract. In common with their vertebrate homologues, the prototypical *Drosophila* TRP channels are activated downstream of phospholipase C (PLC) by unknown mechanism(s). Most recent evidence in *Drosophila* photoreceptors now indicates that excitation is mediated, not by inositol 1,4,5-trisphosphate (IP_3), but by lipid products of PLC action, such as diacylglycerol (DAG), its metabolites (polyunsaturated fatty acids, PUFAs), or the reduction in phosphatidylinositol 4,5-bisphosphate (PIP_2). Compelling evidence for a PKC independent role of DAG comes from mutants of the *rdgA* gene, which encodes DAG kinase. The *rdgA* mutation leads to constitutive activation of both TRP and TRPL channels and dramatically increases sensitivity to light in hypomorphic mutations of PLC or G protein. A role for PIP_2 reduction is suggested by finding that conditions, which lead to acute PIP_2 depletion—monitored by genetically targeted PIP_2-sensitive ion channels—also lead to constitutive activation of TRP channels. Finally, recent data indicate that PUFAs activate TRP channels directly, and independently of PLC or metabolic inhibition. Together with evidence from several mammalian TRP homologues, these results suggest that regulation by lipids may be a defining feature of many TRP channels.

2004 Mammalian TRP channels as molecular targets. Wiley, Chichester (Novartis Foundation Symposium 258) p 160–171

Drosophila photoreceptors respond to light via a G protein-coupled phospholipase C (PLC) transduction cascade, culminating in the opening of cation permeable channels localized in microvilli which form the light-guiding rhabdomere (Fig. 1). Over 30 years ago, a *Drosophila* mutant was isolated in which the response to maintained illumination decayed to baseline over a period of a few seconds (Cosens & Manning 1969). The *trp* gene responsible for this so-called *transient receptor potential* phenotype was subsequently cloned (Montell & Rubin 1989) and later recognized as the defining member of a novel ion channel family when the mutant's light-sensitive conductance was found to have a greatly reduced Ca^{2+} permeability (Hardie & Minke 1992) and when it was realized that both *trp* and a homologue, *trp*-like (*trpl*), showed structural homologies to the superfamily of voltage-gated channel genes (Phillips et al 1992). In common with the vertebrate

TRPC subfamily to which they are most closely related (Montell 2001), *Drosophila* TRP and TRPL channels are activated downstream of PLC. However, whilst it was originally assumed that they were activated downstream of inositol 1,4,5-trisphosphate (IP_3)-induced Ca^{2+} release, extensive experiments have generally failed to support this view (but see Minke & Cook 2002), and the light response is completely unaffected by a null mutation in the only IP_3 receptor gene in the *Drosophila* genome (Raghu et al 2000a). Phosphatidylinositol 4,5-bisphosphate (PIP_2) hydrolysis by PLC potentially produces two further signals: a rise in diacylglycerol (DAG) and a fall in PIP_2. Initially, DAG was ignored as a candidate, since its most familiar role is to activate protein kinase C (PKC), and mutants of the eye-specific PKC (*inaC*) have defects only in response termination leaving excitation unaffected (Smith et al 1991, Hardie et al 1993). However, recent evidence now strongly indicates that DAG and/or other lipid messengers may indeed mediate excitation in *Drosophila* photoreceptors. At the same time, it appears that many vertebrate TRP homologues may be regulated by DAG or PIP_2, suggesting this may be a common feature of many Ca^{2+} influx pathways (Benham et al 2002, Hardie 2003).

Genetic evidence for excitation by DAG: *rdgA* mutants

Some of the most compelling evidence for an excitatory role for DAG comes from recent studies of mutants of the *rdgA* (retinal degeneration A) gene, which encodes a DAG kinase (DGK) expressed in the photoreceptors (Masai et al 1993). Degeneration in *rdgA* mutants is light independent and unusually severe with the retina already degenerated in newly eclosed flies. Although they are blind, whole-cell recordings from *rdgA* mutant photoreceptors revealed that the light-sensitive TRP and TRPL channels were constitutively active (Raghu et al 2000b). Since conversion of DAG to phosphatidic acid by DGK is the major route for inactivating DAG (Topham & Prescott 1999), this would be consistent with an excitatory role for DAG, which should accumulate in the absence of DGK. In addition, the constitutive activity suggested a mechanism for degeneration—namely the cytotoxic effects of uncontrolled Ca^{2+} influx—and indeed degeneration can be rescued by additional mutations in the TRP channel (Raghu et al 2000b) or the PLC (*norpA*) required for channel activation (Hardie et al 2002). Importantly, the light response was also restored in such *rdgA*;*trp* and *norpA*;*rdgA* double mutants allowing insight into the role of DGK in phototransduction.

Sensitivity in severe hypomorphic *norpA* mutants is reduced by many orders of magnitude since quantum bumps (the responses to single activated rhodopsin molecules) are reduced *circa.* 10-fold in amplitude and elicited with reduced quantum efficiency and prolonged latencies. In *norpA*;*rdgA* double mutants however, sensitivity to light is massively enhanced (~100–1000 fold) and

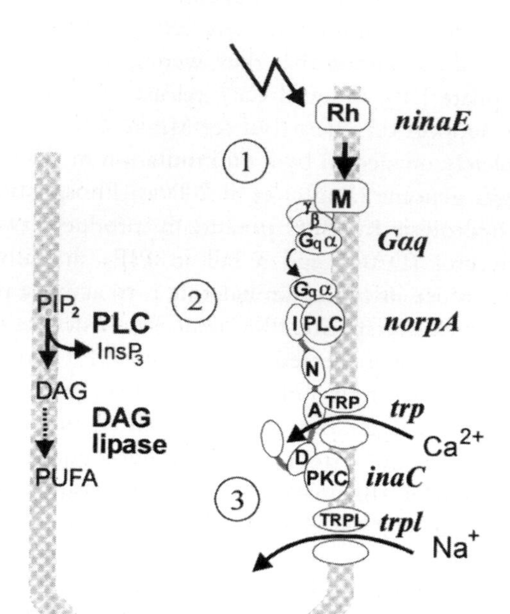
Rh
ninaE
1
M
Gaq
2
PLC
norpA
N
A
TRP
trp
Ca2+
D
PKC
inaC
3
TRPL
trpl
Na+
PIP2
PLC
InsP3
DAG
DAG
lipase
PUFA

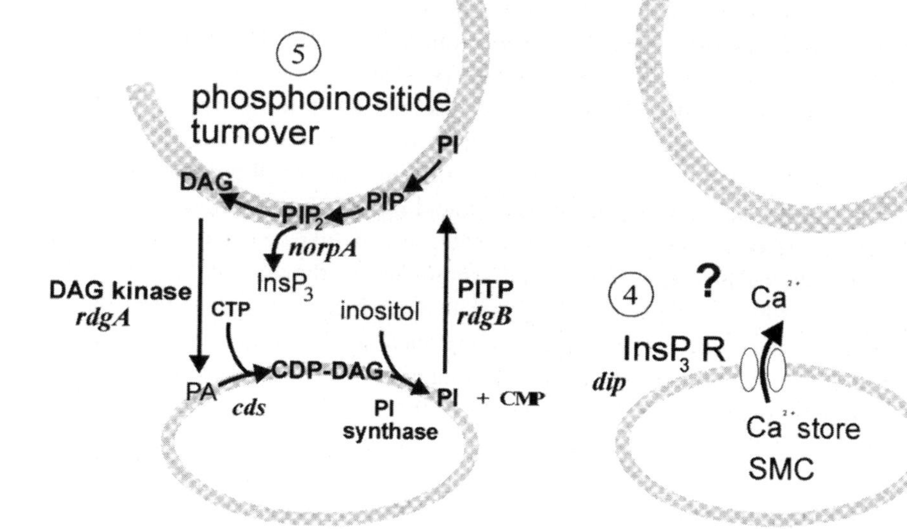
5
phosphoinositide
turnover
PI
DAG
PIP
PIP2
norpA
DAG kinase
rdgA
InsP3
CTP
inositol
PITP
rdgB
CDP-DAG
PA
cds
PI
synthase
PI
+ CMP
4
?
Ca
InsP3 R
dip
Ca store
SMC

quantum bump amplitude is restored to wild-type (WT) levels. Importantly, this effect of the *rdgA* mutation can be very effectively phenocopied simply by restricting the supply of ATP for kinase activity. Thus a similar enhancement of sensitivity and bump amplitude can be achieved in *norpA* simply by omitting ATP from the recording electrode (Hardie et al 2002).

These effects of the *rdgA* mutation clearly point to DAG as an excitatory messenger or intermediate. In *norpA* mutants the rate of generation of DAG by residual PLC activity is presumably too slow to overcome its removal by DGK resulting in near elimination of the response. However, if DGK function is impaired by mutation or ATP deprivation, even a low rate of PLC activity can generate sufficient DAG to overcome the threshold for channel activation.

Activation by polyunsaturated fatty acids

The simplest explanation for these results would be that DAG directly gates the TRP and TRPL channels as has been proposed for several vertebrate TRP homologues (Hofmann et al 1999). But, although DAG has been reported to activate recombinant TRPL channels (Estacion et al 2001), it is unclear if this is a direct action and attempts to activate TRP or TRPL channels by application of exogenous DAG in photoreceptors have thus far proved unsuccessful. Whilst this may represent a technical problem of access in sufficient concentrations, another possibility is that the active ligand is downstream of DAG. In particular DAG may be further metabolised by DAG lipase to release polyunsaturated fatty acids (PUFAs) such as arachidonic and linolenic acid. Indeed, to date PUFAs are

---

FIG. 1. Elements of the *Drosophila* phototransduction cascade shown on a schematic microvillus. (1) Photoisomerization of rhodopsin to metarhodopsin (Rh ▶M, encoded by *ninaE* gene) activates heterotrimeric G_q protein, releasing the $G_q\alpha$ subunit (*Gαq* gene). (2) $G_q\alpha$ activates PLC (*norpA* gene), releasing IP_3 and DAG from PIP_2. DAG may further release polyunsaturated fatty acids (PUFAs) via DAG lipase. (3) Two classes of light-sensitive channels are activated by an unknown mechanism: the current is mediated predominantly by TRP channels, which have a high selectivity for Ca^{2+} (P_{Ca}:P_{Na} *c.* 110:1), with a minor contribution from the less Ca^{2+}-permeable TRPL channels (Niemeyer et al 1996; Reuss et al 1997) and possibly a recently discovered third homologue, TRPγ (Xu et al 2000). The TRP channel, protein kinase C (PKC, *inaC* gene) and PLC are coordinated into a signalling complex by the scaffolding protein, INAD, which contains five PDZ domains. At the base of the microvilli specialized smooth endoplasmic reticulum forms the submicrovillar cisternae (SMC): although these may represent Ca^{2+} stores (4) endowed with IP_3 receptors (IP_3R, *dip* gene), they may be more important for phosphoinositide turnover (5), whereby DAG is converted to phosphatidic acid (PA) via DAG kinase (*rdgA* gene) and to CDP-DAG via CD synthase (*cds* gene) in the SMC. After conversion to phosphatidyl inositol (PI) by PI synthase, PI is transported back to the microvillar membrane by a PI transfer protein (*rdgB* gene) and converted to PIP_2 via serial phosphorylation (PI kinase and PIP kinase).

the only potential ligands found to activate *Drosophila* TRP and TRPL channels *in situ* (Chyb et al 1999).

Although PUFAs activate channels even in excised inside-out patches with doses as low as 2 μM, the suggestion that they act directly on the channels has been challenged. Firstly, although they confirmed that PUFAs were potent activators of TRPL channels expressed in insect Sf9 cells, Schilling and colleagues found that PLC inhibitors suppressed the action of PUFAs, suggesting that activation was indirect, via PUFAs activating PLC (Estacion et al 2001). However, this cannot readily account for their action *in situ*. Firstly, the possibility that PUFAs activate the channels via PLC encoded by *norpA* can be excluded since PUFAs activate TRP channels with undiminished potency in severe *norpA* mutants. Secondly, if an alternative PLC isoform were responsible, the potency of PUFAs should be much greater on an *rdgA* background, since this massively enhances activation by residual PLC activity. However, no difference in the potency of PUFAs was found when applied, e.g. to *norpA* compared to *norpA*;*rdgA* double mutants.

A second objection relates to the tendency for *Drosophila* TRP and TRPL channels to activate spontaneously during whole-cell recordings made without ATP in the electrode (Hardie & Minke 1994). This activation can be accelerated by application of mitcohondrial inhibitors and since such inhibitors were found to be effective in *norpA* mutants, it was concluded that they acted downstream of PLC (Agam et al 2000). Since PUFAs are known to act as mitochondrial uncouplers, it was proposed that this accounted for their ability to activate TRP channels *in situ* (Agam et al 2000). To explore this possibility we compared the actions of PUFAs and mitochondrial inhibitors in *norpA* mutants. To appreciate these results it must be explained that PLC is an obligatory GTPase activating protein required for inactivation of the activated $G_q\alpha$ subunit (Cook et al 2000). This means that in *norpA* hypomorphs containing only trace amounts of PLC, free $G_q\alpha$ subunits can remain active for several minutes before finally encountering a PLC molecule. One consequence of this is that on establishing the whole-cell recording configuration in *norpA* photoreceptors there is a small spontaneous inward current created by the summation of quantum bumps with very long latencies generated by prior illumination (Cook et al 2000, Hardie et al 2002). As discussed above, quantum bumps in *norpA* are reduced in amplitude, but can be greatly enhanced by the *rdgA* mutation or by ATP depletion. Consequently, it seems likely that the activation of inward currents by metabolic inhibitors in *norpA* mutants could be directly attributed to impairment of DGK activity and the resulting amplification of the miniature quantum bumps underlying the spontaneous currents. If this is the case then the inhibitors should no longer be effective if the spontaneous current is eliminated. This can be achieved either by waiting in the dark (the current decays to baseline after *circa*. 20 minutes) or by eliminating the G protein α-subunit (in a *norpA*;*Gαq* double mutant). In either case mitochondrial inhibitors were no longer able to activate any channels at all in the absence of spontaneous currents; however, under the same conditions PUFAs invariably activated TRP channels with undiminished potency. Their action thus is independent of any role as mitochondrial uncouplers, but consistent with a direct action on the channels. Whether PUFAs are the endogenous excitatory messengers, in which case they would presumably be released from DAG by DAG lipase, or whether for example they mimic the effect of DAG, remains to be determined.

PIP_2 depletion

In addition to the mounting evidence for DAG's role as an essential messenger (or intermediate) of excitation, recent studies have also raised the intriguing possibility that PIP_2 reduction may contribute to excitation of TRP and/or TRPL channels. The first indication came from studies showing that recombinant TRPL channels in inside-out patches could be activated by application of exogenous PLCβ and subsequently suppressed by application of PIP_2 (Estacion et al 2001). However, recordings of light-activated TRPL channel activity, isolated in *trp* mutants, suggest a different picture *in vivo*. In the *trp* mutant prolonged illumination leads to complete loss of PIP_2 in the microvillar membrane, because Ca^{2+} influx via the TRP channels appears to be required for efficient and rapid PIP_2 recycling; but under these conditions the remaining TRPL channels rapidly close and remain profoundly inactivated until PIP_2 is resynthesized (Hardie et al 2001). This collapse of the response is entirely consistent with a role for DAG in excitation, since the substrate for its generation is exhausted, but clearly the opposite of what would be expected if PIP_2 depletion directly activated the channels. Interestingly, TRP channels behave rather differently. The Ca^{2+} influx required for maintaining PIP_2 levels can also be blocked by removing extracellular Ca^{2+}: under these conditions flashes of light which deplete a substantial fraction of PIP_2 usually result in a failure of the TRP channels to close after termination of the light flash. A similar failure in response termination can also be observed in mutants of the *rdgB* gene, which encodes an essential component of the PIP_2 recycling pathway (Hardie et al 2001). Since large amounts of DAG are also produced under these conditions, this cannot necessarily be attributed to PIP_2 depletion alone; however, a hypothesis worth further investigation is that channels may be activated by simultaneous generation of DAG and depletion of PIP_2. For example, the TRP/TRPL proteins (or associated proteins) might incorporate domains capable of binding PIP_2 and DAG (or PUFA); binding to PIP_2 would stabilize the closed state, whilst DAG/PUFA would stabilize the open state.

Conclusion

Drosophila phototransduction has long been an influential genetic model for G protein-coupled signalling and PLC signalling in particular. Perhaps foremost amongst the many mutants and genes isolated from this pathway have been the light-sensitive TRP channels, which have emerged as a large family of cation channels responsible, *inter alia*, for most Ca^{2+} influx associated with PLC activation. Although traditionally considered to be activated downstream of the IP_3 receptor, both in *Drosophila* and increasingly in many vertebrate systems, it now appears that many Ca^{2+} influx pathways may instead be regulated by lipid products of PLC activity. Nevertheless, the detailed gating mechanism of TRP channels remains unresolved and will continue so until a number of key questions have been answered. In *Drosophila* these include whether there is a DAG lipase (which has yet to be cloned in any eukaryote) expressed in *Drosophila* photoreceptors and which is required for phototransduction, and whether PUFAs are released in response to illumination. In addition it will also be important to determine whether there are DAG, PUFA or PIP_2 binding domains on the TRP and TRPL channels or associated proteins and to demonstrate directly that these are responsible for channel gating.

Acknowledgements

The author's work described in this review was supported by grants from the Wellcome Trust, the Biotechnology and Biological Sciences Research Council (BBSRC) and the Medical Research Council (MRC).

References

Agam K, von Campenhausen M, Levy S et al 2000 Metabolic stress reversibly activates the *Drosophila* light-sensitive channels TRP and TRPL in vivo. J Neurosci 20:5748–5755

Benham CD, Davis JB, Randall AD 2002 Vanilloid and TRP channels: a family of lipid-gated cation channels. Neuropharmacology 42:873–888

Chyb S, Raghu P, Hardie RC 1999 Polyunsaturated fatty acids activate the *Drosophila* light-sensitive channels TRP and TRPL. Nature 397:255–259

Cook B, Bar-Yaacov M, Cohen BenAmi H et al 2000 Phospholipase C and termination of G-protein-mediated signalling in vivo. Nat Cell Biol 2:296–301

Cosens DJ, Manning A 1969 Abnormal electroretinogram from a *Drosophila* mutant. Nature 224:285–287

Estacion M, Sinkins WG, Schilling WP 2001 Regulation of *Drosophila* transient receptor potential-like (TrpL) channels by phospholipase C-dependent mechanisms. J Physiol 530: 1–19

Hardie RC 2003 Regulation of TRP channels via lipid second messengers. Annu Rev Physiol 65:735–759

Hardie RC, Minke B 1992 The *trp* gene is essential for a light-activated Ca^{2+} channel in Drosophila photoreceptors. Neuron 8:643–651

Hardie RC, Minke B 1994 Spontaneous activation of light-sensitive channels in *Drosophila* photoreceptors. J Gen Physiol 103:389–407
Hardie RC, Peretz A, Suss-Toby E et al 1993 Protein kinase C is required for light adaptation in *Drosophila* photoreceptors. Nature 363:634–637
Hardie RC, Raghu P, Moore S, Juusola M, Baines RA, Sweeney ST 2001 Calcium influx via TRP channels is required to maintain PIP_2 levels in Drosophila photoreceptors. Neuron 30: 149–159
Hardie RC, Martin F, Cochrane GW, Juusola M, Georgiev P, Raghu P 2002 Molecular basis of amplification in Drosophila phototransduction. Roles for G protein, phospholipase C, and diacylglycerol kinase. Neuron 36:689–701
Hofmann T, Obukhov AG, Schaefer M, Harteneck C, Gudermann T, Schultz G 1999 Direct activation of human TRPC6 and TRPC3 channels by diacylglycerol. Nature 397:259–263
Masai I, Okazaki A, Hosoya T, Hotta Y 1993 *Drosophila* retinal degeneration A gene encodes an eye-specific diacylglycerol kinase with cysteine-rich zinc-finger motifs and ankyrin repeats. Proc Natl Acad Sci USA 90:11157–11161
Minke B, Cook B 2002 TRP Channel proteins and signal transduction. Physiol Rev 82:429–472
Montell C 2001 Physiology, phylogeny, and functions of the TRP superfamily of cation channels. Sci STKE 90:RE1
Montell C, Rubin GM 1989 Molecular characterization of Drosophila trp locus, a putative integral membrane protein required for phototransduction. Neuron 2:1313–1323
Niemeyer BA, Suzuki E, Scott K, Jalink K, Zuker CS 1996 The Drosophila light-activated conductance is composed of the two channels TRP and TRPL. Cell 85:651–659
Phillips AM, Bull A, Kelly LE 1992 Identification of a Drosophila gene encoding a calmodulin-binding protein with homology to the trp phototransduction gene. Neuron 8:631–642
Raghu P, Colley NJ, Webel R et al 2000a Normal phototransduction in Drosophila photoreceptors lacking an InsP(3) receptor gene. Mol Cell Neurosci 15:429–445
Raghu P, Usher K, Jonas S, Chyb S, Polyanovsky A, Hardie RC 2000b Constitutive activity of the light-sensitive channels TRP and TRPL in the Drosophila diacylglycerol kinase mutant, rdgA. Neuron 26:169–179
Reuss H, Mojet MH, Chyb S, Hardie RC 1997 In vivo analysis of the Drosophila light-sensitive channels, TRP and TRPL. Neuron 19:1249–1259
Smith DP, Ranganathan R, Hardy RW, Marx J, Tsuchida T, Zuker CS 1991 Photoreceptor deactivation and retinal degeneration mediated by a photoreceptor-specific protein-kinase-C. Science 254:1478–1484
Topham MK, Prescott SM 1999 Mammalian diacylglycerol kinases, a family of lipid kinases with signaling functions. J Biol Chem 274:11447–11450
Xu XZS, Chien F, Butler A, Salkoff L, Montell C 2000 TRP gamma, a Drosophila TRP-related subunit, forms a regulated cation channel with TRPL. Neuron 26:647–657

DISCUSSION

Westwick: I am going to ask you the same question I asked Craig Montell. When you showed your PIP_2 cycles, you never brought in phosphoinositide 3-kinase (PI3K). This is another mechanism for getting rid of PIP_2, because you will get PIP_3 which won't bind to your biosensor. If you block PI3K in your system, what happens?

Hardie: Wortmannin has no obvious effect, but this is just a negative result and otherwise we have not looked at PI3K. As Craig Montell said, mutants affecting phosphatidylinositol 3,4,5-trisphosphate (PIP_3) don't appear to affect activation, and I would doubt PI3K or PIP_3 have any effect on the direct activation cycle. It may be involved in membrane recycling and protein translocation.

Westwick: Does your PIP_2 sensitive ion channel respond to PIP_3?

Hardie: No, it is extremely specific for PIP_2, which is one of the reasons why we chose it. It is the most specific of all the *Kir* channels.

Westwick: If you add PIP_3 to your system, what happens?

Hardie: We haven't done that experiment.

Montell: Related to this, if you eliminate or overexpress the PTEN in fly photoreceptors, there is no effect on activation but such manipulations do affect translocation of arrestin. I was wondering what your views were about the results that Charles Zuker's lab published a few years ago (Scott et al 1997). They concluded that the TRP phenotype was due to the very robust Ca^{2+}-dependent inactivation of TRPL.

Hardie: The TRPL channel does undergo Ca^{2+}-dependent inactivation, but this is unrelated to what is generally considered to be the true 'TRP decay' which is associated with profound prolonged loss of sensitivity. There is actually a clear biphasic decay of the response in *trp* mutants. The first rapid decay is Ca^{2+}-dependent inactivation and the slow decay reflects PIP_2 depletion. If you do these experiments in the absence of any Ca^{2+}, the rapid component is eliminated but the slow component is accelerated. The reason for this phenotype seems to be that the Ca^{2+} influx through the TRP channel is required for two things: first, rapid inhibition of PLC, and secondly for facilitating PIP_2 recycling. We are looking at the moment at what level this occurs; for example the PITP protein might be one candidate.

Montell: I am not defending the work from Zuker's lab, but all they did was mutate the calmodulin binding site of TRPL, and were able to suppress the transient phenotype. It seemed like a straightforward result.

Hardie: Once again, this relates to the biphasic decay and the rapid initial Ca^{2+}-dependent inactivation. In Zuker's paper they only gave rather short bright flashes as stimuli. I think they were therefore probably just looking at the rapid component of inactivation and not at stimuli that actually induced the loss of PIP_2.

Gill: Is there any effect of the RHC lipase inhibitor compound?

Hardie: The lipase inhibitor itself is ineffective. This doesn't surprise me. There are many established pharmacological blockers that don't work well in *Drosophila* photoreceptors. This might be related to the exceptionally high concentration of protein in this system, accessibility to the microvillar membrane or pharmacological non-equivalence.

Penner: I have a more fundamental question. Why do flies resort to lipids which are sticky and probably much more difficult to handle? If they want to have a really fast on–off response, a lipid is a poor choice. I would use something like cGMP.

Hardie: I think the rapid kinetics is due to the localization of the signalling cascade to a single microvillus. These are short distances. There is no problem in getting the kinetics of the fly photoresponse based on a model of protein and lipid diffusion within the dimensions of the microvillus. With very plausible assumptions we can easily reproduce the kinetics in our model (M. Postma, R.C. Hardie, unpublished results).

Penner: I don't doubt this.

Hardie: One of the crucial features in the way the fly works is that excitation from each photon is probably restricted to a single microvillus. This can readily be understood with a lipid messenger; however, a water soluble messenger could diffuse out of the microvillus and excite a number of surrounding microvilli. One of the remarkable properties of fly photoreceptors is that, unlike the rods in our own eyes, not only can they respond to single photons, but they can also respond under bright daylight conditions when they are absorbing about half a million photons a second. The only way I think this can be achieved is if each microvillus acts as a semi-autonomous unit. Each microvillus can be activated, deactivated and recover within 100 ms, and with about 50 000 microvilli, this would allow photons to be processed at *c.* 5×10^5 per second. If you had a soluble messenger, high light intensities would probably simply flood the whole cell. This may be a good reason for using a membrane-delimited system.

Putney: I would like to know what the lateral diffusion rate of DAG is in membranes, compared with a low molecular weight substance in solution.

Hardie: It's much slower of course; there are a variety of figures in the literature, ranging from about 1–20 $\mu m^2/s$. Whether it is at the low or high end of that range will influence how easily the kinetics can be modelled by lipid diffusion. We are intending to measure lipid diffusion coefficients directly by measuring fluorescently tagged lipids in the microvilli.

Ambudkar: I am a little confused. You said that the slow inactivation is because of PIP_2 depletion. At the same time, when you added PIP_2 to the channel when it was active, you blocked it.

Hardie: This reflects the difference between the TRP and the TRPL channel. The slow inactivation is seen in the *trp* mutant which has just TRPL channels. When PIP_2 is depleted those channels close and won't open again until the PIP_2 is resynthesized and given another flash of light. If we deplete PIP_2 when TRP channels are present—and we can do this by making use of mutants which have defects in the PIP_2 recycling pathway—then when PIP_2 is depleted the TRP channels stay open. They can be closed by re-adding PIP_2. I was distinguishing quite strictly between TRP and TRPL under those circumstances. The TRP channels appear to stay open when PIP_2 is depleted; the TRPL channels stay closed.

Putney: Does this behaviour speak to the possibility of a PIP_2 control mechanism regulating TRPL combinations? That is, can loss of PIP_2 itself activate TRP?

Hardie: I'm suggesting that after the channels have been activated, they will remain open if the PIP_2 is depleted. I think they need the DAG or a DAG metabolite for activation in the first place. But PIP_2 may be required for reclosing. Just to give a specific model, perhaps the TRP channel is bound to PIP_2 in the closed state, activation involves DAG binding possibly to an overlapping binding site, displacing the PIP_2, and then when the DAG disappears the PIP_2 rebinds. If it doesn't, the channel stays open even though the DAG is gone.

Putney: The *rdgA* phenotype could be explained by the fact that without rephosphorylating DAG you can't remake PI and subsequently PIP_2.

Hardie: I think we can exclude that, because the phenotype of the *rdgA* in enhancing sensitivity in a *norpA* or a $G\alpha q$ background, for example, is specific for *rdgA*, in that mutants in other, subsequent steps in the PIP_2 recycling pathway won't work. E.g. if you take the next enzyme, CDP DAG synthase, and make a double mutant for this with $G\alpha q$, it actually further reduces the sensitivity. So we are confident that it is really the DAG and not the PIP_2 that is required to explain that phenotype. You can, however, it seems, get a situation where the channels are maintained in an open state due to the absence of PIP_2. We wanted to get a way of depleting PIP_2 without generating DAG to see whether this could be sufficient, but it has proved very difficult. We have tried expressing PIP_2 phosphatases in the cells, but this doesn't get rid of the PIP_2 (P. Raghu and R. C. Hardie, unpublished results).

Fleig: Perhaps an experiment that might be worth doing is to try to keep PIP_2 levels in the plasma membrane and see whether you can still get activation of the channel by DAG.

Hardie: In a heterologous expression system, we do.

Fleig: This would argue that PIP_2 depletion has nothing to do with the activation of the channel.

Hardie: I don't think so. The channels would be maintained in an open state in the absence of DAG or fatty acids, simply because PIP_2 levels are depleted.

Fleig: You said that the TRP channel is permeable to other divalent cations. Have you looked with zinc and nickel?

Hardie: No, we have used barium, manganese and magnesium. They are all more permeable than any monovalent cations.

Scharenberg: Is the situation with the other TRPs similar?

Fleig: Ca^{2+} channels are thought to permeate manganese, barium and strontium. Cadmium usually blocks at micromolar levels.

Montell: I don't think there is any conflict with what you were describing about the activation mechanism and anything I said about the signalplex. There is currently no evidence that the signalplex has a specific role in activation.

Hardie: We agree on that.

Montell: But it definitely has a role in the termination of the photoresponse. You mentioned that the signalplex is necessary for getting the proteins into the rhabdomeres. INAD definitely doesn't have a role in targeting the proteins to the rhabdomeres. Instead, it has a role in retaining the proteins there once they are targeted to the rhabdomeres through mechanisms independent of INAD. It was our expectation that the signalplex would have a role in activation, but if you disrupt the direct interaction between TRP and INAD the activation kinetics don't seem to be profoundly affected.

Hardie: My understanding is that TRP is targeted without INAD, but that PLC and PKC need to be pre-assembled with INAD to get into the microvilli (Tsunoda et al 1997), but in any case, my point is that it is the microvillar organization that may be the most critical for rapid and efficient activation rather than the signalplex *per se*.

Gill: This is an important point. Many of us in the mammalian field have been led to think that organization within the signalplex was the paradigm for how things were being turned on. Now we are thinking that perhaps it is not so important for rapid turning on. Could it be that the organization in the eye is also different? That because there is rapid turnover of these proteins, they have to be kept in a particular location?

Montell: The main caveat to this is that we can't exclude that indirect interactions with the signalplex are required for the rapid activation. The mutation in TRP that prevented direct binding to INAD did not disrupt indirect interactions between TRP and INAD.

Hardie: To summarize, the signaplex **does** appear to be required for rapid response termination, but whilst we can't exclude a direct role for the signalplex in activation, there isn't any compelling and specific evidence for this. On the other hand it seems clear that microvillar localization of TRP, G protein and PLC in appropriate numbers is essential for rapid activation.

References

Scott K, Sun YM, Beckingham K, Zuker CS 1997 Calmodulin regulation of Drosophila light-activated channels and receptor function mediates termination of the light response in vivo. Cell 91:375–383

Tsunoda S, Sierralta J, Sun YM et al 1997 A multivalent PDZ-domain protein assembles signalling complexes in a G-protein-coupled cascade. Nature 388:243–249

Control of TRPC and store-operated channels by protein kinase C

Kartik Venkatachalam, Fei Zheng and Donald L. Gill[1]

Department of Biochemistry and Molecular Biology, University of Maryland School of Medicine, 108 North Greene Street, Baltimore, MD 21201, USA

Abstract. TRPC channels are widely expressed among cells and are believed to play important roles in receptor-mediated Ca^{2+} signalling. We determined that the function of TRPC channels is highly regulated by protein kinase C (PKC). Application of diacylglycerol (DAG) or elevated endogenous DAG resulting from either DAG-lipase or DAG-kinase inhibition, completely prevented TRPC5 or TRPC4 activation in both HEK293 cells and DT40 cells. This inhibitory action of DAG on TRPC5 and TRPC4 channels was clearly mediated by PKC, in distinction to the stimulatory action of DAG on TRPC3 which was PKC-independent. PKC activation totally blocked TRPC3 channel-activated in response to OAG, and was restored by PKC-blockade. PKC-inhibition resulted in decreased TRPC3 channel deactivation. Store-operated Ca^{2+} entry in response to PLC-coupled receptor activation but not store-depletion *per se*, was substantially reduced by OAG or DAG-lipase inhibition in a PKC-dependent manner. The results reveal that each TRPC subtype is strongly inhibited by DAG-induced PKC activation reflecting a likely universal feedback control on TRPCs. The profound yet distinct control by PKC and DAG on the activation of TRPC channel subtypes may be the basis of a spectrum of regulatory phenotypes of expressed TRPC channels.

2004 Mammalian TRP channels as molecular targets. Wiley, Chichester (Novartis Foundation Symposium 258) p 172–188

The TRPC family of channels are ubiquitously expressed in vertebrate cells and are the products of at least seven genes coding for cation channels that appear to be activated primarily in response to phospholipase C (PLC)-coupled receptors (Montell et al 2002a, Venkatachalam et al 2002). TRPC channels are related closely in structure and function to the group of TRP channel proteins first identified in *Drosophila* that mediate the PLC-dependent light-induced current in retinal cells (Montell 2001). Interest has focused on the vertebrate TRPC subfamily since these channels have been implicated as important mediators of Ca^{2+} entry

[1]This paper was presented at the symposium by Donald L. Gill to whom correspondence should be addressed.

(Venkatachalam et al 2002, Zitt et al 2002). Evidence indicates that they may function as 'store-operated' channels (SOCs) (Venkatachalam et al 2002, Putney Jr et al 2001, Birnbaumer et al 1996, Philipp et al 1996, Kiselyov et al 1998, Philipp et al 1998, Vannier et al 1999, Philipp et al 2000, Liu et al 2000a) mediating the process of capacitative Ca^{2+} entry—essential for longer term Ca^{2+} signals and replenishment of Ca^{2+} stores (Putney Jr et al 2001, Venkatachalam et al 2002). Studies on the coupling between TRPC channels and intracellular inositol 1,4,5-trisphosphate receptors (IP_3Rs) (Kiselyov et al 1998, Ma et al 2000, Boulay et al 1999, Zhang et al 2001, Tang et al 2001) have suggested that TRPC channels can receive information directly from Ca^{2+} stores. However, there is also considerable evidence that TRPC channels can function independently of stores (Venkatachalam et al 2002, Zitt et al 2002, Hofmann et al 2000, Harteneck et al 2000, Putney Jr et al 2001). Our analyses reveal that TRPC3 channels are activated in response to PLC-coupled receptors and mimicked by the application of exogenous diacylglycerol (DAG) (Venkatachalam et al 2001), consistent with the earlier report from Hofmann and colleagues indicating that members of the closely related subgroup of TRPC3, TRPC6 and TRPC7 channels, can each be activated in response to DAG through a mechanism independent of PKC (Hofmann et al 1999). Other members of the TRPC channel family appear to behave differently. Thus, the subgroup represented by the closely related TRPC4 and TRPC5 channel proteins are reported to respond to store-depletion (Philipp et al 1996, 1998, 2000) and to have an essential requirement for the IP_3R (Kanki et al 2001). Moreover, both TRPC4 and TRPC5 channels are reported to be unresponsive to application of DAG (Hofmann et al 1999). We therefore considered it important to investigate the role of store-emptying and IP_3Rs in the activation of TRPC4 and TRPC5 channels utilizing the DT40 knockout cell lines, and to assess how the activation of these channels in response to PLC-coupled receptors compares with the activation of TRPC3 channels. Our results indicate some important differences in the role of DAG as a mediator of TRPC channel activation, and reveal that each TRPC subtype is strongly inhibited by DAG-induced protein kinase C (PKC) activation reflecting a likely universal feedback control mechanism for TRPC channels.

Using either the DT40 B cell line or HEK 293 cells, we found that the functional phenotype of expressed TRPC5 channels appears almost identical to that of the TRPC3 channel we studied earlier (Venkatachalam et al 2001). Both channels can be activated in response to G protein-coupled receptor (GPCR)-induced activation of $PLC\beta$ or receptor-induced tyrosine kinase-mediated activation of $PLC\gamma$. And the activation of both channels does not require the presence of IP_3Rs or store-depletion. Since PLC activation is required and since TRPC3 channels can be activated by exogenously applied DAG, we concluded that DAG is the mediator through which TRPC3 channels are stimulated in response to receptors

(Venkatachalam et al 2001). However, with respect to TRPC5, it was earlier shown by Hofmann et al (1999) that, in contrast to its stimulation of TRPC3 channels, DAG does not activate TRPC5 channels. It was important therefore to ascertain whether a similar differential effect of DAG applied to the function of TRPC3 and TRPC5 channels in our systems, or whether our expression conditions had somehow rendered the TRPC5 channel sensitive to DAG.

We transiently transfected the HEK 293 cells with TRPC5 or TRPC3 and analysed channel activation in response to both CCh-mediated PLCβ activation and treatment with the cell permeant analogue of DAG, 1-oleoyl-2-acetyl-*sn*-glycerol (OAG) (Hofmann et al 1999, Ma et al 2000). It is clear from the data in Fig. 1A that TRPC5-mediated Sr^{2+} entry was activated in response to carbachol (CCh)-stimulation, however, after cessation of entry following removal of Sr^{2+} and CCh, subsequent addition of 100 μM OAG with Sr^{2+} resulted in no entry. On the other hand, in exactly analogous experiments on TRPC3-transfected cells (Fig. 1B), the final addition of OAG caused a robust entry of Sr^{2+} 100 μM OAG (Fig. 1B). Using TRPC5-transfected DT40 cells, the addition of OAG in the presence of Ba^{2+} induced no entry, even though subsequent BCR cross-linking by anti-IgM induced a substantial entry of Ba^{2+} (Fig. 1C). In this experiment, however, the increased F_{340}/F_{380} ratiometric signal also has a substantial component from the B cell receptor-induced release of stores. Therefore, we undertook the same experiment using the DT40 triple-$IP_3R^{-/-}$ cells devoid of IP_3Rs (Fig. 1D). In this case, the Ca^{2+} store-release component was eliminated, and while OAG again had no effect on Ba^{2+} entry, B cell receptor cross-linking resulted in TRPC5-mediated Ba^{2+} entry. The lag of approximately 1 min before the start of Ba^{2+} entry was consistently observed. This appears to reflect the slow B cell receptor-induced activation of PLCγ2.

A further question was whether the lack of effect of OAG on TRPC5 channels might reflect some divergence in the function of the permeant DAG analogue from

FIG. 1. OAG activates TRPC3 channels but not TRPC5 channels in both HEK 293 and DT40 cells. Standard conditions included Ca^{2+}-free medium; *bars* indicate replacement of Ca^{2+}-free media with media containing Sr^{2+} or Ba^{2+}. (A) In HEK 293 cells co-transfected with TRPC5 and eYFP, rapid IP_3-mediated Ca^{2+} release was induced by the addition of 100 μM CCh (*bar*). Addition of 1 mM Sr^{2+} (*bar*) caused entry of Sr^{2+} via activated TRPC5. Subsequent addition of 100 μM OAG (*arrow*) did not activate the channel. (B) In HEK 293 cells cotransfected with TRPC3 and eYFP, rapid IP_3-mediated Ca^{2+} release was induced by addition of 100 μM CCh (*bar*). Addition of 1 mM Sr^{2+} (*bar*) caused entry of Sr^{2+} via activated TRPC3. Subsequent addition of 100 μM OAG also caused activation of the channel leading to Sr^{2+} entry. (C) In DT40 *wt* cells co-transfected with TRPC5 and eYFP, addition of 3 mM Ba^{2+} (*bar*) did not lead to any constitutive entry via TRPC5. 100 μM OAG (*bar*) added in the presence of Ba^{2+} did not activate TRPC5. Subsequent addition of 3 μg/ml anti-IgM caused IP_3-mediated Ca^{2+} release due to activation of PLCγ and Ba^{2+} entry due to TRPC5 activation. (D) Same as in *C*, but in the DT40 *$IP_3R^{-/-}$* cells.

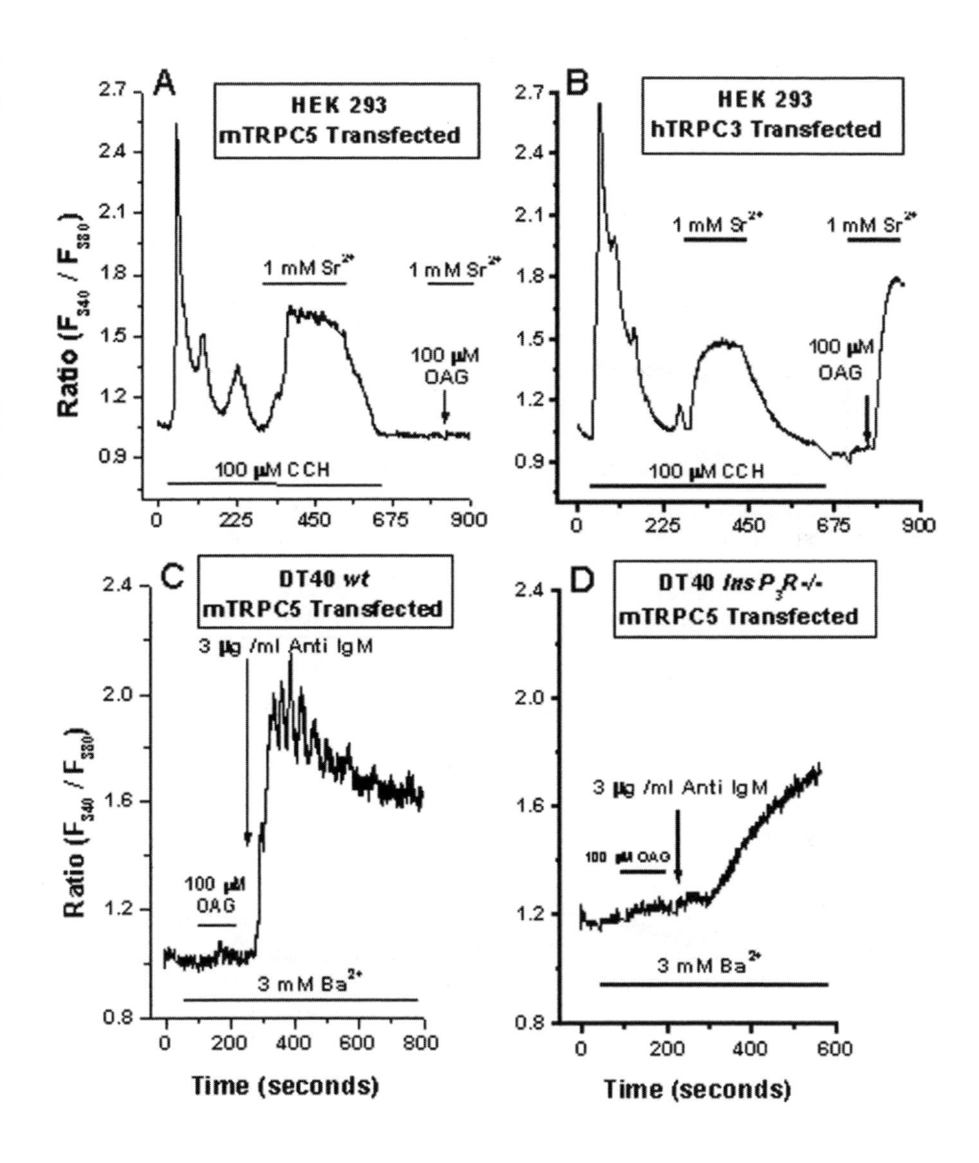
A
HEK 293
mTRPC5 Transfected
Ratio (F340 / F380)
1 mM Sr2+
1 mM Sr2+
100 µM
OAG
100 µM CCH
B
HEK 293
hTRPC3 Transfected
1 mM Sr2+
1 mM Sr2+
100 µM
OAG
100 µM CCH
C
DT40 wt
mTRPC5 Transfected
3 µg /ml Anti IgM
100 µM
OAG
3 mM Ba2+
Time (seconds)
D
DT40 Ins P3R -/-
mTRPC5 Transfected
3 µg /ml Anti IgM
100 µM OAG
3 mM Ba2+
Time (seconds)

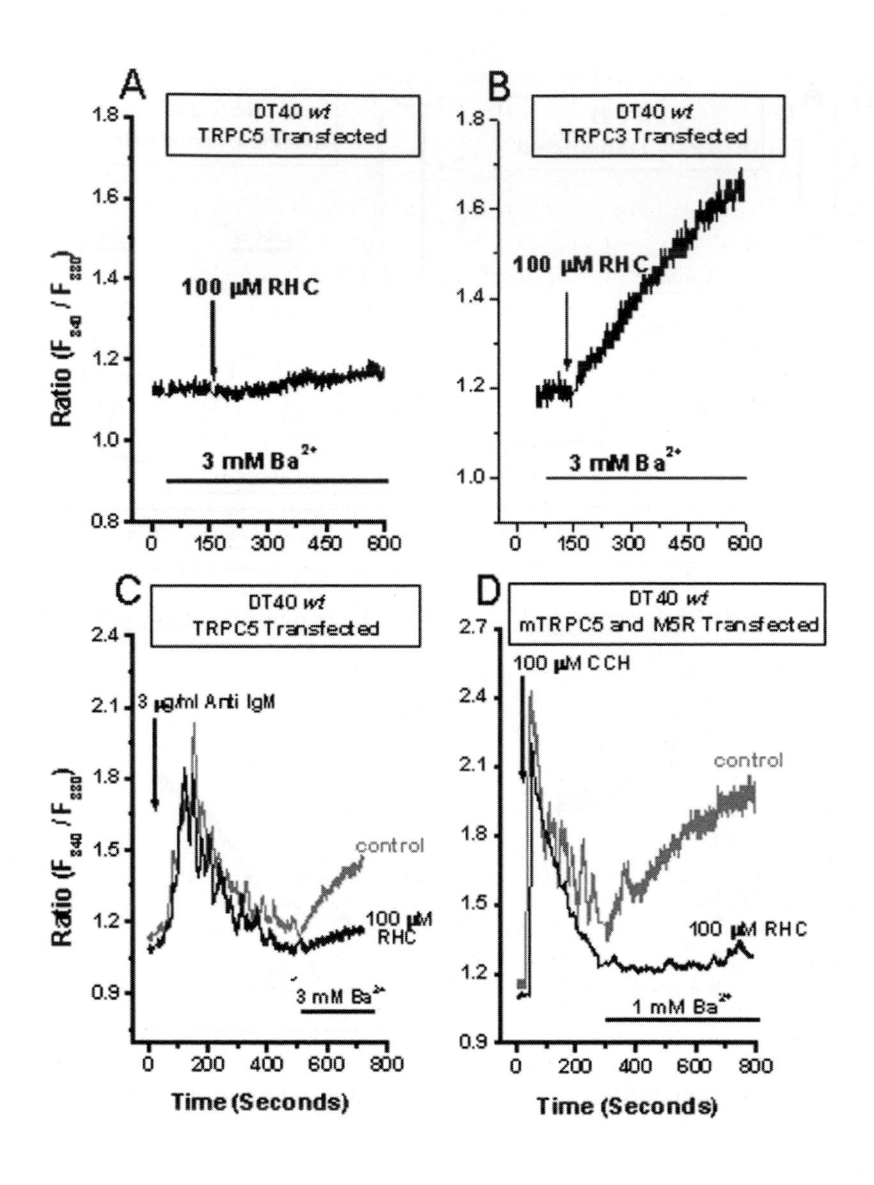
A
DT40 wt
TRPC5 Transfected
100 μM RHC
3 mM Ba2+
Ratio (F340 / F380)
B
DT40 wt
TRPC3 Transfected
100 μM RHC
3 mM Ba2+
C
DT40 wt
TRPC5 Transfected
3 μg/ml Anti IgM
control
100 μM RHC
3 mM Ba2+
Time (Seconds)
D
DT40 wt
mTRPC5 and M5R Transfected
100 μM CCH
control
100 μM RHC
1 mM Ba2+
Time (Seconds)

the function of authentic, endogenously generated DAG. Endogenous DAG undergoes continual turnover through the combined actions of DAG kinase and DAG lipase, and the latter can be effectively eliminated by the DAG lipase inhibitor, RHC-80267, resulting in a rapid elevation in the steady-state level of endogenous DAG sufficient to activate TRPC3 channels (Ma et al 2000, Hofmann et al 1999). We examined the action of RHC-80267 on DT40-*wt* cells transiently transfected with TRPC5 and found that it had no effect on Ba^{2+} entry (Fig. 2A), whereas it clearly activated Ba^{2+} entry in TRPC3-transfected DT40-*wt* cells (Fig. 2B), confirming previous observations on DAG activation of this channel (Ma et al 2000). Thus, despite the many similarities in function of TRPC3 and TRPC5 channels, it appears the TRPC5 channel differs in being insensitive to either exogenously added or endogenously generated DAG. With these observations in mind, we sought to evaluate whether DAG was playing a role in agonist-mediated activation of TRPC5, that is, whether agonist-mediated activation of TRPC5 was also independent of elevated levels of DAG. Therefore, we undertook experiments to assess whether increasing levels of DAG with RHC-80267 would have any permissive effect on receptor-induced TRPC5 activation. To our surprise we found that RHC-80267 completely blocked TRPC5 activation. Thus, as shown in Fig. 2C, using TRPC5-transfected DT40 cells, we abolished the activation of Ba^{2+} entry in response to B cell receptor cross-linking in the presence of RHC-80267. Likewise, the DAG lipase blocker completely prevented TRPC5 activation in response to CCh in DT40 cells co-transfected with TRPC5 and the M5 muscarinic receptor (M5R) (Fig. 2D). Therefore, it appears that RHC-80267-mediated elevation of DAG blocks activation of TRPC5 whereas it activates TRPC3.

We also assessed the actions of modifying DAG on TRPC5 channels by examining the effects of exogenous OAG added directly to the cells, and by

FIG. 2. Inhibition of DAG-lipase causes activation of TRPC3 channels and inhibits TRPC5 channels. Standard conditions included Ca^{2+}-free medium; *bars* indicate replacement of Ca^{2+}-free media with media containing Ba^{2+}. (A) In DT40 *wt* cells co-transfected with TRPC5 and eYFP, addition of 3 mM Ba^{2+} (*bar*) did not cause constitutive entry via TRPC5. Subsequent addition of 100 μM RHC-80267 (RHC) (*arrow*) did not activate TRPC5. (B) In DT40 *wt* cells co-transfected with TRPC3 and eYFP, addition of 3 mM Ba^{2+} (*bar*) did not result in constitutive entry via TRPC3. However, subsequent addition of 100 μM RHC (*arrow*) activated TRPC3 and led to a rapid entry of Ba^{2+}. (C) In DT40 *wt* cells co-transfected with TRPC5 and eYFP, addition of 3 μg/ml anti-IgM led to IP_3-mediated Ca^{2+} release and subsequent TRPC5-mediated Ba^{2+} entry upon addition of 3 mM Ba^{2+} (*bar*) (*grey trace*). In the presence of 100 M RHC (*black trace*), IP_3-mediated Ca^{2+} release was intact but TRPC5 mediated Ba^{2+} entry was absent. (D) Using DT40 *wt* cells co-transfected with TRPC5, M5R and eYFP, addition of 100 μM CCh led to rapid IP_3-mediated Ca^{2+} release and subsequent TRPC5-mediated Ba^{2+} entry upon addition of 1 mM Ba^{2+} (*bar*) (*grey trace*). In the presence of 100 μM RHC (*black trace*), IP_3-mediated Ca^{2+} release was intact but TRPC5-mediated Ba^{2+} entry was absent.

modifying the function of the DAG-kinase. In HEK293 cells, the presence of RHC-80267 added together with CCh completely abolished the activation of TRPC5 channels, confirming the results obtained using DT40 cells and revealing that the action of RHC-80267 is not to alter the function or production of IP_3 through modification of PLC. Application of 100 μM OAG together with CCh also completely prevented the activation of the TRPC5 channel, consistent with the conclusion that DAG itself is mediating the inhibitory action on TRPC5 activation. Moreover, the DAG kinase inhibitor R59949 also completely blocked the TRPC5 channel activation providing further verification that increased DAG results in the deactivation of TRPC5 channel activity. The results provide compelling evidence that elevation of endogenous DAG or exogenous addition of OAG, lead to a complete inhibition of the TRPC5 channel. We observed essentially the same inhibitory action of OAG on the TRPC4 channel, which is a close structural and functional relative of the TRPC5 channel (Montell 2001). Crucial to ascertain was whether the novel inhibitory action of DAG on TRPC5 and TRPC4 channels was related to PKC.

To evaluate a role for PKC on the inhibitory action of DAG on TRPC5 channels, we utilized the aminoalkyl bisindolylmaleimide, GF 109203X, which is recognized as a highly selective and potent inhibitor of multiple PKC subtypes (Toullec et al 1991). We examined the action of this PKC-modifier on DT40-*wt* cells co-transfected with the M5R and TRPC5 channel, determining its effect on the actions of the DAG lipase and DAG kinase inhibitors and exogenously added OAG. As shown in Fig. 3A, the activation of Ba^{2+} entry through TRPC5 channels in response to CCh was blocked by the DAG lipase inhibitor, RHC-80267. Importantly, when GF 109203X was present with the DAG-lipase inhibitor, TRPC5 activation was exactly as without inhibitors. Thus, the PKC blocker prevented the inhibition of TRPC5 channels resulting from DAG lipase inhibition. We next assessed the effect of the PKC blocker on the action of directly added OAG (Fig. 3B), the results clearly indicating that the inhibitory action of OAG was also prevented by the simultaneous presence of GF 109203X. Lastly, TRPC5 channel inhibition by the DAG kinase blocker, R59949, was also reversed by the PKC inhibitor (Fig. 3C). The results provide rather compelling evidence that the effects of each of these different means to induce increased DAG levels can be reversed by inhibition of PKC. The fact that we have a 'return' of function induced by inhibition of PKC provides evidence that the function of the TRPC5 channel *per se* is not directly modified by any of the agents used. Instead, the results indicate that PKC has an important modulatory role in the receptor-induced coupling process that leads to TRPC5 channel activation.

Our question next was whether other members of the TRPC channel family might be similarly PKC regulated. We turned our attention to the TRPC3 channel which we have studied in detail (Ma et al 2000, Venkatachalam et al

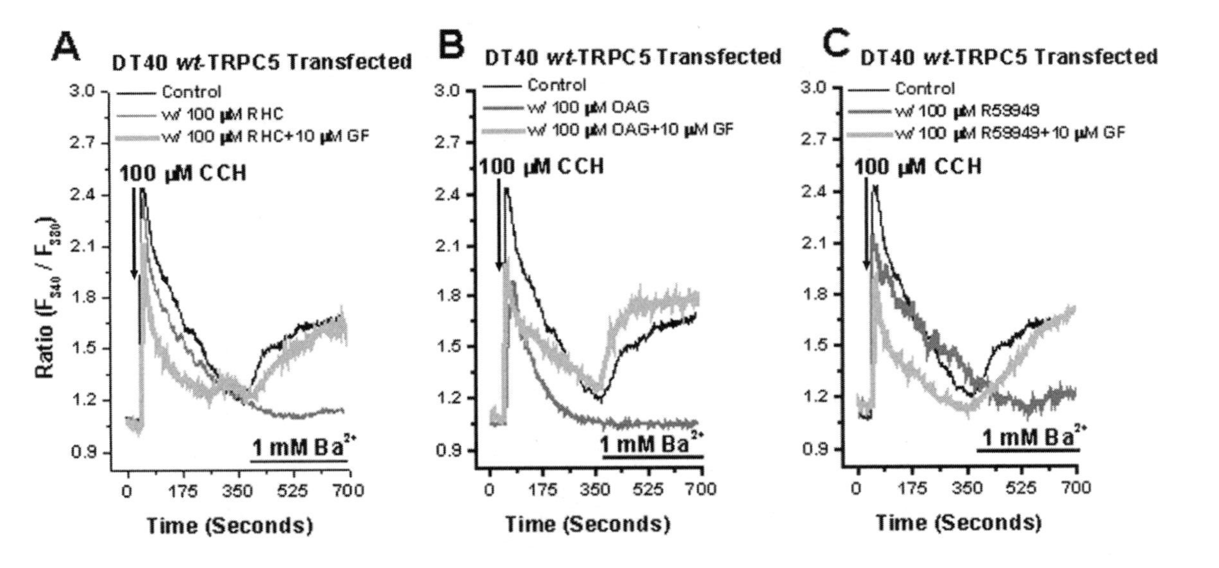

FIG. 3. DAG-mediated inhibition of TRPC5 channels is dependent on PKC. Experiments used DT40 *wt* cells co-transfected with TRPC5, M5R and eYFP. Standard conditions included Ca^{2+}-free medium; *bars* indicate replacement of Ca^{2+}-free medium with medium containing Ba^{2+}. (A) Addition of 100 μM CCh (*arrow*) caused a rapid IP_3-mediated Ca^{2+} release. Subsequent addition of 1 mM Ba^{2+} caused TRPC5-mediated Ba^{2+} entry in the control cells (*black trace*). When the same trace was performed with 100 μM RHC added with CCh (*arrow*) (*grey trace*), TRPC5-mediated Ba^{2+} entry was absent without affecting CCh-mediated Ca^{2+} release. When both 10 μM GF 109203X (GF) and 100 μM RHC were added with CCh (*arrow*) (*light grey trace*), TRPC5-mediated Ba^{2+} entry was recovered. (B) Addition of 100 μM CCh (*arrow*) caused rapid IP_3-mediated Ca^{2+} release. Subsequent addition of 1 mM Ba^{2+} caused TRPC5-mediated Ba^{2+} entry in the control cells (*black trace*). When the same trace was performed with 100 μM OAG added with CCh (*arrow*) (*grey trace*), TRPC5-mediated Ba^{2+} entry was absent without affecting CCh-mediated Ca^{2+} release. When both 10 μM GF and 100 μM OAG were added with CCh (*arrow*) (*light grey trace*), TRPC5-mediated Ba^{2+} entry was recovered. (C) Addition of 100 μM CCh (*arrow*) caused rapid IP_3-mediated Ca^{2+} release. Subsequent addition of 1 mM Ba^{2+} caused TRPC5-mediated Ba^{2+} entry in control cells (*black trace*). Performed with 100 μM R59949 added with CCh (*arrow*) (*grey trace*), TRPC5-mediated Ba^{2+} entry was absent without affecting CCh-mediated Ca^{2+} release. When both 10 μM GF and 100 μM R59949 were added with CCh (*arrow*) (*light grey trace*), TRPC5-mediated Ba^{2+} entry was recovered.

2001). We needed a means to activate PKC that was independent of DAG which is clearly an activator of the TRPC3 channel (Hofmann et al 1999, Venkatachalam et al 2001). We therefore utilized the powerful PKC-activator, phorbol myristate acetate (PMA), which causes pronounced PKC-mediated phosphorylation of targets at nanomolar levels (Quest et al 1997). Using the stably TRPC3-transfected HEK 293 T3-65 cell line used in earlier studies (Ma et al 2000), we found that OAG-mediated activation of TRPC3 is totally abolished by a 5 min pretreatment with 1 μM PMA (Fig. 4A). This inhibition of TRPC3 activity was completely reversed when cells were pretreated with PMA together with 10 μM GF 109203X (Fig. 4A). This provides compelling evidence that the TRPC3 channel is also PKC-modulated. Thus, it appears that DAG is inducing a potentially crucial bimodal regulation of TRPC3 channels. Moreover, closer examination of the data in Fig. 4A reveals that whereas TRPC3 channel activity following OAG addition is transient (the activity deactivates in the continued presence of 100 μM OAG), in the presence of the PKC inhibitor, this deactivation is clearly retarded. In contrast, the *rate* of OAG-induced activation of TRPC3 is identical in the presence or absence of GF 109203X. In other words, it appears that DAG rapidly activates TRPC3 prior to a slower PKC-mediated deactivation of the channel.

So far we have addressed the function and regulation of exogenously expressed TRPCs. Although controversial, much recent work provides evidence that endogenous store-operated Ca^{2+} channels involve the function of TRPC channels (Venkatachalam et al 2002). Since the action of PKC may be a useful and hitherto unrecognized signature of TRPC channel function, we examined the effects of PKC modification on endogenous store-operated Ca^{2+} entry. The data in Fig. 4B reveal that the potent PKC activator, PMA, had no effect on the rate or duration of Ca^{2+} entry induced in HEK 293 cells in response to complete emptying of stores induced by thapsigargin (TG). However, an interesting finding was that the Ca^{2+} entry induced in response to activation of the endogenous muscarinic receptor was prevented by almost 70% in the presence of the DAG lipase inhibitor, RHC-80267, whereas Ca^{2+} release from stores was unaffected (Fig. 4C). In this case, the more potent direct PKC activator, PMA, completely prevented receptor-induced store-emptying (likely as a result of direct actions on PLC), and hence could not be used to examine effects on entry. However, the inhibitory action of RHC-80267 on receptor-induced Ca^{2+} entry was exactly mimicked by addition of exogenous OAG (not shown). Significantly, the inhibitory action of DAG lipase blockade on Ca^{2+} entry in response to CCh was completely reversed by the PKC blocker, GF 109203X (Fig. 4C). Likewise, the inhibitory action of OAG was completely reversed by the PKC blocker (data not shown). These results provide evidence for a potential link between the endogenous entry of Ca^{2+} induced by a receptor and the activity of exogenously

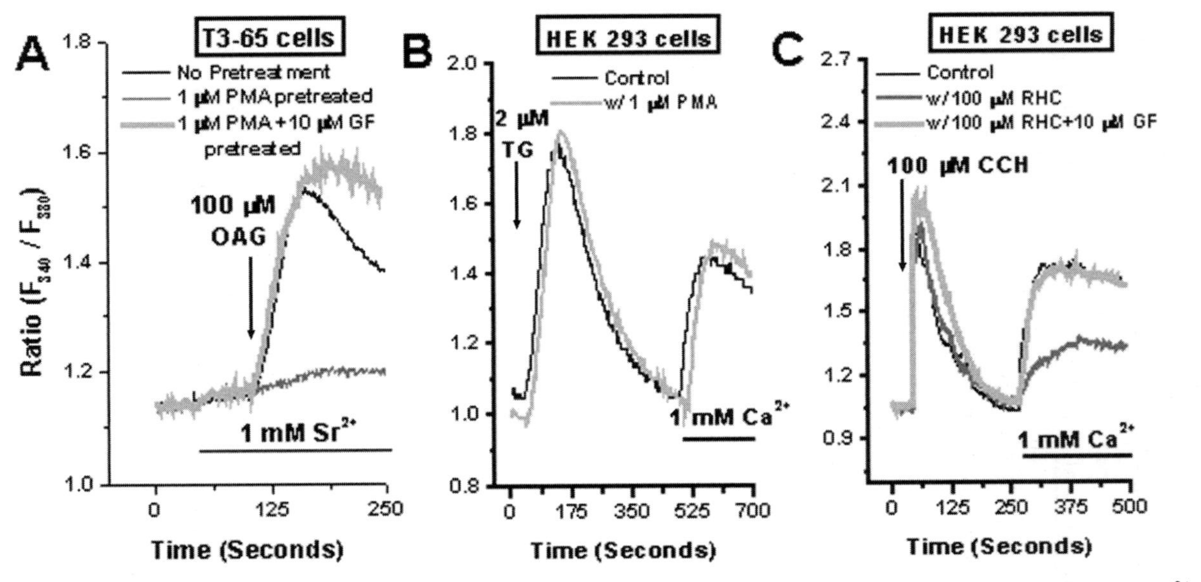

FIG. 4. PKC activation inhibits both OAG-mediated TRPC3 channel activation and development of endogenous GPCR-mediated Ca^{2+} entry without affecting Ca^{2+} entry due to TG-mediated store depletion. Standard conditions included Ca^{2+}-free medium; *bars* indicate replacement of Ca^{2+}-free media with media containing either Sr^{2+} or Ca^{2+}. (A) In the TRPC3 stably expressing T3-65 clone of HEK 293 cells, addition of 1 mM Sr^{2+} (*bar*) led to minimal constitutive entry via TRPC3. Subsequent addition of 100 M OAG (*arrow*) in control cells (*black trace*) led to TRPC3 activation and Sr^{2+} entry. 5 min pretreatment with 1 μM PMA completely prevented OAG-mediated TRPC3 activation (*grey trace*). 5 min pretreatment with 1 μM PMA and 10 μM GF rescued the effect of PMA on OAG-mediated TRPC3 activation and led to substantial Sr^{2+} entry (*light grey trace*). (B) In HEK 293 cells, passive store depletion with 2 μM TG (*arrow*) led to activation of store-operated Ca^{2+} entry upon the readdition of 1 mM Ca^{2+} (*bar*) (*black trace*). Pretreatment with 1 μM PMA did not affect either TG-mediated Ca^{2+} release or subsequent Ca^{2+} entry (*light grey trace*). (C) In HEK 293 cells, addition of 100 μM CCh (*arrow*) led to rapid IP_3-mediated Ca^{2+} release in control cells (*black trace*). Subsequent addition of 1 mM Ca^{2+} (*bar*) resulted in Ca^{2+} entry due to GPCR-mediated Ca^{2+} release. When 100 μM RHC was added with CCh (*arrow*), IP_3-mediated Ca^{2+} release was the same size as in control cells, but subsequent Ca^{2+} entry was reduced by about 60% (*grey trace*). When both 100 μM RHC and 10 M GF were added with CCh (*arrow*), both IP_3-mediated Ca^{2+} release and subsequent Ca^{2+} entry were the same as in control cells (*light grey trace*).

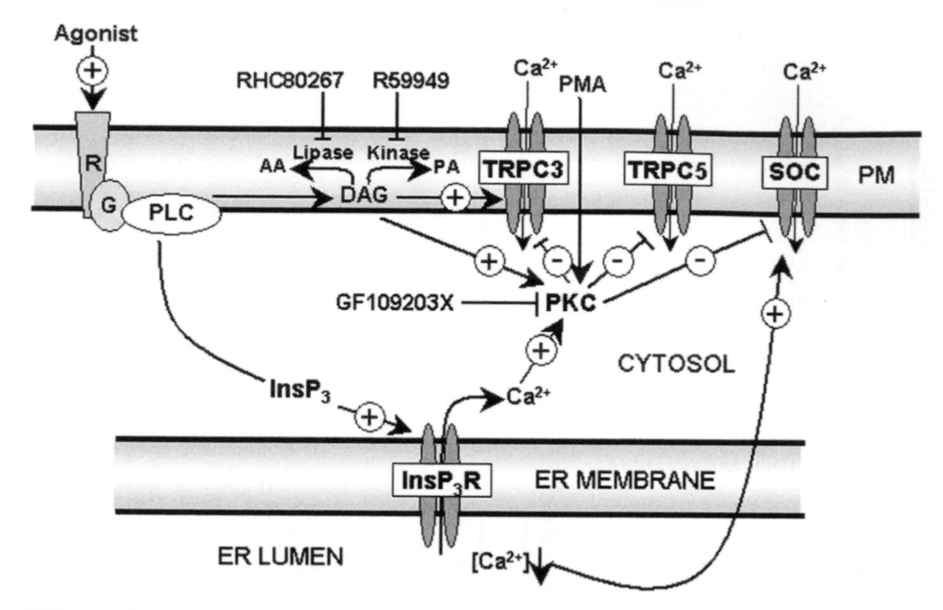

FIG. 5. Diagram to summarize the modifications and actions of DAG and PKC on the activation of TRPC channels and SOCs. Established and potentially significant stimulatory and inhibitory regulatory pathways are indicated by circled plus-signs and minus-signs, respectively. Details of these pathways are given in the text.

expressed TRPC channels. However, in contrast to the complete inhibition of TRPC channels by PKC, the partial effect on receptor-induced endogenous Ca^{2+} entry may reflect heterogeneity of channel subtypes involved in this process.

Overall, our results indicate that three members of the TRPC family of channels are negatively regulated by PKC. A summary of results and modification is given in Fig. 5. The TRPC4 and TRPC5 channels form a structurally closely related subgroup (Montell 2001). The TRPC3 channel is part of a structurally distinct subgroup of closely related channels including TRPC6 and TRPC7 (Montell et al 2002b, Montell 2001). This latter subgroup is distinguished functionally by being activated by DAG through a non-PKC mechanism (Hofmann et al 1999). Although we have not examined the actions of PKC on TRPC6 and TRPC7, given their structural and functional similarity to TRPC3, it would be surprising if they were distinct with respect to the PKC effects. Therefore, the actions of PKC on TRPC channels from different subgroups may signify a universal and important component in the feedback regulation of TRPC channels following PLC-dependent activation. Somewhat enigmatically, while we may have shed light on a potentially important turn-off mechanism for TRPC channels, the mediation of the turning-on of TRPC channels is still a mystery. Thus, whereas TRPC channels seem to be universally activated by receptor-induced PLC activation, only the TRPC3/6/7 subgroup appear to respond to DAG (Zakim 1996). What accounts for activation of TRPC4/5 channels that are unresponsive to DAG? Certainly, a large body of evidence has pointed to the other PLC product, IP_3, functioning through IP_3Rs to activate TRPC channels (Kiselyov et al 1998, 1999, Ma et al 2000, Boulay et al 1999, Birnbaumer et al 2000, Zhang et al 2001, Tang et al 2001, Venkatachalam et al 2002). Indeed, while there are conflicting reports, evidence suggests that IP_3Rs can exert a direct conformational-coupling role in the activation of TRPC channels in addition to activation resulting from store-depletion (reviewed in Venkatachalam et al 2002). However, our studies reveal that receptors, either G protein-coupled through PLCβ, or tyrosine kinase-coupled through PLCγ2, can activate TRPC5 channels in DT40 cells devoid of any IP_3Rs, a conclusion mirroring activation of TRPC3 channels (Venkatachalam et al 2001). The enigma of TRPC activation extends to the prototypic *Drosophila* TRP channel which also is dependent on receptor-induced PLC activation even though PLC products have no obvious mediating action (Montell 1999, 2001). Closely resembling vertebrate TRPC channels, the *Drosophila* TRP channel exists in a functional complex containing photoreceptor, PLC, PKC and calmodulin, held within the PDZ domain-containing INAD scaffold protein (Montell 1999, 2001). Indeed there is evidence that the PKC within this complex directly phosphorylates and inhibits the TRP channel in a negative feedback loop controlling phototransduction (Liu et al 2000b, Huber et al 1998). In vertebrate systems, TRPC channels may be organized within similar regulatory complexes via PDZ domain-containing proteins such as NHERF which is shown to interact with and organize TRPC4 and TRPC5 channels and PLCβ isoforms (Tang et al 2000). It is also well known that PKC-mediated inhibition of receptor-induced PLC provides an important feedback loop mediated by DAG and Ca^{2+} on the PLC enzyme (Ryu et al 1990, Yue et al 2000, Filtz et al 1999, Litosch 1997). Notable in the current studies is that induction of DAG by DAG lipase or DAG kinase activation, or the application of OAG, fully inhibits TRPC5 channel activation through a PKC-dependent mechanism, but only slightly reduces receptor-induced PLC activation revealed by the IP_3-mediated release of Ca^{2+} from stores. In contrast, application of the potent PKC-activator, PMA, prevents both PLC and TRPC activation. Thus, it may be that the PKC affecting TRPC channels is not the same as that which controls PLC. Indeed, it is possible that a subpopulation of PKC, perhaps that tightly associated with the TRPC-containing complex, is highly responsive to changes in DAG induced within the membrane. This may contrast with a more globally distributed subfraction of PKC exerting actions on PLC, which is less sensitive to membrane DAG changes but nevertheless highly activated by PMA. The function of a closely coupled PKC moiety within a local TRPC channel complex which is highly-sensitive to local DAG levels, provides an intriguing control process for the entry channels. Indeed, control of the assembly of the complex with respect to the relative amounts of PKC in combination with TRPC channel subunits that are DAG-sensitive (such as TRPC3) or insensitive (such as TRPC5), may provide functional channel assemblies that have profoundly different responsiveness to receptor activation.

Acknowledgement

This work was supported by NIH grant HL55426.

References

Birnbaumer L, Zhu X, Jiang M et al 1996 On the molecular basis and regulation of cellular capacitative calcium entry: roles for Trp proteins. Proc Natl Acad Sci USA 93:15195–15202

Birnbaumer L, Boulay G, Brown D et al 2000 Mechanism of capacitative Ca^{2+} entry (CCE): interaction between IP_3 receptor and TRP links the internal calcium storage compartment to plasma membrane CCE channels. Recent Prog Horm Res 55:127–161

Boulay G, Brown DM, Qin N et al 1999 Modulation of Ca^{2+} entry by polypeptides of the inositol 1,4, 5-trisphosphate receptor (IP_3R) that bind transient receptor potential (TRP): evidence for roles of TRP and IP_3R in store depletion-activated Ca^{2+} entry. Proc Natl Acad Sci USA 96:14955–14960

Filtz TM, Cunningham ML, Stanig KJ, Paterson A, Harden TK 1999 Phosphorylation by protein kinase C decreases catalytic activity of avian phospholipase C-beta. Biochem J 338:257–264

Harteneck C, Plant TD, Schultz G 2000 From worm to man: three subfamilies of TRP channels. Trends Neurosci 23:159–166

Hofmann T, Obukhov AG, Schaefer M, Harteneck C, Gudermann T, Schultz G 1999 Direct activation of human TRP6 and TRP3 channels by diacylglycerol. Nature 397:259–263

Hofmann T, Schaefer M, Schultz G, Gudermann T 2000 Transient receptor potential channels as molecular substrates of receptor-mediated cation entry. J Mol Med 78:14–25

Huber A, Sander P, Bahner M, Paulsen R 1998 The TRP Ca2+ channel assembled in a signaling complex by the PDZ domain protein INAD is phosphorylated through the interaction with protein kinase C (ePKC). FEBS Lett 425:317–322

Kanki H, Kinoshita M, Akaike A, Satoh M, Mori Y, Kaneko S 2001 Activation of inositol 1,4,5-trisphosphate receptor is essential for the opening of mouse TRP5 channels. Mol Pharmacol 60:989–998

Kiselyov KI, Xu X, Mohayeva G, Kuo T et al 1998 Functional interaction between $InsP_3$ receptors and store-operated Htrp3 channels. Nature 396:478–482

Kiselyov KI, Mignery GA, Zhu MX, Muallem S 1999 The N-terminal domain of the IP_3 receptor gates store-operated hTrp3 channels. Mol Cell 4:423–429

Litosch I 1997 G-protein betagamma subunits antagonize protein kinase C-dependent phosphorylation and inhibition of phospholipase C-beta1. Biochem J 326:701–707

Liu X, Wang W, Singh BB, Lockwich T, Jadlowiec J et al 2000a Trp1, a candidate protein for the store-operated Ca^{2+} influx mechanism in salivary gland cells. J Biol Chem 275:3403–3411

Liu M, Parker LL, Wadzinski BE, Shieh BH 2000b Reversible phosphorylation of the signal transduction complex in Drosophila photoreceptors. J Biol Chem 275:12194–12199

Ma H-T, Patterson RL, van Rossum DB, Birnbaumer L, Mikoshiba K, Gill DL 2000 Requirement of the inositol trisphosphate receptor for activation of store-operated Ca^{2+} channels. Science 287:1647–1651

Montell C 1999 Visual transduction in Drosophila. Annu Rev Cell Dev Biol 15:231–268

Montell C 2001 Physiology, phylogeny, and functions of the TRP superfamily of cation channels. Sci STKE 90:RE1

Montell C, Birnbaumer L, Flockerzi V 2002a The TRP channels, a remarkably functional family. Cell 108:595–598

Montell C, Birnbaumer L, Flockerzi V et al 2002b A unified nomenclature for the superfamily of TRP cation channels. Mol Cell 9:229–231

Philipp S, Cavalié A, Freichel M et al 1996 A mammalian capacitative calcium entry channel homologous to *Drosophila* TRP and TRPL. EMBO J 15:6166–6171

Philipp S, Hambrecht J, Braslavski L et al 1998 A novel capacitative calcium entry channel expressed in excitable cells. EMBO J 17:4274–4282

Philipp S, Trost C, Warnat J et al 2000 TRP4(CCE1)is part of native Ca^{2+} release-activated Ca^{2+}-like channels in adrenal cells. J Biol Chem 275:23965–23972

Putney JW Jr, Broad LM, Braun FJ, Lievremont JP, Bird GS 2001 Mechanisms of capacitative calcium entry. J Cell Sci 114:2223–2229

Quest AF, Ghosh S, Xie WQ, Bell RM 1997 DAG second messengers: molecular switches and growth control. Adv Exp Med Biol 400A:297–303

Ryu SH, Kim UH, Wahl MI et al 1990 Feedback regulation of phospholipase C-beta by protein kinase C. J Biol Chem 265:17941–17945

Tang Y, Tang J, Chen Z et al 2000 Association of mammalian trp4 and phospholipase C isozymes with a PDZ domain-containing protein, NHERF. J Biol Chem 275:37559–37564

Tang J, Lin Y, Zhang Z, Tikunova S, Birnbaumer L, Zhu MX 2001 Identification of common binding sites for calmodulin and IP_3 receptors on the carboxyl-termini of Trp channels. J Biol Chem 276:21303–21310

Toullec D, Pianetti P, Coste H et al 1991 The bisindolylmaleimide GF 109203X is a potent and selective inhibitor of protein kinase C. J Biol Chem 266:15771–15781

Vannier B, Peyton M, Boulay G et al 1999 Mouse trp2, the homologue of the human trpc2 pseudogene, encodes mTrp2, a store depletion-activated capacitative Ca^{2+} entry channel. Proc Natl Acad Sci USA 96:2060–2064

Venkatachalam K, Ma HT, Ford DL, Gill DL 2001 Expression of functional receptor-coupled TRPC3 channels in DT40 triple $InsP_3$ receptor-knockout cells. J Biol Chem 276:33980–33985

Venkatachalam K, van Rossum DB, Patterson RL, Ma HT, Gill DL 2002 The cellular and molecular basis of store-operated calcium entry. Nat Cell Biol 4:E263–E272

Yue C, Ku CY, Liu M, Simon MI, Sanborn BM 2000 Molecular mechanism of the inhibition of phospholipase C beta 3 by protein kinase C. J Biol Chem 275:30220–30225

Zakim D 1996 Fatty acids enter cells by simple diffusion. Proc Soc Exp Biol Med 212:5–14

Zhang Z, Tang J, Tikunova S et al 2001 Activation of Trp3 by inositol 1,4,5-trisphosphate receptors through displacement of inhibitory calmodulin from a common binding domain. Proc Natl Acad Sci USA 98:3168–3173

Zitt C, Halaszovich CR, Luckhoff A 2002 The TRP family of cation channels: probing and advancing the concepts on receptor-activated calcium entry. Prog Neurobiol 66:243–264

DISCUSSION

Nilius: I have a general comment. I would like to say something that might be helpful for the whole TRP channel field. We have now seen hundreds of pictures showing a store depletion followed by a reapplication of Ca^{2+}. We don't care about what might be the driving force. What really bothers me is that we treat these Ca^{2+} reapplication signals like Ca^{2+} entry. But when you do this, please do the first time derivative. You have to take into account the change in the rate of rise. No one is showing this. I have serious problems understanding many of these data because of this.

Gill: That's a reasonable comment. To say that when we look at changes in the steady-state level of Ca^{2+} after emptying stores we are not necessarily looking at the kinetics of the activation of the channel. But if there are changes that are reflected by some pharmacologically modified mechanism, I don't think these are invalid, either. They need to be validated by channel work as well.

Penner: Let me add one comment. Christina Fasolato is a PC12 expert, and she came to our lab because she wanted to find store-operated Ca^{2+} entry mechanisms in the PC12 cells. She brought with her more than a dozen different clones of PC12 cells. She has looked unsuccessfully for over eight months for a SOC entry mechanism in PC12 cells. She did over 900 cells.

Gill: Did she do fura though?

Penner: She did fura. With fura you see the typical thapsigargin response when readmitting Ca^{2+}, but she wanted to find the electrophysiological correlate of that response. She gave up because there was nothing that we could see in terms of an electrical current that was not voltage dependent. There is voltage-dependent current.

Muallem: What does Martha Nowycky see?

Penner: I don't know what she sees. I have never seen any current in chromaffin cells either.

Gill: That's why we didn't do it in PC12 cells alone: we did it in every cell we could get our hands on.

Penner: What is the basis for that response?

Nilius: Instead of measuring the amplitude of these Ca^{2+} signals, why does nobody look at the first-time derivative?

Putney: We do with barium. I agree that it is better to measure current rather than fluorescence, but we can't discount all experiments that don't do this.

Penner: I'd like to know what the basis of this is. Don Gill has said that it is store-operated Ca^{2+} entry, and the Ca^{2+} entry is normally carried by an ion channel. I would like to know how we can look at this at the molecular level.

Putney: Maybe we can. Let me give an example. We have done an experiment in RBL cells in which we voltage clamp but don't buffer the Ca^{2+}. We measure the Ca^{2+} and put on carbachol, thapsigargin or ionomycin and see the normal fura response such as Don Gill shows, but there is no current that we can detect. If we do the maths we can easily show that it is possible to obtain this size of Ca^{2+} signal with a current that one cannot detect. It is below the level of detection, but that doesn't mean that it is not there.

Nilius: Even if you don't measure currents, the more sensitive signal for Ca^{2+} entry is the first-time derivative. That is simple biophysics. Why is no one doing it?

Putney: I have a question that has to do with mechanism. There were two points that you made in your talk. There was a dissociation between receptor activation and the entry which you think is store operated, but there was a disconnect between receptor activation and thapsigargin activation. I think you can explain both of those by changes in PLC activity. I think it is possible that in the absence of PLCγ the PLCβ activity is not maintained with time, but in fact desensitizes more rapidly through some mechanism. This would give the result you obtained. It is also possible that even though RHC does not activate PKC enough to inhibit the PLC activity when you first put on carbachol, the combination of carbachol and RHC produces much more PKC activation so that by the time you add Ca^{2+} back, you have again greatly down-regulated your PLC activity. To put it another way, I would say that one would like to see the time course of IP_3 under those two experimental conditions.

Gill: We should check that out.

Putney: I would add that the phenotype that you see in the PLCγ knockout DT40 cells, we can see frequently in wild-type cells. We have seen it in a number of different cell lines where the receptor happens to be rapidly desensitized. The receptor desensitizes, the IP_3 goes away and the store-operated channels are still open. Then, when you add Ca^{2+} back when IP_3 is gone, the stores refill, but with little or no cytoplasmic signal.

Gill: Are you talking about PLCγ2 knockout cells?

Putney: We can see that phenotype in wild-type DT40 cells and in other cell lines. Years ago when we saw that kind of phenotype—release without entry—with one agonist, and then we would see release and entry with a different agonist, when we measured the time course of IP_3 this explained the difference. There is a precedent for explaining this kind of behaviour in the terms I have described.

Muallem: It is not true that almost all the results can be explained by different stimulation intensity. The LIM mutant cannot be explained by different stimulation intensity.

Putney: I am not saying that the PLCγ is providing the PLC activity, but it is regulating the PLCβ activity in some way.

Gill: If it is a change in PLCβ, it is still dependent on the PLCγ that is present. We don't know what the mechanism is, and we have stated this.

Scharenberg: But you are connecting the receptor to the channel.

Putney: The way you draw the model does make a difference. What you are implying in your model is some parallel pathway for regulation of the channel which is independent of store depletion. My explanation is that it could all be through store depletion.

Gill: It could be that there is a more efficient store depletion, or a more efficient coupling process that is occurring.

Scharenberg: I have done measurements where we have had differences similar to what Don Gill has observed. If we use just a receptor binding assay to measure IP_3 accumulation, within 15–20 s we see a rapid large accumulation. Then, very small differences in accumulation over time will produce a digital difference. This fits beautifully the idea that there is a threshold: if you are below it the stores will refill; if you are above it they stay depleted and you have entry.

Montell: One possible explanation for your data demonstrating that the PLCγ LIM derivative rescues TCR signalling is that PLCγ and PLCβ could be interacting. Do the two proteins co-immunoprecipitate (co-IP)?

Gill: They do interact.

Montell: Do they still co-IP with the LIM mutant?

Gill: I don't know that. It is one single mutation so you would expect them to.

Westwick: There are data out there in human B cells showing that activation of the human B cell receptor is wortmannin sensitive. Have you put wortmannin in your system?

Gill: No.

Westwick: I think that the difference you got with the SH3-deleted domain is because you are interfering with the PI3K orientation with PLC also. I suspect that if you IP you will find you have lost PI3K association.

Gill: That is a good point.

Barritt: I wanted to take a slightly different angle on your work with PC12 cells and PLC gamma. Using an 'Australian clone' of PC12 cells we have shown that acetylcholine and thapsigargin stimulated Ca^{2+} inflow (Tesfai et al 2001). We interpreted this — rightly or wrongly — to be store-activated inflow. Separately, along with DAG-stimulated Ca^{2+} inflow, we demonstrated the presence of TRPC3 and TRPC6 proteins in those cells. Are the two activation pathways that you are using, thapsigargin and agonist, actually activating the same channel?

Gill: They are very selective for Ca^{2+}. We would imagine it is the same channel, but we don't know this for sure. We think there is input from the receptor in one way as opposed to only input from the stores in the other way.

Reference

Tesfai Y, Brereton HM, Barritt GJ 2001 A diacylglycerol-activated Ca^{2+} channel in PC12 cells (an adrenal chromaffin cell line) correlates with expression of the TRP-6 (transient receptor potential) protein. Biochem J 358:717–726

TRPC4 and TRPC4-deficient mice

Marc Freichel, Stephan Philipp, Adolfo Cavalié and Veit Flockerzi[1]

Experimentelle und Klinische Pharmakologie und Toxikologie, Medizinische Fakultät der Universität des Saarlandes, D 66421 Homburg, Germany

Abstract. TRP proteins, in most cases, provide localized Ca^{2+} increases for spatially defined signal transduction processes. They are activated by as yet unclear mechanisms, many involving the complex phospholipase C and phosphatidylinositol pathways. In mouse endothelial cells at least seven TRPs are expressed, including TRPC1, TRPC2, TRPC3, TRPC4, TRPC6, TRPV4 and TRPM4. As shown previously, TRPC4 is an indispensable component of agonist-induced Ca^{2+} entry channels in native endothelial cells which essentially contributes to agonist-induced vessel relaxation and microvascular endothelial permeability, although, it is still open, whether TRPC4 acts as channel-forming subunit and/or essential constituent for channel activation. Utilizing the mouse model is one way to address this question and to provide novel insights for the biological functions of TRPC4. Here we review recent results on heterologously expressed TRPC4 and summarize what is known on the phenotype of the TRPC4$^{-/-}$ mice generated in our laboratory.

2004 Mammalian TRP channels as molecular targets. Wiley, Chichester (Novartis Foundation Symposium 258) p 189–203

TRPC4 gene and protein

The mouse *Trpc4* gene is localized on chromosome 3.D (Ensembl gene ID ENSMUSG00000027748) and extends over a region of 163.3 Kb (Fig. 1A). Analysis of its organization using the DNA sequence obtained from sequencing and subcloning of P1 clones isolated from a murine 129 SvJ genomic library (Genome Systems) and a database search in the mouse genome database (Ensembl) identified 11 exons (Fig. 1A). Comparing the region of mouse chromosome 3 harbouring the *Trpc4* gene with corresponding regions of genomes from other species using the Ensembl syntenyview feature (Clamp et al 2003) reveals that the mouse region is syntenic to a human 4.9 Mb region on chromosome 13.q13.3 where the human Trpc4 gene is localized. The mouse *Trpc4* gene gives rise to two TRPC4 variants (GenBank accession numbers U50922, U50921), TRPC4 (relative molecular mass $M_r \sim 102\,000$) and

[1]This paper was presented at the symposium by Marc Freichel, although correspondence should be addressed to Veit Flockerzi.

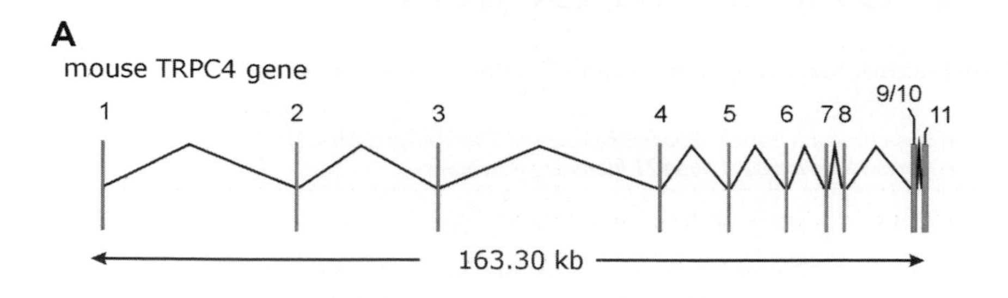
A
mouse TRPC4 gene
1
2
3
4
5
6
7
8
9/10
11
163.30 kb

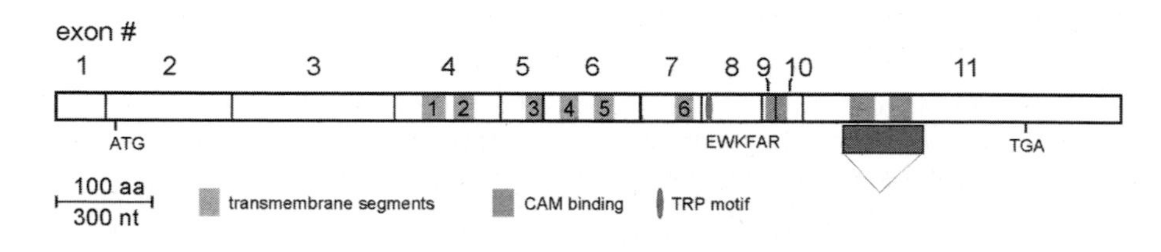
exon #
1
2
3
4
5
6
7
8
9
10
11
ATG
EWKFAR
TGA
100 aa
300 nt
transmembrane segments
CAM binding
TRP motif

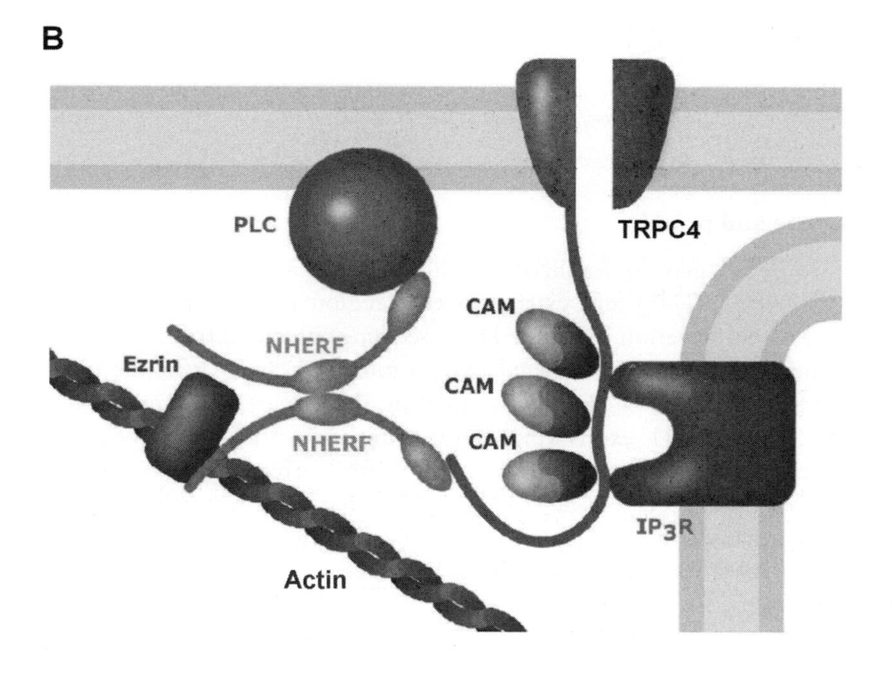
B
PLC
TRPC4
CAM
CAM
CAM
NHERF
NHERF
Ezrin
IP3R
Actin

TABLE 1 TRPC4 contains three calmodulin binding sites

Binding site	*Position*	*Ca^{2+}-dependent binding (EC_{50} [μM])*	*Binding at 1 mM Ca^{2+} (K_D[μM])*
1*	694–728	27.9 ± 4.5 (6)	0.094 ± 0.006 (3)
2**	787–812	1.0	0.6
3*	829–853	16.6 ± 2.3 (6)	0.189 ± 0.007 (3)

± SEM (*n*); data from Trost et al (2001) (*) and Tang et al (2001) (**).

$TRPC4_{\triangle 781-864}$ ($M_r \sim 94\,000$), which are both identified by TRPC4-specific antibodies in homogenates and microsomes from certain mouse tissues (Freichel et al 2001). The $TRPC4_{\triangle 781-864}$ variant is accomplished by additional splicing of 252 bp within exon 11 of the gene using splice donor and acceptor consensus sites within this exon. Interestingly, alternative splicing leading to two protein variants occurs also in the corresponding sequence of the human *TRPC4* gene (Mery et al 2001) but not in the bovine *TRPC4* gene (S. Philipp, V. Flockerzi, unpublished).

The TRPC4 protein complex

TRPC4 like other members of the TRPC subfamily has a Ca^{2+}-dependent calmodulin binding domain (Fig. 1B) in its C terminus which has been implied in feedback inhibition of channel activity by calcium (Tang et al 2001). Most intriguingly TRPC4 comprises three calmodulin binding sites within its C-terminus (Tang et al 2001, Trost et al 2001) compared to only one in the C-termini of the other TRPCs (Table 1). Two of the three binding sites are localized within the 84 amino acid residues absent in the $TRPC4_{\triangle 781-864}$ variant.

FIG. 1. TRPC4 gene (A) and TRPC4 containing protein complex (B). (A) Intron–exon distribution (exons shown as grey vertical lines in the upper part of the figure) along the mouse chromosome 3.D. Exons 9 and 10 appear as a single line because the intron in between consists of 106 bp only. The positions of the predicted transmembrane segments (grey numbered boxes), calmodulin binding sites (grey unnumbered boxes) and the TRP motif are indicated. The dark grey rectangular box indicates the 252 nucleotides within exon 11 that are spliced out in $TRPC4_{\triangle 781-864}$. (B) Scheme of the TRPC4 containing protein complex summarizing recent experimental data: TRPC4 resides within the plasma membrane; its C-terminal region comprises three binding sites for calmodulin (CAM), two of which overlap with binding sites for the inositol trisphosphate receptor (IP_3R). The very C-terminus of TRPC4 interacts with the first PDZ domain of EBP 50 or NHERF, which is known to be able to dimerize apparently via its second PDZ domain and which binds via its C-terminus to ezrin, linking the protein complex to the actin cytoskeleton (Shenolikar & Weinman 2001, all other references in the text). The first PDZ domain of NHERF also binds phospholipase Cβ1 (PLC).

Interestingly, the amino acids representing calmodulin binding sites 1 and 3 have also been identified as parts of inositol trisphosphate receptor-binding domains (Tang et al 2001, Zhang et al 2001) that have been implied in mechanisms for channel activation associated with Ca^{2+}-store depletion.

The highest sequence similarity among the members of the TRPC subfamily lies within their transmembrane regions and an immediately following group of six amino acids, EWKFAR, referred to as the TRP motif (Fig. 1A), which is present not only in the C-type but, in a slightly degenerated form, also in the V- and M-type TRPs. C-terminal to the TRP motif, TRPC channels share very limited sequence identity but for the presence of a group of five amino acids, VTTRL, present at the very C-terminus of TRPC4 and TRPC5. The C-terminal 'TRL' motif is known to bind with high affinity to the first PDZ domain of the scaffold protein ezrin/moesin/radixin-binding phosphoprotein (EBP)50 also known as regulatory factor of the Na^+/H^+ exchanger (NHERF) and it could be demonstrated that both, TRPC4 and TRPC5 do bind to EBP50 (Tang et al 2000). While the first PDZ domain of EBP50 also binds to PLCβ1 (Tang et al 2000), its C-terminus binds to ezrin, linking the protein complex to the actin cytoskeleton. This also implies that TRPC4 is organised in a signalling complex (Fig. 1B) near the plasma membrane (Mery et al 2002).

Is TRPC4 an ion channel?

Given that *Drosophila* TRP requires phospholipase C (PLC) for activity *in vivo*, mammalian TRPCs were predicted to be PLC-dependent ion channels. Activation of PLC could be coupled to TRP channel activation via relief of phosphatidylinositol 4,5-bisphosphate (PIP_2)-mediated channel repression and/or production of inositol 1,4,5-trisphosphate (IP_3) and diacylglycerol (DAG). According to one mechanism, referred to as store-operated Ca^{2+} entry (Putney 1986), IP_3-induced release of Ca^{2+} from intracellular stores induces sustained Ca^{2+} influx by activation of a plasma-membrane Ca^{2+} entry channel. Leading contenders for channels activated by the latter mechanism are TRPC1, TRPC3, TRPC4 and TRPC5, although none of the published reports have unequivocal evidence for such a mechanism. We have shown that application of common store-depletion protocols activates cation currents across the plasma membrane of HEK 293 cells and CHO cells transiently and stably expressing the bovine TRPC4 protein which shares 97.4% sequence identity with the TRPC4 mouse variant containing all three calmodulin binding sites (Philipp et al 1996, Warnat et al 1999). In an extension of this work, we have provided evidence that TRPC4 contributes to a Ca^{2+} current in adrenal cells which is activated by common store depletion protocols (Philipp et al 2000). These adrenal cells do express endogenous TRPC4 and expression of TRPC4 cDNA in antisense orientation reduced the Ca^{2+}

current and the level of the TRPC4 protein by 50 to 60%. Expression of the corresponding human TRPC4 variant in CHO cells leads to a constitutively active non-selective cation current which could not be further increased by PLC-linked receptor activation, by store-depletion protocols or in the presence of the synthetic diacylglycerol OAG (McKay et al 2000). When co-expressed with mouse TRPC1 in HEK 293 cells mouse TRPC4 resulted in a novel non-selective cation channel with a voltage-dependence similar to NMDA receptor channels, but unlike that of any reported TRPC channel (Strübing et al 2001). Expression of the mouse $TRPC4_{\triangle 781-864}$ variant, which lacks two of the three calmodulin binding sites yields currents which seemingly were activated by yet unidentified components of the PLC pathway other than diacylglycerol or store depletion. In some of the latter studies a mouse $TRPC4_{\triangle 781-864}$ variant was used, which additionally lacks the 21 C-terminal amino acid residues (Schaefer et al 2000) including the EBP 50 interaction domain. Apparently, heterologous expression of TRPC4 leads to markedly different results which, in part, could be explained by channels which are formed by various TRPC heteromers.

Essential molecular determinants of ion permeation through TRP channels should reside within a protein domain, which participates in the formation of the ion permeable pathway of the channel. This domain is therefore called the pore loop. So far, this region has only been characterized for TRP channels formed by members of the TRPV group. Here, TRPV1 (Garciá-Martínez et al 2000), TRPV2, TRPV3 and TRPV4 (Voets et al 2002) channels only poorly discriminate between monovalent and divalent cations, especially Na^+ and Ca^{2+}, whereas TRPV5 and TRPV6 form highly selective Ca^{2+} channels. Thus the relative permeability to Ca^{2+} (P_{Ca}) is more than 100-fold higher than to Na^+ (P_{Na}) in these two channels (Fig. 2). A single negatively charged aspartic acid residue (D, shown in bold in Fig. 2) in the pore loop of TRPV5, which is conserved in TRPV6, but not in the other members of the TRPV subfamily, has been shown to be responsible for the high Ca^{2+} selectivity of channels formed by TRPV5 (Nilius et al 2001). TRPC4 like the other members of the TRPC subfamily does not fit in this scheme (Fig. 2) and experimental evidence for the presence of an ion conducting pore is still pending.

TRPC4-deficient mice

A clear limitation of studies on TRPC channels is the lack of specific channel blockers. Organic compounds (e.g. ruthenium red, econazole, miconazole, SK&F 96365) and inorganic blockers (e.g. La^{3+}, Gd^{3+}) have generally been found to be of insufficient potency and specificity. To investigate the contributions of TRPC4 upon Ca^{2+}-entry into native cells and its biological significance for organ or systemic functions, we therefore inactivated the TRPC4 gene in embryonic stem cells by homologous recombination to ablate endogenous

		pCa/pNa
TRPV 4	SETFSAFLLDLFKLTIGMGDLEML	<10
TRPV 1	YNSLYSTCLELFKFTIGMGDLEFT	<10
TRPV 2	YRGILEASLELFKFTIGMGELAFQ	<10
TRPV 5	YPTALFSTFELF-LTII**D**GPANYS	>100
TRPV 6	YPMALFSTFELF-LTII**D**GPANYD	>100
TRPC 4	LQSLFWSIFGLINLYVTNVKAQHE	?*

*current absent in TRPC 4 knockout: *pCa/pNa > 100

FIG. 2. Does TRPC4 contribute to a channel pore? Alignment of the TRPV pore regions and the putative TRPC4 pore region and selectivity of the channels for Ca^{2+} compared to Na^{+} (pCa/pNa). So far molecular determinants of the ion permeation have been identified only for TRPV4 (Voets et al 2002), TRPV1 (Garciá-Martínez et al 2000) and TRPV5 (Nilius et al 2001). pCa/pNa of currents absent in TRPC4 knockout are from Freichel et al (2001).

TRPC4 expression in native cells of the mouse. In primary endothelial cells isolated from mouse aorta by a nonenzymatic explantation method (Fig. 3B and Suh et al 1999) 3.9 kb TRPC4 transcripts are readily identified (Fig. 3A). By comparison of endothelial cells isolated from wild-type and TRPC4-deficient ($TRPC4^{-/-}$) mice, TRPC4 proteins were identified as indispensable components of store-operated Ca^{2+} channels in mouse aortic endothelial cells either as channel-forming subunits or as essential constituents for channel activation (Freichel et al 2001). Additionally, Ca^{2+} entry into $TRPC4^{-/-}$ endothelial cells induced by agonists such as acetylcholine and ATP is drastically decreased leading to a markedly impaired endothelium-dependent relaxation of blood vessels. Following stimulation with thrombin, which activates G protein-coupled proteinase-activated receptor 1 (PAR-1) and plays an important role in the pathogenesis of vascular injury and tissue inflammation, the Ca^{2+} influx in endothelial cells isolated from the lung of neonatal $TRPC4^{-/-}$ mice was also markedly reduced (Tiruppathi et al 2002). Because thrombin-induced Ca^{2+} entry is a critical determinant of endothelial permeability (Sandoval et al 2001, Lum et al 1992) thrombin-induced changes in transendothelial electrical resistance of cultured endothelial monolayers were measured to assess endothelial cell retraction. Thrombin-induced decreases of transendothelial resistance in $TRPC4^{-/-}$ confluent monolayers were significantly smaller than in wild-type controls. These results suggest a role of TRPC4 channel activation in thrombin-induced increase of

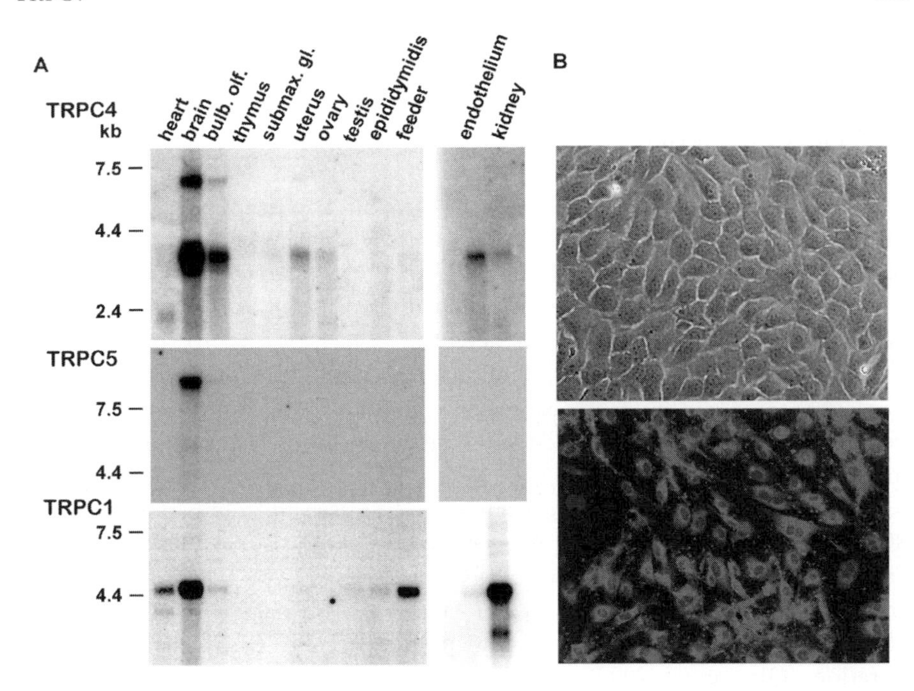

FIG. 3. TRPC4 expression (A) in mouse tissues and (B) characterization of explanted murine macrovascular endothelium. (A) Autoradiogram of RNA blot following hybridization with [α^{32}P] dCTP-labelled probes. Isolation of poly(A)$^+$ RNA, gel electrophoresis, blotting and washing procedures were performed as described (Freichel et al 1996). Following cDNA probes were used: TRPC4, nucleotides 892–3039 (Accession number U50922); TRPC5, nucleotides 199–2897 (Accession number AJ006204), TRPC1, nucleotides 4–358 (Accession number U40980). Filters were exposed to X-ray films for up to 21 days. (B) Morphology of murine aortic endothelial cells (MAECs) at confluence by phase contrast image (upper panel, 200) at passage 4 (P4). MAECs were isolated using a non-enzymatic method (Suh et al 1999) and characterized using a LDL-uptake assay. Epifluorescence photograph (lower panel) of subconfluent MAECs after incubation for four hours in the presence of acetylated human low density lipoprotein (DiI-Ac-LDL, 10 μg/ml in culture medium, Cat No 4003, Cell Systems) labelled with a fluorescent dye 1,1′-dioactadecyl -3,3,3′,3′-tetramethylindocarbocyanine perchlorate (DiI). Following DiI-Ac-LDL uptake cells were washed twice with PBS and incubated with Hoechst 33258 (300 nM in PBS, H-1398, MoBiTec) for visualization of cell nuclei (these show as dark grey within the lighter grey stained cells).

endothelial permeability and could be corroborated by studying the permeability changes of microvessels in intact lungs. In these *ex vivo* preparations the agonist-activated permeability was reduced by about 50% (Tiruppathi et al 2002). Clearly, these results support the conclusion that TRPC4 is or is part of a Ca^{2+} entry pathway in macrovascular endothelium. In addition to TRPC4, endothelial cells do also express TRPC1 (Fig. 3A), TRPC2, TRPC3, TRPC6, TRPM4 (Freichel et al 1999, M. Freichel, V. Flockerzi, unpublished) and TRPV4 (Wissenbach et al 2000) which raises the possibility that Ca^{2+}-entry channels in these cells may be formed by TRP heteromers (Hofmann et al 2002, Goel et al 2002). However, only TRPV4 has been demonstrated to contain an ion conducting pore (Voets et al 2002).

TRPC4 is not only expressed in endothelium, but also in a variety of other cells and tissues and the 3.9 kb TRPC4 transcripts are readily identified in brain, uterus, ovary and kidney. (Fig. 3A). In brain, TRPC4 is expressed in the olfactory bulb, septal nuclei, hippocampus, cortex and cerebellum. Overall its expression does not match the expression pattern of TRPC5, which seems to be expressed predominantly in brain (Fig. 3; Philipp et al 1998), and of TRPC1. TRPC1 transcripts are most abundantly expressed in kidney, brain, embryonic fibroblasts (feeder) and heart, but also present in olfactory bulb, testis, epididymis and endothelial cells (Fig. 3A).

TRPC4 was shown to be expressed in a cell line derived from pancreatic β cells (Roe et al 1998, Qian et al 2002). We analysed the expression of TRPC4 in mouse pancreatic islets and could detect TRPC4 transcripts by RT-PCR and TRPC4 proteins (C. Trost, M. Freichel, V. Flockerzi, unpublished). TRPC4-mediated Ca^{2+}-entry might be involved in regulating excitability of β cells and insulin secretion (Qian et al 2002). To test this hypothesis *in vivo* blood glucose homeostasis was analysed in wild-type and TRPC4-deficient mice by comparing glucose tolerance after intraperitoneal injection of D-glucose (2 g/kg body weight). Both basal glucose levels under fasting conditions as well as their increase following the glucose challenge were not altered in female or male $TRPC4^{-/-}$ mice compared to litter-matched control animals (Fig. 4). Similar results were obtained when other anaesthesia regimens such as ketamine (100 mg/kg)/diazepam (5 mg/kg) or tribromoethanol (Avertin, 450 mg/kg) were used (data not shown). However, to elucidate the functional role of TRPC4 in β cells, studies investigating Ca^{2+}-entry pathways in β cells and insulin secretion from isolated islets are required.

In summary, the TRPC4 gene product is involved in agonist-induced and PLC-dependent Ca^{2+} entry. This is supported by the results of three independent experimental approaches pursued in our laboratory, including

- heterologous expression of TRPC4 in HEK 293 cells and CHO cells (Philipp et al 1996, Warnat et al 1999)
- targeting of the TRPC4 gene product in an adrenal cell line, that endogenously expresses TRPC4 proteins by an antisense strategy and, finally
- by targeting the TRPC4 gene in the mouse. Further efforts to characterize the TRPC4 containing protein complex, the functional significance of the 'long' TRPC4 and the 'short' $TRPC4_{\triangle 781-864}$ variant, respectively, and the roles of

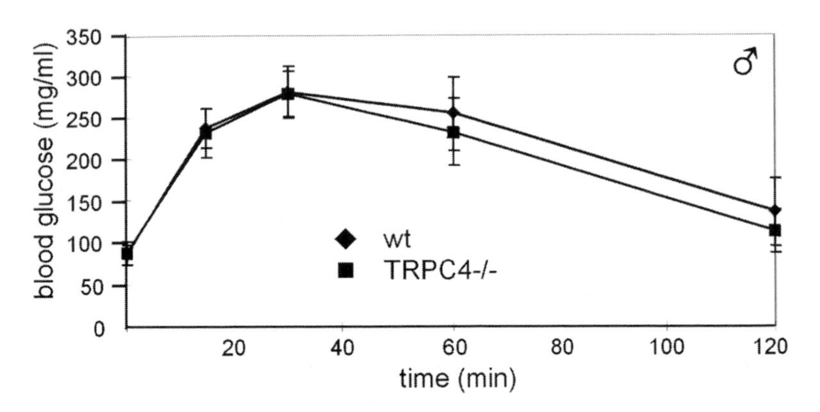

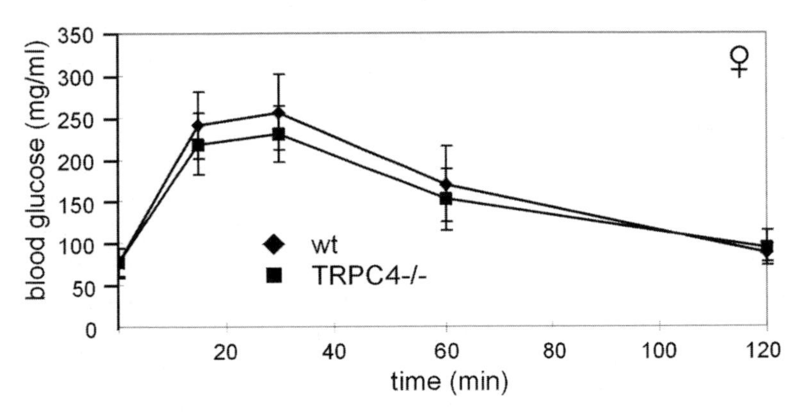

FIG. 4. Glucose tolerance test in TRPC4 deficient mice. Litter-matched 4 months old male (upper panel) or female mice (lower panel) of mixed genetic background (C57Bl6/J 129 SvJ) were fasted over night (14–16 h) and anaesthesized (60 mg/kg pentobarbital i.p.). Blood glucose levels were measured using an Accutrend sensor (Boehringer Mannheim) from tail bleedings before and 15, 30, 60 and 120 min after intraperitoneal injection of D-glucose (2 g/kg body weight).

TRPC4 in other native cells, especially in brain should bring us more complete understanding of the TRPC4 function and physiological roles.

Acknowledgements

We thank Stefanie Buchholz, Kerstin Fischer and Susanne Stolz for excellent technical assistance. This work was supported in part by the Deutsche Forschungsgemeinschaft.

References

Clamp M, Andrews D, Barker D et al 2003 Ensembl 2002: accommodating comparative genomics. Nucleic Acids Res 31:38–42

Freichel M, Zink-Lorenz A, Holloschi A, Hafner M, Flockerzi V, Raue F 1996 Expression of a calcium-sensing receptor in a human medullary thyroid carcinoma cell line and its contribution to calcitonin secretion. Endocrinology 137:3842–3848

Freichel M, Schweig U, Stauffenberger S, Freise D, Schorb W, Flockerzi V 1999 Store-operated cation channels in the heart and cells of the cardiovascular system. Cell Physiol Biochem 9:270–283

Freichel M, Suh SH, Pfeifer A et al 2001 Lack of an endothelial store-operated Ca^{2+} current impairs agonist-dependent vasorelaxation in $TRP4^{-/-}$ mice. Nat Cell Biol 3: 121–127

Garcia-Martinez C, Morenilla-Palao C, Planells-Cases R, Merino JM, Ferrer-Montiel A 2000 Identification of an aspartic residue in the P-loop of the vanilloid receptor that modulates pore properties. J Biol Chem 275:32552–32558

Goel M, Sinkins WG, Schilling WP 2002 Selective association of TRPC channel subunits in rat brain synaptosomes. J Biol Chem 277:48303–48310

Hofmann T, Schaefer M, Schultz G, Gudermann T 2002 Subunit composition of mammalian transient receptor potential channels in living cells. Proc Natl Acad Sci USA 99:7461–7466

Lum H, Aschner JL, Phillips PG, Fletcher PW, Malik AB 1992 Time course of thrombin-induced increase in endothelial permeability: relationship to Ca^{2+}i and inositol polyphosphates. Am J Physiol 263:L219–L225

McKay RR, Szymeczek-Seay CL, Lievremont JP et al 2000 Cloning and expression of the human transient receptor potential 4 (TRP4) gene: localization and functional expression of human TRP4 and TRP3. Biochem J 351:735–746

Mery L, Magnino F, Schmidt K, Krause KH, Dufour JF 2001 Alternative splice variants of hTrp4 differentially interact with the C-terminal portion of the inositol 1,4,5-trisphosphate receptors. FEBS Lett 487:377–383

Mery L, Strauss B, Dufour JF, Krause KH, Hoth M 2002 The PDZ-interacting domain of TRPC4 controls its localization and surface expression in HEK293 cells. J Cell Sci 115:3497–3508

Nilius B, Vennekens R, Prenen J, Hoenderop JG, Droogmans G, Bindels RJ 2001 The single pore residue Asp542 determines Ca^{2+} permeation and Mg^{2+} block of the epithelial Ca^{2+} channel. J Biol Chem 276:1020–1025

Philipp S, Cavalie A, Freichel M et al 1996 A mammalian capacitative calcium entry channel homologous to Drosophila TRP and TRPL. EMBO J 15:6166–6171

Philipp S, Hambrecht J, Braslavski L et al 1998 A novel capacitative calcium entry channel expressed in excitable cells. EMBO J 17:4274–4282

Philipp S, Trost C, Warnat J et al 2000 TRP4 (CCE1) protein is part of native calcium release-activated Ca^{2+}-like channels in adrenal cells. J Biol Chem 275:23965–23972

Putney JW Jr 1986 A model for receptor-regulated calcium entry. Cell Calcium 7:1–12

Qian F, Huang P, Ma L, Kuznetsov A, Tamarina N, Philipson LH 2002 TRP genes: candidates for nonselective cation channels and store-operated channels in insulin-secreting cells. Diabetes 51:S183–S189

Roe MW, Worley JF 3rd, Qian F et al 1998 Characterization of a Ca^{2+} release-activated nonselective cation current regulating membrane potential and $[Ca^{2+}]$i oscillations in transgenically derived beta-cells. J Biol Chem 273:10402–10410

Sandoval R, Malik AB, Naqvi T, Mehta D, Tiruppathi C 2001 Requirement for Ca^{2+} signaling in the mechanism of thrombin-induced increase in endothelial permeability. Am J Physiol Lung Cell Physiol 280:L239–L247

Schaefer M, Plant TD, Obukhov AG, Hofmann T, Gudermann T, Schultz G 2000 Receptor-mediated regulation of the nonselective cation channels TRPC4 and TRPC5. J Biol Chem 275:17517–17526

Shenolikar S, Weinman EJ 2001 NHERF: targeting and trafficking membrane proteins. Am J Physiol Renal Physiol 280:F389–F395

Suh SH, Vennekens R, Manolopoulos VG et al 1999 Characterisation of explanted endothelial cells from mouse aorta: electrophysiology and Ca^{2+} signalling. Pflugers Arch 438:612–620

Tang Y, Tang J, Chen Z et al 2000 Association of mammalian trp4 and phospholipase C isozymes with a PDZ domain-containing protein, NHERF. J Biol Chem 275:37559–37564

Tang J, Lin Y, Zhang Z, Tikunova S, Birnbaumer L, Zhu MX 2001 Identification of common binding sites for calmodulin and inositol 1,4,5-trisphosphate receptors on the carboxyl termini of trp channels. J Biol Chem 276:21303–21310

Tiruppathi C, Freichel M, Vogel SM et al 2002 Impairment of store-operated Ca^{2+} entry in TRPC4($^{-/-}$) mice interferes with increase in lung microvascular permeability. Circ Res 91:70–76

Trost C, Bergs C, Himmerkus N, Flockerzi V 2001 The transient receptor potential, TRP4, cation channel is a novel member of the family of calmodulin binding proteins. Biochem J 355:663–670

Voets T, Prenen J, Vriens J et al 2002 Molecular determinants of permeation through the cation channel TRPV4. J Biol Chem 277:33704–33710

Warnat J, Philipp S, Zimmer S, Flockerzi V, Cavalie A 1999 Phenotype of a recombinant store-operated channel: highly selective permeation of Ca^{2+}. J Physiol 518:631–638

Wissenbach U, Bodding M, Freichel M, Flockerzi V 2000 Trp12, a novel Trp related protein from kidney. FEBBS Lett 485:127–134

Zhang Z, Tang J, Tikunova S et al 2001 Activation of Trp3 by inositol 1,4,5-trisphosphate receptors through displacement of inhibitory calmodulin from a common binding domain. Proc Natl Acad Sci USA 98:3168–3173

DISCUSSION

Montell: In light of the results that Thomas Gudermann showed, demonstrating that *Trpc3* mRNA was elevated in *Trpc6* knockout mice, was there an effect on *Trpc5* mRNA levels in your *Trpc4* knockout? Even better, did you see an effect on the protein levels?

Freichel: I looked in the brain with northern blots. I didn't take a truly quantitative approach like that of Thomas Gudermann, but I didn't see a massive increase or decrease of any of the other TRPs I looked at. This is all I have done so far.

Muallem: Did I understand correctly that these mutant mice have tachycardia but normal blood pressure?

Freichel: That is correct.

Muallem: How does this happen?

Freichel: They don't have a defect in smooth muscle. Perhaps they have decreased peripheral resistance and therefore they compensate for this by tachycardia. But there is no evidence for this. Maybe this TRP is important in neurons of the autonomic nervous system: there might be a defect there. When we did that study we started with the hypothesis that they have lower blood pressure. There are many components influencing blood pressure. We looked with all the drugs that we applied. This was a blinded study, so we didn't know which mice we were working with. We focused on vascular resistance and things like that. We should have given a β blocker, for example, and then it should go down to the same level. We haven't done this yet. Or we could isolate atria from the mice and let them beat spontaneously, and these should be normal if there is no cardiac defect. If there is increased cardiac pacing we should see this with isolated hearts.

Muallem: If the amount of blood coming out of the heart is the same, if stroke volume is the same, then you have to get it. I still can't understand this.

Freichel: We don't know what the stroke volume is.

Benham: Cardiac output could be down.

Scharenberg: If peripheral resistance is lower then stroke volume could still be the same.

Muallem: He said peripheral resistance is the same.

Freichel: What I showed is just the response to α1-adrenergic agonists. This is the same.

Scharenberg: Stroke volume will be affected by filling pressure. There are a lot of variables here.

Kunze: A classical way to look at the baroreceptor reflex control of the heart rate is to infuse different levels of phenylephrine or nitroprusside to manipulate the blood pressure. This has no direct effect on the heart, but produces a reflex change in heart rate through the vagal and sympathetic innervation of the heart.

Freichel: We got an increase of the mean arterial pressure.

Kunze: And the heart rate went down. If you did this in graded fashion you could look to see whether or not the reflex curve for the vagal control of heart rate was still intact. I recall seeing a decrease which made me think that the vagal reflex was still working.

Freichel: We also injected i.v. sodium nitroprusside. The drop in blood pressure was the same. For the vagal reflex you can plot changes of heart rate against changes in blood pressure. There is a difference in the absolute level of the heart rate because the *Trpc4* knockout mice are tachycardic, but the slope of this plot is the same. The response to a change in blood pressure looks the same.

Cox: I don't know whether you have looked at this, but would you expect the mice to show resistance to various epileptic kindling modes? This is the induction of epilepsy in the animal by chemical or electrical means.

Freichel: We haven't done that.

Cox: It would be a good validation for an anti-epileptogenic use for modulators of this (these) target(s).

Freichel: We have looked at motor coordination with a rotarod-test because of the expression in the cerebellum. There was no difference. We also did a test which gives an idea about anxiety, and we saw no difference there either.

Montell: You looked at a lot of phenotypes based on the expression pattern. One tissue that you didn't look at was olfactory tissue. You mentioned that TRPC4 was expressed in the olfactory bulb. It is probably not involved in olfactory transduction because cGMP channels are involved. Perhaps it has a role in adaptation. Have you looked to see whether olfactory adaptation might be disrupted in the knockouts?

Freichel: No. We looked at electroretinograms and saw no difference.

Montell: Is TRPC4 expressed in the inner or outer segments of the photoreceptor cells?

Freichel: We have looked just by RT-PCR. There was no difference in the cones and rods.

Gudermann: Could you explain why the response to L-NAME is blunted in these mice? Is it because they are already used to a defective relaxation system?

Freichel: From the aortic ring studies we think that NO formation in the endothelium is diminished. There is a basal release of NO which contributes to the maintenance of a basal blood pressure. When you acutely give this NO blocker you take away that NO when it is just released and contribute to that basal blood pressure, and then it goes up because this NO cannot be released anymore. This results in a massive increase. If you now block the basal NO release in knockouts, you just see an initial peak and then they go back to normal very quickly, whereas the wild-types stay elevated. This means in the normal mouse, this NO contributes to that setting where you get that basal blood pressure. If you now block this NO formation in the *Trpc4* knockout mouse you don't see this effect. This suggests that in the knockout there is a drastic reduction of that NO component in the maintenance of basal blood pressure.

Penner: You mentioned the renin-secreting cells. Are you expecting a stop of renin release?

Freichel: We went through the physiological tests by the tissues and cells where *Trpc4* is expressed. We haven't really seen expression of *Trpc4* in juxtaglomerular cells by *in situ* hybridization, but we just did these experiments because of the expression of *Trpc4* in the kidney. Where else in the kidney is there good agonist or store-operated Ca^{2+} entry?

Putney: The epithelial cells have store-operated entry by criteria such as thapsigargin or ionomycin treatment, although it is not that clear what regulates them physiologically.

Penner: The reason I am asking is because the renin-secreting cells behave the other way round. An increase in Ca^{2+} actually stops renin release.

Kunze: Did you look at the respiratory changes? I am asking about cardiovascular and respiratory changes because the sensory reflex pathways that control these functions express TRPC4?

Freichel: No.

Hardie: Did you mention circadian rhythms?

Freichel: I mentioned the circadian rhythm of the heart rate. We measure continuously over 24 h and then we looked at heart rates for each time of the day. We averaged heart rate after this recovery over 3 or 4 days and saw that at almost all times of the day there is this tachycardia.

Hardie: It would be interesting to look at photic entrainment. There is a subclass of light-sensitive retinal ganglion cells which depolarize, possibly through a TRP channel. Melanopsin is probably the photopigment that activates them, since melanopsin knockouts have specific defects in photic entrainment.

Montell: Other features and molecules, which participate in the photoresponses in retinal ganglion cells and fly photoreceptor cells are similar. Both types of cells depolarize in response to light. Furthermore, the fly rhodopsins are more similar to melanopsin than to the mammalian rhodopsins or the colour pigments expressed in cones. There is some evidence that the photoresponse in retinal ganglion cells is a PLC-mediated pathway, as is the case in fly phototransduction.

Putney: I have a more general question. On the basis of all of the experiments with the *Trpc4* knockout mouse, what is the overall feeling of your laboratory about TRPC4 and store-operated channels?

Freichel: The overexpression studies that were published many years ago in HEK and CHO cells demonstrated a store-operated current (Philipp et al 1996, Warnat et al 1999). This work hasn't been reproduced by other labs. Then Stefan Philipp did antisense work in adrenal cells and Markus Hoth measured store-operated Ca^{2+}-selective current which looked like I_{CRAC}; this current was reduced and the reduction in current corresponded to the reduction of the protein (Philipp et al 2000). In the mice we now also have evidence for TRPC4 as an essential component of a store-operated channel in endothelial cells. But this does not necessarily have to mean that TRPC4 mediates a store-operated current in all cell types.

Nilius: When we express TRPC4 in HEK cells or other endothelial cells, we see a non-selective cation current. There is no doubt that if we measure endothelial cells with this kind of protocol, there is a kind of store-depletion-activated channel. But since we have done so many things with pores, I am very sceptical that it is a Ca^{2+}-selective channel. I think it is a necessary component of a complex. I don't believe that a TRPC4 pore has anything to do with a highly Ca^{2+}-selective pore.

Putney: If I understand you correctly, one issue is that we understand to some degree the pore properties of expressed TRPC4, which is a non-selective cation channel that isn't regulated by store depletion in most laboratories, at least when expressed in cell lines. Even in the case of the mouse where there is an endothelial cell current that is reduced, that current doesn't match what you would expect just from expressed TRPC4.

Nilius: Yes, it is a different situation.

Fleig: Have you looked at immune cells such as mast cells and lymphocytes?

Freichel: We are looking in mast cells. TRPC4 is expressed in mast cells. We have looked at peritoneal mast cells with Markus Hoth and he recorded a CRAC-like current that was reduced by about a third in the knockout. We looked for expression in these cells with RT-PCR, and currently we are looking at mast cells derived from bone marrow where we find that TRPC4 is also expressed.

Benham: Bernd Nilius, can we say that there is a similarity between the *Trpc4* knockout and the *Trpc6* knockout, in that there is an increase in non-selective current? This suggests that if you take something away from the complex that it degrades the channel in the sense of making it less Ca^{2+} selective.

Nilius: That is a good point. There is certainly more going on.

Freichel: Jim Putney, do you agree with the conclusion that TRPC4 has something to do with the pore or is an activating component of the channel in the mouse endothelial cells?

Putney: I was much more encouraged that this was the case until I heard Bernd Nilius' testimony about the list of currents that are changed in the knockouts. When you are faced with a list of currents you can't just pick the one that you thought TRPC4 was before you did the experiment. It means all bets are off.

Nilius: For me, the most exciting result in this knockout is the macroscopic signals, such as the measurement of agonist-induced changes in intracellular Ca^{2+} in clamped endothelial cells. This knockout is a very good example for asking the question about whether TRPs are, like CFTR, channel regulators. This is something that we haven't addressed at all.

Scharenberg: I would say that this is also a very good argument for doing genetics in somatic cells, where you don't have developmental issues. In mice you have up-regulation, down-regulation, developmental problems and compensation. For the most part, those problems are not nearly as prominent if you do your knockouts in a cell line.

References

Philipp S, Cavalie A, Freichel M et al 1996 A mammalian capacitative calcium entry channel homologous to Drosophila TRP and TRPL. EMBO J 15:6166–6171

Philipp S, Trost C, Warnat J et al 2000 TRP4 (CCE1) protein is part of native calcium release-activated Ca^{2+}-like channels in adrenal cells. J Biol Chem 275:23965–23972

Warnat J, Philipp S, Zimmer S, Flockerzi V, Cavalie A 1999 Phenotype of a recombinant store-operated channel: highly selective permeation of Ca^{2+}. J Physiol 518:631–638

TRP channels as drug targets

Su Li, John Westwick, Brian Cox and Chris T. Poll

Novartis Respiratory Research Centre, Wimblehurst Road, Horsham, West Sussex, RH12 5AB, UK

Abstract. Ca^{2+} channel antagonists acting on electrically-excitable cells have proved to be valuable therapeutic agents. The discovery of such agents and the identification of their molecular target resulted from the investigation of unexpected actions of known pharmacological agents. Ca^{2+} influx through receptor-operated channels in electrically non-excitable cells such as leukocytes is also functionally important, but to date the channels involved have not been successfully exploited as drug targets for anti-inflammatory therapy. Until recently, research in this area has been hindered by the lack of obvious molecular identity, but the emergence of the transient receptor potential (TRP) cation family has yielded promising candidates which may underpin the different receptor-operated Ca^{2+} influx pathways present in leukocytes. In addition, receptor-operated Ca^{2+} influx channels are also expressed in electrically-excitable cells suggesting that receptor-operated Ca^{2+} entry pathways are likely to be of wider significance and emphasizes the breadth of their potential as novel, and as yet, unexplored and unexploited drug targets.

2004 Mammalian TRP channels as molecular targets. Wiley, Chichester (Novartis Foundation Symposium 258) p 204–221

Ca^{2+} influx through plasma membrane channels in all cell types is of fundamental importance in both physiology and pathophysiology, e.g. contraction of muscle cells, neurotransmitter release from nerve terminals, secretion by epithelial cells and leukocyte activation. Therefore modulation of cell function by targeting these channels represents a potentially effective approach for therapeutic intervention.

The two main Ca^{2+} influx pathways which have been under investigation are those Ca^{2+} channels which are voltage-activated and those which are not. Although their function is similar — they regulate the source of one of the most important cytosolic signalling messengers in the body, Ca^{2+} — the drug discovery path trodden by researchers is distinctly different.

Discovery of L-type voltage-operated Ca^{2+} channel blockers

Blockers of L-type voltage-gated Ca^{2+} channels (L-VOCCs) present in vascular smooth muscle are widely used and very effective drugs for the treatment of cardiovascular diseases such as hypertension and, in particular, angina. The very earliest compounds, verapamil and prenylamine (Fig 1), from Knoll and Hoechst respectively, were developed as coronary vasodilators working as β-adrenoceptor antagonists. However, both had unexpected cardio-suppressant side effects, which Fleckenstein investigated, and he discovered by chance that they were in fact Ca^{2+} channel blockers (Fleckenstein 1983). He demonstrated that these compounds mimicked the cardiac effect of withdrawal of Ca^{2+} ions, which inhibits the cardiac excitation-coupling resulting in diminished contractile force. These effects could be rapidly overcome by addition of Ca^{2+} ions, β catecholamines or cardiac glycosides. Hence the initial concept of Ca^{2+} antagonism was coined disproving the initial assumption that both of these compounds were β-adrenoceptor antagonists (Fleckenstein 1983). More Ca^{2+} antagonists were subsequently identified such as D600, a methoxy derivative of verapamil (Fig. 1), which had a similar profile to verapamil although it was more potent, and in 1969, the dihydropyridines nifedipine and niludipine (Fig. 1) which were also strong coronary vasodilators and had negative ionotropic effects showed a similar mechanism to verapamil and D600. These observations led Fleckenstein to propose a new distinct pharmacological class of potent inhibitors of contraction-excitation coupling existed, the 'Ca^{2+} antagonists'. The biochemical isolation and identification for the site of action of the compounds came from the availability of highly radioactive ^{3}H-labelled nitrendipine (Fig. 1) which demonstrated high affinity binding to protein extracts from heart, coronary arteries and aorta leading to the identification of the α subunit of the L-VOCC and subsequently the additional accessory subunits of the native channel. The results of these biochemical experiments also suggested that the dihydropyridines, which are chemically different Ca^{2+} antagonists to verapamil or diltiazem (Fig. 1), did not have identical binding sites on the channel nor shared a common mechanism of action, e.g. state-dependence of inhibition by dihydropyridines. The differential binding affinities of the dihydropyridines measured in a range of tissues also supported existence of different channel subtypes in different tissues.

Molecular cloning of the L-VOCC subunits came in the early 1990s and functional data obtained with these cloned proteins confirmed the biochemical, pharmacological and electrophysiological characteristics of the native channel.

Hence, the unexpected effect of pharmacological agents in functional cardiovascular assays led to the discovery of a new class of cardiovascular drug and the identification of their molecular target.

Overview of receptor-activated Ca^{2+} channel blockers

Receptor-activated Ca^{2+} influx channels (ROCCs) are not voltage-activated and have a broader cellular distribution than their voltage-operated counterparts.

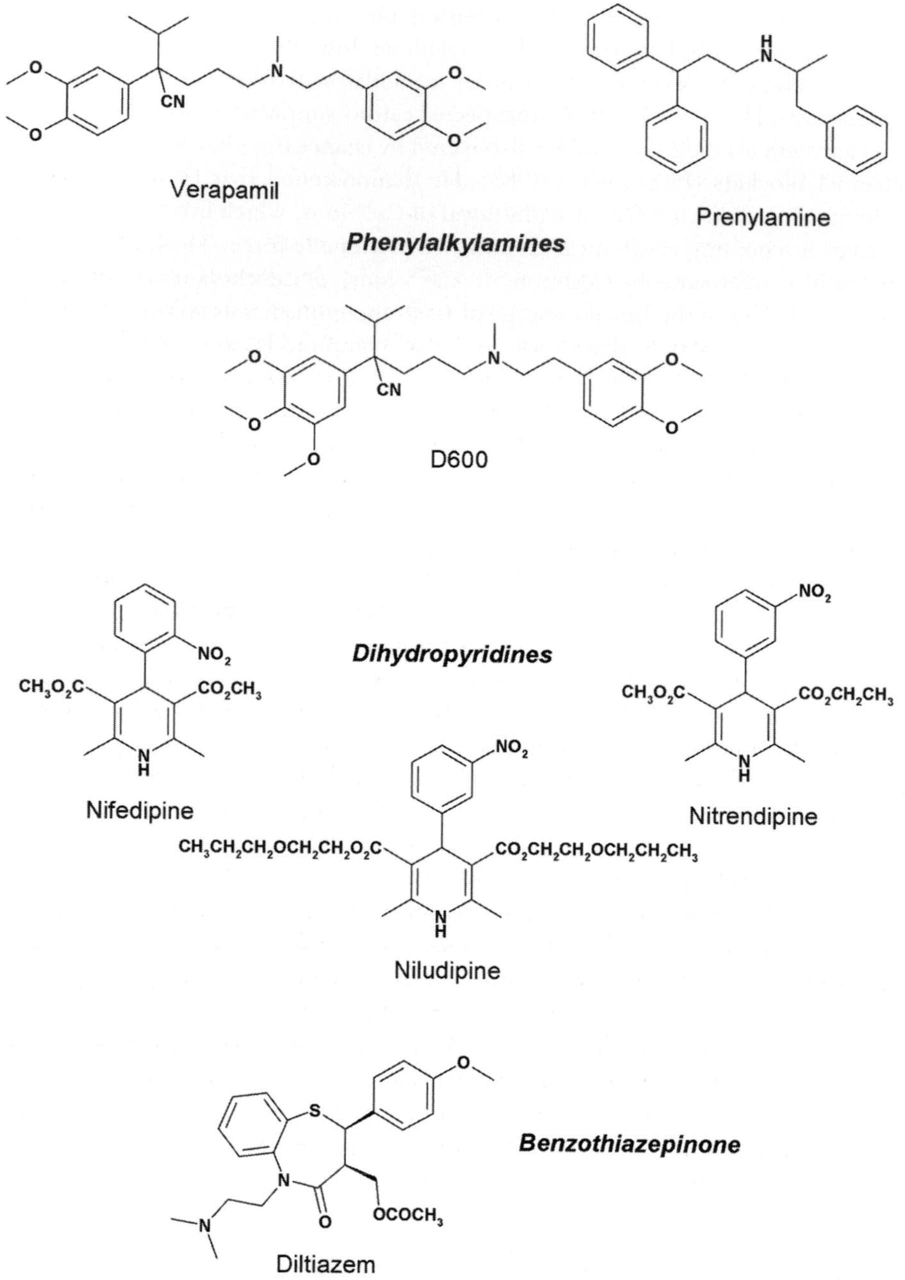

Verapamil

Prenylamine

Phenylalkylamines

D600

Dihydropyridines

Nifedipine

Nitrendipine

Niludipine

Benzothiazepinone

Diltiazem

FIG 1. Structures of L-type voltage-operated Ca^{2+} channel blockers.

They are present in both electrically-excitable cells such as muscle and nerve cells and electrically-unexcitable cells such as leukocytes and endothelial cells. Ca^{2+} entry through ROCCs has been most-intensively studied *in vitro* in leukocytes and leukocytic cell lines, both electrophysiologically and by using fluorescent Ca^{2+} indicators to monitor intracellular Ca^{2+} levels (Grynkiewicz et al 1985). These Ca^{2+} influx pathways include the highly Ca^{2+} selective current I_{CRAC} (Ca^{2+}-release activated Ca^{2+} current) in mast cells and lymphocytes (Hoth & Penner 1992, Zweifach & Lewis 1995) and Ca^{2+}-permeable non-selective cation channels in mast cells, (Franzius et al 1994), promyelocytic cell line HL60 (Krautwurst et al 1993), human neutrophils (von Tscharner et al 1986), monocytes and macrophages (Malayev & Nelson 1995, Naumov et al 1995).

As expected, the realization of the functional importance of receptor-activated Ca^{2+} influx in leukocytes was not only of intense academic interest but was also of potential therapeutic interest in the development of anti-inflammatory drugs with novel mechanisms of action. After fluorescent intracellular Ca^{2+} indicators became available, several compounds were developed and receptor-activated/receptor-mediated/receptor-operated-store-operated Ca^{2+} channel/influx blockers—identified on the basis of their ability to inhibit Ca^{2+} influx in native cells using these Ca^{2+} indicators. These include SK&F 96365 and LOE 908 (Fig. 2). Their use as pharmacological tools has provided evidence of the heterogeneity of receptor-operated Ca^{2+} influx channels. However, unlike the situation with the L-VOCCs, none of these tools have been potent or selective enough to be used to isolate and identify the molecular target.

SK&F 96365, probably the best known synthetic ROCC blocker compound, inhibits receptor-activated Ca^{2+} influx and activation of platelets (Merritt et al 1990), neutrophils (Merritt et al 1990) and lymphocytes (Chung et al 1994). However, the molecular identity of the channels responsible for Ca^{2+} influx in the target cells was, and is still, unknown. The impact of not being able to identify this vital piece of information was to severely limit the ability of drug companies to develop therapeutically useful blockers of receptor-operated channels. SK&F 96365 was an optimized compound with improved potency over the prototypical compound SC38249 (Fig. 2) which was originally designed as a thromboxane synthetase inhibitor (Howson et al 1990). However the level of potency and selectivity vs. other channels e.g. L-VOCCs and Cl^- channels that could be achieved has not been sufficient to enable the progression from useful experimental pharmacological tool to therapeutic drug. The fundamental problem has been that, unlike the voltage-gated Ca^{2+} channel blockers, the molecular identity of the target channel was unknown and therefore no clear structure-activity relationship has been demonstrated.

LOE 908, developed as an inhibitor of human neutrophilic ROCCs (Krautwurst et al 1993) has also proved to be a useful experimental tool. For

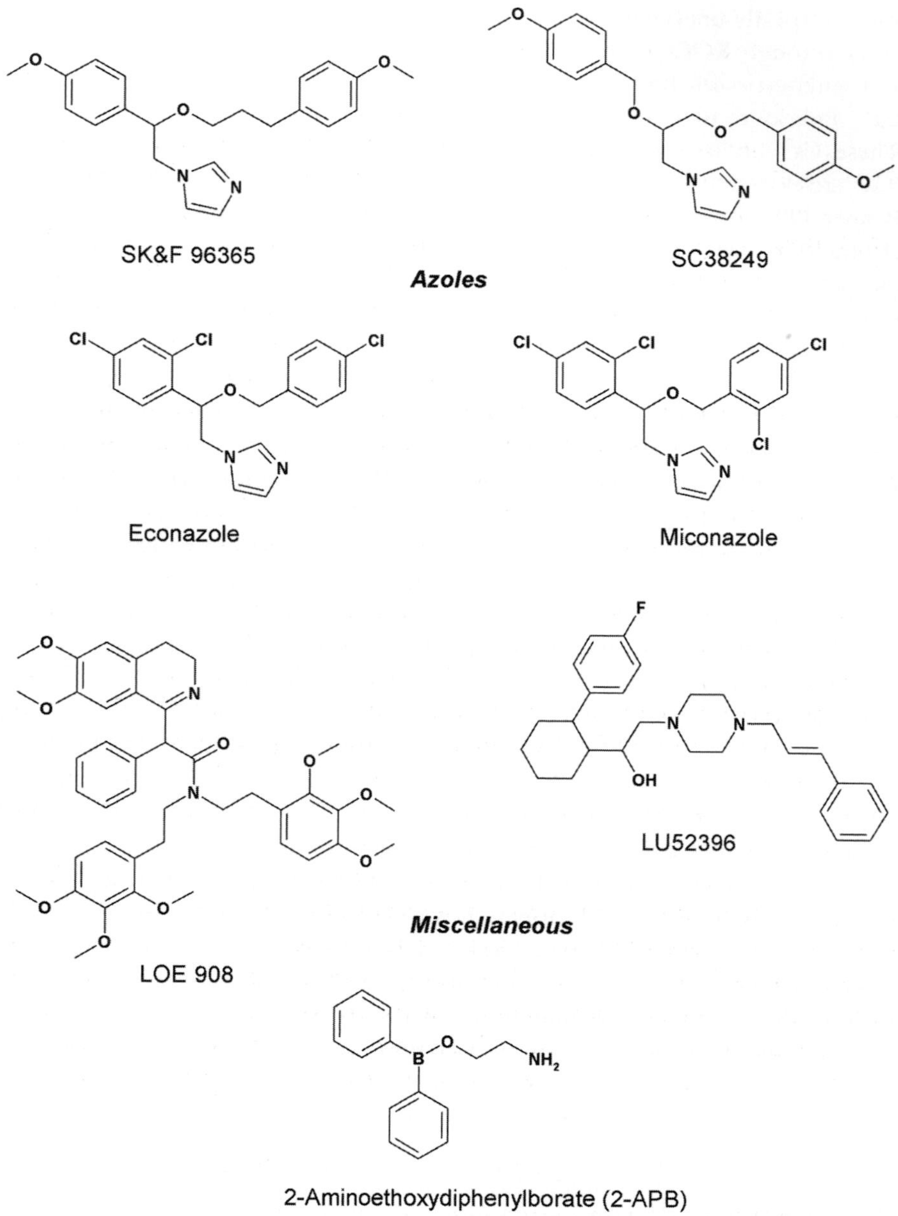

SK&F 96365

SC38249

Azoles

Econazole

Miconazole

LOE 908

LU52396

Miscellaneous

2-Aminoethoxydiphenylborate (2-APB)

FIG. 2. Receptor-activated Ca^{2+} channel blockers; azoles and miscellaneous compounds.

instance, LOE 908 was used in combination with SK&F 96365 to demonstrate the presence of at least pharmacologically distinct Ca^{2+}-influx pathways in rat aortic smooth muscle cells (Iwamuro et al 1999).

A diverse range of compounds have also been found to inhibit receptor-operated Ca^{2+} influx. These include cytochrome P450 inhibitors, such as econazole and miconazole (Fig. 2) which are imidazoles like SK&F 96365 and which inhibit receptor-operated Ca^{2+} influx in a wide variety of cell types (e.g. neutrophils and platelets). There is evidence to suggest that the mechanism by which these compounds inhibit Ca^{2+} influx does not involve cytochrome P450 (reviewed by Clementi & Meldolesi 1996). LU52396 (Fig. 2), like SC38249, was designed as a thromboxane synthetase inhibitor and found to inhibit thromboxane-independent agonist-stimulated Ca^{2+} influx in platelets (Clementi et al 1995). More recently, 2-aminoethoxydiphenyl borate (2-APB) (Fig. 2) originally described as an IP_3 receptor antagonist has also been postulated to be a ROCC blocker (reviewed by Bootman et al 2002). However, as is the case with the more chemically complex organic channel blockers already mentioned above, 2-APB also shows little selectivity at the concentrations which block Ca^{2+} influx and therefore these effects can be difficult to attribute to its ROCC-blocking activity.

Inorganic divalent and trivalent cations are also widely used in the study of ROCCs. There are channel blockers e.g. Ni^{2+}, and the lanthanides La^{3+} and Gd^{3+} which can discriminate between ROCC Ca^{2+}-entry pathways (Itagaki et al 2002) and channel-permeant cations which are used as surrogates for Ca^{2+} influx as they are either not substrates for Ca^{2+}-efflux pathways (e.g. Ba^{2+}, Sr^{2+}) or interact differentially with fluorescent Ca^{2+} indicators (e.g. Mn^{2+}). The heterogeneity of native ROCCs is also supported by the differential permeability to Mn^{2+} (Demaurex et al 1992) and Sr^{2+} (Itagaki et al 2002).

In summary, there is much circumstantial evidence to suggest that each cell type can possess multiple Ca^{2+} influx pathways and that there is heterogeneity between cell types on the basis of the data obtained using the pharmacological tools available to date. To enable the full exploration and exploitation of the therapeutic potential of ROCCs requires identification of the molecular components of these channels and the development of more discriminating pharmacological tools with properties which allow their development as therapeutically useful agents.

Evolution of drug discovery approaches: implications for ROCCs

The completion of the sequencing of the human genome (Venter et al 2001) has provided molecular information about the candidate genes encoding these channel proteins and the escalation and miniaturization of drug discovery technologies in the last decade suggests that never before has there been a better chance for identifying novel ROCC channel blockers. The combination of molecular information and increase in compound screening capabilities available now allows the use of a rational approach to systematically identify these elusive channels in the disease-relevant cells to identify modulators of these channels which could fully exploit their therapeutic potential.

TRP channels: targets with real potential for drug discovery?

The discovery of human homologues of transient receptor potential (TRP) channels in the middle of the last decade has provided to date the most promising molecular candidates for these elusive ROCCs. Within this gene family there are now three main subfamilies encoding approximately 20 cation channels. Most are non-selective cation channels, whilst there are a few which show selective permeability for certain divalent cations, e.g. Ca^{2+} or Mg^{2+} (Clapham et al 2001, Montell et al 2002).

The canonical TRP channels, the TRPC subfamily, are the original members which were discovered and are homologous to the *Drosophila trp* channel — the founder member of the TRP gene family and shown as a receptor-activated Ca^{2+}-permeable channel. Members of the TRPC family still represent some of the best candidates for ROCCs by virtue of their close similarity of electrophysiological fingerprint, pharmacological sensitivity to certain native ROCCs e.g. the α1-adrenoceptor-activated non-selective cation channel (α1AR-NSCC) in portal vein smooth muscle (Inoue et al 2001) and the functional importance as gene-specific knockdown of TRPC6 expression reduces α1AR-mediated NSCC current and Ca^{2+} influx. A role of TRPC6 in vascular smooth muscle function has also been suggested in rat cerebral artery myogenic tone (Welsh et al 2002) and also in pulmonary artery smooth muscle proliferation which is a component of pulmonary hypertension (Yu et al 2003).

TRPC6 expression is particularly high in mouse and human lung (Boulay et al 1997, Riccio et al 2002). More detailed analysis of specific tissues of the lung have indicated the presence of TRPC6 mRNA and protein expression in human airway smooth muscle (Corteling et al 2003) suggesting that in addition to having a role in vascular smooth muscle function, TRPC6 may also be involved in bronchial smooth muscle contraction. Therefore TRPC6 channel blockers may represent a novel mechanism for the development of bronchodilator therapy. Gorenne et al (1998) demonstrated that ROCC-mediated Ca^{2+} influx contributes a major source of Ca^{2+} required for bronchoconstriction of human bronchioles stimulated by a wide range of spasmogenic agents, e.g. LTD_4, allergen, acetylcholine and histamine. The epithelial cells of the lung lining the airway lumen and submucosal glands both strongly express TRPC6-immunoreactive staining (Corteling et al 2003). As epithelial cell Cl^- ion secretion can involve increases in cytosolic Ca^{2+} levels (Chiyotani et al 1994, Ko et al 1997) and receptor-stimulated Ca^{2+} influx has been demonstrated in well-differentiated human bronchial epithelial cells (Mesher et al 2003), TRPC6 may also be implicated in epithelial cell function.

Leukocytes are one of the cell types in which ROCCs have been most intensely studied and ROCC-mediated Ca^{2+} most widely documented, and it is not surprising that many TRP channels have been shown to be expressed in these cells (reviewed by Li et al 2002). Because many leukocyte responses such as release of inflammatory mediators and degranulation which cause tissue damage have a ROCC-mediated Ca^{2+}-influx-dependent component, modulation of Ca^{2+} influx in these leukocytes is a potential anti-inflammatory approach. With respect to inflammatory diseases such as chronic obstructive pulmonary disease (COPD) and asthma, the main orchestrative leukocytes involved are the lung macrophage (Barnes 1998) and the $CD4^+$ lymphocyte, respectively (Hamid et al 1991), whilst the main infiltrating granulocytes are neutrophils in COPD and eosinophils in asthma. Amongst the TRPC members, TRPC6 is the mRNA and protein for which there are clearest data to suggest its presence in a range of leukocytes such as the lymphocyte (Gamberucci et al 2002, S. Li, unpublished data), the neutrophil (Heiner et al 2003, S. Li, unpublished data) and the lung macrophage (S. Li, unpublished data). However, it remains to be determined what the functional importance of TRPC6 is in these leukocytes.

Other TRP channels which may also be of relevance to COPD and asthma include TRPM2 which appears to be selectively highly expressed in granulocytic and monocytic cells (Sano et al 2001, Heiner et al 2003). It is activated by hydrogen peroxide and has been suggested to play a role in oxidant stress (Wehage et al 2002). The presence of TRPV4 protein has also been demonstrated in lung macrophages, epithelium and in tissue neutrophils (Delany et al 2001).

Concluding remarks

Antagonists of L-VOCCs in electrically-excitable cells are valuable therapeutic agents for the treatment of cardiovascular diseases. It is also clear that Ca^{2+}-permeable channels are functionally important in electrically non-excitable cells such as leukocytes and potential targets for therapeutic intervention. Previously, progress in this area has been hindered by the lack of information about the molecular identity of these targets. However, members of the recently discovered TRP gene family represent promising molecular candidates for these channels. The challenge now is to identify which TRPs are functionally important in disease-relevant cell types and to fully explore and exploit them as drug targets. Due to the molecular information and tools now available, we are in a better position to meet this challenge.

References

Barnes PJ 1998 Chronic obstructive pulmonary disease: new opportunities for drug development. Trends Pharmacol Sci 19:415–423

Bootman MD, Collins TJ, Mackenzie L, Roderick HL, Berridge MJ, Peppiatt CM 2002 2-aminoethoxydiphenyl borate (2-APB) is a reliable blocker of store-operated Ca^{2+} entry but an inconsistent inhibitor of InsP3-induced Ca^{2+} release. FASEB J 16:1145–1150

Boulay G, Zhu X, Peyton M et al 1997 Cloning and expression of a novel mammalian homolog of Drosophila transient receptor potential (Trp) involved in calcium entry secondary to activation of receptors coupled by the Gq class of G protein. J Biol Chem 272: 29672–29680

Chiyotani A, Tamaoki J, Takeuchi S, Kondo M, Isono K, Konno K 1994 Stimulation by menthol of Cl secretion via a Ca^{2+}-dependent mechanism in canine airway epithelium. Br J Pharmacol 112:571–575

Chung SC, McDonald TV, Gardner P 1994 Inhibition by SK&F 96365 of Ca^{2+} current, IL-2 production and activation in T lymphocytes. Br J Pharmacol 113:861–868

Clapham DE, Runnels LW, Strubing C 2001 The TRP ion channel family. Nat Rev Neurosci 2:387–396

Clementi E, Martini A, Stefani G, Meldolesi J, Volpe P 1995 LU52396, an inhibitor of the store-dependent (capacitative) Ca^{2+} influx. Eur J Pharmacol 289:23–31

Clementi E and Meldolesi J 1996 Pharmacological and functional properties of voltage-independent Ca^{2+} channels. Cell Calcium 19:269–279

Corteling RL, Li SW, Giddings J, Westwick J, Poll C, Hall JP 2003 Expression of TRPC6 and related TRP family members in human airway smooth muscle and lung tissue. Am J Respir Cell Mol Biol 30:145–154

Delany NS, Hurle M, Facer P et al 2001 Identification and characterization of a novel human vanilloid receptor-like protein, VRL-2. Physiol Genomics 4:165–174

Demaurex N, Lew DP, Krause KH 1992 Cyclopiazonic acid depletes intracellular Ca^{2+} stores and activates an influx pathway for divalent cations in HL-60 cells. J Biol Chem 267: 2318–2324

Fleckenstein A 1983 History of calcium antagonists. Circ Res 52:I3–I16

Franzius D, Hoth M, Penner R 1994 Non-specific effects of calcium entry antagonists in mast cells. Pflugers Arch 428:433–438

Gamberucci A, Giurisato E, Pizzo P et al 2002 Diacylglycerol activates the influx of extracellular cations in T-lymphocytes independently of intracellular calcium-store depletion and possibly involving endogenous TRP6 gene products. Biochem J 364:245–254

Gorenne I, Labat C, Gascard JP, Norel X, Nashashibi N, Brink C 1998 Leukotriene D_4 contractions in human airways are blocked by SK&F 96365, an inhibitor of receptor-mediated calcium entry. J Pharmacol Exp Ther 284:549–552

Grynkiewicz G, Poenie M, Tsien RY 1985 A new generation of Ca^{2+} indicators with greatly improved fluorescence properties. J Biol Chem 260:3440–3450

Hamid Q, Azzawi M, Ying S et al 1991 Expression of mRNA for interleukin-5 in mucosal bronchial biopsies from asthma. J Clin Invest 87:1541–1546

Heiner I, Eisfeld J, Halaszovich CR et al 2003 Expression profile of the transient receptor potential (TRP) family in neutrophil granulocytes: evidence for currents through long TRP channel 2 induced by ADP-ribose and NAD. Biochem J 371:1045–1053

Hoth M, Penner R 1992 Depletion of intracellular calcium stores activates a calcium current in mast cells. Nature 355:353–356

Howson W, Armstrong WP, Cassidy K et al 1990 Design and synthesis of a series of glycerol-derived receptor mediated calcium entry (RCME) blockers. Eur J Med Chem 25: 595–602

Inoue R, Okada T, Onoue H et al 2001 The transient receptor potential protein homologue TRP6 is the essential component of vascular alpha(1)-adrenoceptor-activated Ca^{2+}-permeable cation channel. Circ Res 88:325–332

Itagaki K, Kannan KB, Livingston DH et al 2002 Store-operated calcium entry in human neutrophils reflects multiple contributions from independently regulated pathways. J Immunol 168:4063–4069

Iwamuro Y, Miwa S, Zhang XF et al 1999 Activation of three types of voltage-independent Ca^{2+} channel in A7r5 cells by endothelin-1 as revealed by a novel Ca^{2+} channel blocker LOE 908. Br J Pharmacol 126:1107–1114

Ko WH, Wilson SM, Wong PY 1997 Purine and pyrimidine nucleotide receptors in the apical membranes of equine cultured epithelia. Br J Pharmacol 121:150–156

Krautwurst D, Hescheler J, Arndts D, Losel W, Hammer R, Schultz G 1993 Novel potent inhibitor of receptor-activated nonselective cation currents in HL-60 cells. Mol Pharmacol 43:655–659

Li SW, Westwick J, Poll CT 2002 Receptor-operated Ca^{2+} influx channels in leukocytes: a therapeutic target? Trends Pharmacol Sci 23:63–70

Malayev A, Nelson DJ 1995 Extracellular pH modulates the Ca^{2+} current activated by depletion of intracellular Ca^{2+} stores in human macrophages. J Membr Biol 146:101–111

Merritt JE, Armstrong WP, Benham CD et al 1990 SK&F 96365, a novel inhibitor of receptor-mediated calcium entry. Biochem J 271:515–522

Mesher J, Li SW, Poll CT, Danahay H 2003 Preliminary pharmacological characterisation of UTP-induced Ca^{2+} influx in well-differentiated human bronchial epithelial cells. Br J Pharmacol 138:115P

Montell C, Birnbaumer L, Flockerzi V et al 2002 A unified nomenclature for the superfamily of TRP cation channels. Mol Cell 9:229–231

Naumov AP, Kiselyov KI, Mamin AG, Kaznacheyeva EV, Kuryshev YA, Mozhayeva GN 1995 ATP-operated calcium-permeable channels activated via a guanine nucleotide-dependent mechanism in rat macrophages. J Physiol 486:339–347

Riccio A, Medhurst AD, Mattei C et al 2002 mRNA distribution analysis of human TRPC family in CNS and peripheral tissues. Brain Res Mol Brain Res 109:95–104

Sano Y, Inamura K, Miyake A et al 2001 Immunocyte Ca^{2+} influx system mediated by LTRPC2. Science 293:1327–1330

Venter JC, Adams MD, Myers EW et al 2001 The sequence of the human genome. Science 291:1304–1351

von Tscharner V, Prod'hom B, Baggiolini M, Reuter H 1986 Ion channels in human neutrophils activated by a rise in free cytosolic calcium concentration. Nature 324:369–372

Wehage E, Eisfeld J, Heiner I et al 2002 Activation of the cation channel long transient receptor potential channel 2 (LTRPC2) by hydrogen peroxide. A splice variant reveals a mode of activation independent of ADP-ribose. J Biol Chem 277:23150–23156

Welsh DG, Morielli AD, Nelson MT, Brayden JE 2002 Transient receptor potential channels regulate myogenic tone of resistance arteries. Circ Res 90:248–250

Yu Y, Sweeney M, Zhang S et al 2003 PDGF stimulates pulmonary vascular smooth muscle cell proliferation by upregulating TRPC6 expression. Am J Physiol Cell Physiol 284:C316–C330

Zweifach A, Lewis RS 1995 Rapid inactivation of depletion-activated calcium current (I_{CRAC}) due to local calcium feedback. J Gen Physiol 105:209–226

DISCUSSION

Ambudkar: I liked those experiments in which you were looking at polarity to see which direction the Ca^{2+} was coming in. It was nice to see basolateral Ca^{2+}

influx: this is quite consistent with previous studies in polarized epithelial cells. You found TRPC6 in the apical membrane, however. Would you therefore conclude that the UTP-mediated Ca^{2+} influx is not coming through TRPC6?

Li: I don't think we can conclude that without knocking out TRPC6. It could be due to the sensitivity with which the antibody can detect low levels of TRPC6 located intracellularly within the epithelium. Although TRPC6 looks to be mainly apical, there is some staining deeper down in the epithelial layer. But how many functional TRP channels are needed in order to see a functional effect?

Ambudkar: Have you looked at other TRPCs? Are there any in the basolateral membrane? We find TRPC1 in basolateral membrane in submandibular gland acinar cells, which are also nicely polarized.

Li: We found TRPC1 message, and this was actually higher in the differentiated human bronchial epithelial cells (HBECs). But we haven't stained with anti-TRPC1 antibody to see where this is localized.

Muallem: This experiment is pretty controversial. It has now been done by two different prominent groups, and they get completely opposite results (Paradiso et al 1995, Gordjani et al 1997). Paradiso et al (1995) got influx from both membranes and Gordjani et al (1997) have similar results to yours. The story might depend on culture conditions and so on.

Li: There could be regional differences depending on the location within the lung from which the epithelium was removed. It could also be agonist dependent: we have only used one stimulus. It might be interesting to use other stimuli such as methacholine.

Putney: The conclusion in the Boucher paper (Paradiso et al 1995) was that the agonist activation showed sidedness, but thapsigargin activation did not. Assuming that these are store-operated channels (SOCs), the suggestion is that there could be spatial restriction on where IP_3 would act. Up until that point, a lot of the work on diffusion constants for IP_3 had always suggested that IP_3 should act globally.

Muallem: With the work from Gordjani et al (1997), no matter how they tried to deplete the stores they could never get Ca^{2+} influx across the apical membrane.

Putney: Have you tried using thapsigargin on this preparation?

Li: Yes, but we had a practical problem of it sticking to everything. We will try doing the experiment with cyclopiazonic acid (CPA).

Nilius: I found it intriguing that in your polarized cells TRPV4 is by far the most prominent channel. It fits marvellously to our data: EETs are involved, you get a signalling pathway with bradykinin. These data are all converging on TRPV4 activation. So why do you still look for TRPC6?

Li: We started with TRPC6, but we are moving on to look at other TRPs including TRPV4.

Nilius: From your data, TRPV4 is a marvellous target.

Li: I think it looks very interesting. I would like to see whether TRPV4 protein is expressed in these differentiated HBECs too. At the moment I don't know where it is located, and I would be interested to see whether it has differentially localized apically and basolaterally.

Scharenberg: In your quantitative PCR surveys did you see TRPM2 expression in any subsets of lymphocytes, besides mouse cells?

Li: We have only just started looking at TRPM2, so we haven't done profiling in a huge range of cell types. We haven't looked at $CD4^+$ and $CD8^+$ T cells derived from human blood yet, but I think our preliminary data on a sample of mixed lymphocytes suggest that TRPM2 is expressed at quite low levels.

Authi: Do the TRPC6 variants you have detected have any relevance? Certainly the one you pointed out is lacking a transmembrane domain.

Li: I don't know what the functional relevance is of these splice variants.

Authi: Have you looked for these in any disease states?

Li: No.

Putney: Was your conclusion that the SK&F compound doesn't distinguish between the store-operated and receptor-operated channels?

Li: I wasn't trying to differentiate between the two types of channels.

Putney: But LOE does to a large extent?

Poll: That influx pathway in human bronchial epithelial cells is relatively insensitive to both econazole and SK&F 96365.

Li: We were going to look at LOE 908 but we have been using fura-2, and LOE 908 interferes with fura-2 measurements so we would have to change the fluorophore. This is something we are planning to do.

Putney: One observation we made is that in HEK cells when we compare the store operated entry and TRP3, the SK&F compound has some selectivity for TRP3, with a difference of around 1–1.5 orders of magnitude. In the same way that we can use gadolinium to selectively remove the store operated entry and leave the TRPC3, we can do the opposite experiment with SK&F. I think Colin Taylor has some evidence that for real receptor-operated channels, whatever the signal is, it also has that selectivity.

Taylor: It is a very similar story in that low gadolinium and 2-APB will take out capacitative entry and not touch the receptor-activated entry. LOE and SK&F will do the opposite.

Authi: There was a compound recently described called TRIM, which is reported to block SOC activity in anococcygeus smooth muscle (Gibson et al 2001). Have you tried this?

Li: No.

Authi: It was originally described as an inhibitor of NOS activity. This compound has no effect on voltage-gated channels.

Muallem: I have a question for the drug company people. How are you going about drug development with all these completely non-selective drugs?

Li: We have to prioritize and start with the channels that we are most interested in first and target those.

Poll: Presently, TRP pharmacology is sparse and the tools are rather blunt, as we have been discussing. However, running 20 or so TRP high throughput screens would not be easily possible! One approach is to obtain functional data in cell types of interest, for example, by using a gene knockdown approach such as siRNA. This in addition to the evidence we have for cellular and tissue distribution and molecular epidemiology should help us to select a rather more limited panel of TRP channels to run high-throughput screens against to try to get more potent selective tools.

Putney: There are two strategies. You can start with physiological phenomenology and screen libraries, or you can start with a molecule and screen in that direction. I don't know which is best. To me, there is a lesson from the dihydropyridines: the original guess was that these couldn't possibly be useful because Ca^{2+} channels are everywhere. The physiology came later, and they established the subtypes of channels and the specificities. I wish you guys would take your libraries and throw them at the SOCs and ROCs and see what happens.

Penner: I admit that I am amazed that you guys are targeting TRPCs. Don't you do target validations?

Westwick: It all depends on what you mean by 'target validation'. This doesn't just apply to TRPs. You can take the reverse approach: look at the story of how the dihydropyridines came about. They were used later to validate that certain VOCs were involved in certain cells, and not the other way round. We can do it by trying to set up screens where we believe what we are measuring is due to the channels we put in as opposed to those that were already there, or we can just work our way through with siRNA, depending on what function you want to look at. You demonstrate at the same time that you are specifically reducing the protein, the current and what we are more interested in, the downstream functional effect. It is still a pretty long-winded approach.

Penner: Pharmaceutical companies have usually taken the 'FLIPR' (Fluorometric Imaging Plate Reader) approach. They screen for a Ca^{2+}-dependent process, and come up with a compound. Then they work out which target the compound acts on.

Westwick: That is what is referred to as the 'black box' system. You have a readout, you find a compound, but you have no idea of what it is hitting. It takes you a long time to find what that compound hits, which you need to do if you want to optimize it.

Gill: That is one approach.

Westwick: It is, but it is terribly difficult to identify the target once you find some functional effect of a compound.

Gill: As long as you have some function that can be measured, and you have a large enough throughput, you can screen enormous numbers of compounds.

Li: The problem is, how can you actually optimize the compounds?

Westwick: Both approaches are being taken, but I would favour working with a system where you know that a particular TRP or combination of TRPs is actually responsible for your output. If you want to set up a binding assay with the GPCRs you don't need an intact system, but for anything with ion channels you will need an intact system, which by the nature of it is very complex and your compound can hit a range of sites and molecules.

Barritt: Could I try to clarify in my mind your arguments. Are you saying that if, for example, you could transfect cell lines stably with each TRP protein and you could search your library of potential inhibitors, you might find some interesting compounds? However, you would not know what the possible physiological consequences were? Are you saying that you wouldn't do these screens on singly-expressed TRP channels because you might come up with an inhibitor, but you don't know what its physiological action is? I think you are saying that you need to know what the physiological actions of a given TRP channel are before you go to look for inhibitors.

Westwick: Yes. The other problem with the TRPs concerns some of the issues we have been discussing today. We have no direct activators of them. That's a difficulty. If we go back to the vanilloid receptor, the VR1s were screened just like you say and compounds were produced as a result of that.

Li: But then there was an actual ligand.

Penner: They are all capsaicin antagonists; they are not VR1 antagonists.

Westwick: That was based on capsaicin binding. There are compounds which block VR1 which are not capsaicin inhibitors.

Poll: As long as you do some level of molecular validation, that is OK. You don't need to fundamentally understand the physiological role of molecular targets before you do a high throughput screen. But we need to know target cell types of interest. This is the sort of level of evidence to stimulate us to run a high throughput screen to get the tools that will give us further clues as to the function. It is a balance between these two approaches.

Kunze: What I suspect you are saying is that you are going to target a disease or problem that you think is good for the drug company to pursue. For instance, the VR1 is an important one with its role in pain. So, are you going to look for a function and then see how the TRPs fit into that function?

Westwick: You can do it either way. In the beginning, whether it is a TRP, a kinase, a GPCR or a phosphatase, you want to see whether there is a change between the disease and normal states. I am still not convinced this is the case with the TRPs, particularly if we are interested in inflammation. There are very few data in terms of any change in expression of TRPs. This is why we are interested in the differentiated epithelial cells. If we can back this up with the antibodies we are generating to these and show differences in protein and electrophysiology between the undifferentiated and differentiated cells, this is something that is worth targeting.

Putney: A particular molecule doesn't have to be part of the disease to be a part of the cure.

Westwick: But it would enhance the selectivity of your compound if it is only expressed in disease.

Putney: In many ways it seems like this is the toughest thing to do. If you have a molecule that is mutated or under regulated, it is very hard to reverse that. On the other hand, it seems it would be much easier to choose another pathway.

Westwick: Again, I am not arguing that that is the only way. You can just demonstrate that you think it is physiologically relevant, you develop an assay and find a compound. You can find a compound that works in the assay but then to optimize that compound that you want to give orally, you really have to understand your assay and the target which it is working against. The curious thing is, if you go back to the dihydropyridines there was very little known about what was operating Ca^{2+} channels when Bayer found nifedipine. They actually used the dihydropyridines to characterize the existence of the different VOCCs.

Putney: It is also an example of where the molecule we treat is not the cause of the disease.

Westwick: But it is only by good fortune that those particular dihydropyridine-sensitive VOCCs are only in those tissues that have an effect on hypertension. Otherwise we would have had major gut and airway effects, for example.

Poll: That is an interesting point. If we were to discover L-type VOCCs now, knowing their distribution we might predict that there would be rather serious consequences of using a VOCC blocker. This sort of information could inhibit us. The fact that VOCC blockers were discovered to be Ca^{2+} antagonists by looking at functional cardiovascular changes led us subsequently to realise that you could have some selectivity with such an agent.

Putney: I won't beat this to death. I was pushing this idea of trying to get SOC blockers years ago with companies, and the reaction I got is that they are everywhere so this would be a disastrous approach. We now know that they are certainly not the same everywhere, and when everything is worked out at the molecular level, everyone can jump in at that point.

Benham: The thing that the folks who aren't used to working with chemists optimizing compounds probably don't appreciate is how robust the assays must be before chemists are happy with them. You don't make improvements in compounds in great leaps. Everything tends to be done in minute increments. It is a question of optimizing the compound at half a dozen or more different sites. To be able to do this you need an assay sufficiently robust to do large numbers of compounds and that is reproducible. We get into trouble if our assays go out by less than half a log unit.

Penner: If you use a heterologous expression system you can optimize it.

Benham: I did this 15 years ago with Tim Rink and Trevor Hallam, and ended up producing SK&F 96365. This was about as far as we got. We got that far relatively quickly, and then we hit a brick wall because we were dealing with a black box. We had no idea what the molecular basis of the channels was. We didn't know whether the effects we saw in neutrophils were due to the action upon the same proteins as in the platelets, and we didn't understand the difference between store-operated and receptor-mediated Ca^{2+} influx. So we ended up with very confusing structure activity relationships because we were probably looking at multiple molecular pathways contributing to our Ca signals. In retrospect, its not surprising that SKF is non-selective as non-specific compounds are the most effective blockers in these complex systems.

Cox: If you had done the black box approach now you would have come up with the same molecule, and you wouldn't be able to take it anywhere because it hits everything at a low concentration. A chemist isn't going to be able to optimize it any further. Where do you go with it?

Benham: Another problem that is relevant to channels generally is that we are competing for resources against folks who have enzymes as targets that they can crystallize and get good structures for both the bare enzyme and the enzyme with their lead molecules bound. This enables them quickly to generate molecular models of the pharmacophore. Coupled with rapid high throughput screening this means that they can make phenomenal progress in getting from micromolar to picomolar affinities.

Cox: With the SK&F compound, now you have perhaps an insight into the various channels you could go back and think of that compound and all the analogues to see whether you can tease out any structure–activity relationships, albeit very small ones. This wouldn't have been possible before because you didn't have the separate channels to play with.

Penner: SK&F is so non-specific. It is not even specific among the TRPs. It will even block Cl^- channels.

Cox: Presumably there is a pool of analogues. You could look for a fingerprint for each molecule and try to build up a direction to send you down from that particular lead i.e. from a diverse set of analogues, differences in activity across a range of the molecular targets can be exploited to make new compounds with further differential activity (selectivity) or even specificity for a given single target.

Benham: If we got SK&F as one of our hits today, we would probably discard it as a way forward. You would hope to get much better hits than this now when there are a million compounds running through a screen with a lot more diversity in the molecules.

Montell: If you are committed to doing screens, a potentially useful approach is to look for effects resulting from overexpression of random genes obtained from expressed sequence tag (EST) libraries. This would be analogous to screens that are done in flies using the GAL4/UAS system.

Westwick: This has been used in zebrafish as well.

Gill: What about the knockouts? Is anyone making a TRPC3 knockout?

Westwick: Don't we need conditional knockouts?

Putney: People either have already, or are trying to make knockout mice for all the TRPCs, and I assume this also applies to TRPVs and TRPMs.

Muallem: Cre would be a better way to go.

Scharenberg: Developing Cre promoter mice takes time and money. It is non-trivial.

Westwick: From the work we have seen today about the up-regulation of TRPC3 in the TRPC6 knockout, was that due to a developmental change? If you did siRNA in a particular cell type would you also see the same result? Would you get more TRPC3 when you lose TRPC6? There are a lot of people doing genome-wide siRNA: they are taking all 40 000 or so genes, targeting with siRNA and then looking for a functional output. They work back to find out which enzymes or channels the genes are directed against. But it is clear from what we have heard here that this is a risky business. Even if you might knock one down, you might up-regulate half a dozen at the same time.

Scharenberg: It is more a problem in mouse development. There are more opportunities for changes than there are in culture. Culture has its own problems. Choose your poison.

Putney: Whenever you make a mutant mouse you are very fortunate if it doesn't turn you into a developmental biologist!

Gudermann: Every approach has its own advantages and disadvantages. The way we try to find the solution is that we are now constructing transgenic mice that inducibly express dominant negative channel subunits whose expression can hopefully be switched on in the animals fast enough. But if you talk to people who work on PKC they tell us that even within a couple of hours you see down-regulation of one and up-regulation of the other isoform.

Putney: The strong conclusions come about when you try many different approaches and they all lead to the same answer.

Scharenberg: The question is, is it easier to make the mice, or to do the screen, make the drug and then come back and check?

Gill: I don't think it is either or.

Hardie: In *Drosophila* we also have temperature-sensitive alleles which give a reversible knockout in a timescale of seconds to minutes. This is a very nice tool.

Zhu: Say, for example, that we want to know whether there is a drug that specifically targets TRPC3. Is this something that should be screened for now, or should we wait until we find the specific composition of the channel?

Putney: That is a good example. You could set up a screen tomorrow for TRPC3, 6 and 7. It would be pretty robust. But the concern here is that if we got the world's very best TRPC3 blocker, it wouldn't be good for anything if we couldn't optimize it.

Westwick: There are likely a number of companies that have set up screens like this and do have compounds. All I was indicating was caution in terms of what happens with them.

References

Gibson A, Fernandes F, Wallace P, McFadzean I 2001 Selective inhibition of thapsigargin-induced contraction and capacitative calcium entry in mouse anococcygeus by trifluoromethylphenylimidazole (TRIM). Br J Pharmacol 134:233–236

Gordjani N, Nitschke R, Greger R, Leipziger J 1997 Capacitative Ca^{2+} entry (CCE) induced by luminal and basolateral ATP in polarised MDCK-C7 cells is restricted to the basolateral membrane. Cell Calcium 22:121–128

Paradiso AM, Mason SJ, Lazarowski ER, Boucher RC 1995 Membrane-restricted regulation of Ca^{2+} release and influx in polarized epithelia. Nature 377:643–646

Role of TRP channels in oxidative stress

Klaus Groschner, Christian Rosker and Michael Lukas

Department of Pharmacology and Toxicology, Karl-Franzens-University, University of Graz, Universitaetsplatz 2, A-8010 Graz, Austria

Abstract. Increasing evidence suggests a pivotal role of reactive oxygen species (ROS) as well as reactive nitrogen species (RNS) in human pathophysiology. A typical target of ROS/RNS signalling is Ca^{2+} channels which mediate both long-term as well as acute cellular responses to oxidative stress. We have previously reported that cation channels related to the *Drosophila* transient receptor potential gene product (TRPC proteins) are likely to serve as redox sensors in the vascular endothelium, and demonstrated that TRPC3 expression is a determinant of the nitric oxide sensitivity of store-operated Ca^{2+} signalling. Experiments with TRPC species overexpressed in HEK293 cells confirmed that TRPC3 and TRPC4 are able to form redox sensitive cation channels. A key mechanism involved in redox activation of TRPC3 appears to be ROS-induced promotion of protein tyrosine phosphorylation and stimulation of phospholipase C activity. In addition, oxidative stress-induced disruption of caveolin 1-rich lipid raft domains, which interfere with functional TRPC channels, is likely to contribute to redox modulation of TRP proteins and to oxidative stress-induced changes in cellular Ca^{2+} signalling. Taken together, our data suggest TRPC species serve as a link between cellular redox state and Ca^{2+} homeostasis. Thus, modulation of these cellular redox sensors may offer unique opportunities for therapeutic interventions.

2004 Mammalian TRP channels as molecular targets. Wiley, Chichester (Novartis Foundation Symposium 258) p 222–235

The redox status of a cell has been recognized as a factor that determines cellular functions and serves as a switch between apoptotic cell death, differentiation or proliferation (Finkel & Holbrook 2000, Martindale & Holbrook 2002). The biological response to oxidative stress depends not only on the nature of the involved reactive oxygen species but most importantly on the intensity of the redox signal corresponding to a specific level of oxidative stress (Finkel & Holbrook 2000, Martindale & Holbrook 2002). Primary transduction of redox signals is incompletely understood. Nonetheless, a number of redox-sensitive proteins have been identified including protein tyrosine phosphatases (Chiarugi et al 2001) as well as ion channels (Chiamvimonvat et al 1995, Koliwad et al 1996, Li et al 1998, Feng et al 2000), and the modification of critical protein cysteine residues has been proposed as a key mechanism of redox sensing by these signalling molecules (Cooper et al 2002). Redox regulation of Ca^{2+}-entry pathways appears as a crucial determinant of the cellular response to oxidative stress (Kamata & Hirata 1999), and some Ca^{2+}-channel proteins such as the ryanodine receptor (Wang et al 2001) and voltage-gated Ca^{2+} channels (Chiamvimonvat et al 1995) have been suggested as primary redox sensors. A variety of other Ca^{2+} transport systems are expected to sense changes in the redox status indirectly via redox modification of one or more input signalling pathways. Pivotal signals for the control of Ca^{2+} channels are generated by phospholipase C, and activation of the γ isoform of phospholipase C is closely related to redox signalling (Kamata & Hirata 1999, Wang et al 2001). Since members of the TRP protein family have been identified as essential components of phospholipase C (PLC)-regulated, Ca^{2+}-permeable cation conductances (Vennekens et al 2002), we investigated the redox sensitivity of TRPC3, one of the TRPC species tightly regulated by PLC. Initial work suggested TRPC3 channels as the basis of a non-selective cation conductance which depolarizes vascular endothelial cells in the presence of oxidants (Balzer et al 1999). We now provide evidence for a role of oxidant-induced disruption of cholesterol-rich lipid rafts as a mechanism which links cellular redox state to TRPC channel function.

Experimental procedures

Cell culture

HEK293 cells stably expressing the coding region of TRPC3 (T3–9) and TRPC4 (kindly provided by M. X. Zhu), a C-terminal deletion mutant of TRPC3 (truncated at aa725; T3ΔC-4), N-terminal YFP-fusion of TRPC3 (YT3–1), or neomycin resistance only (vector controls) were cultured in DMEM supplemented with 10% FCS and 0.25 g/l geneticin. Vascular endothelial cells were cultured from bovine aorta as described previously (Berk 1999).

DNA constructs and cell transfection

Full length TRPC3 (U47050) was subcloned into the MCS of pEYFP-C1(Clontech), full-length caveolin 1 was subcloned into the MCS of pECFP-C1(Clontech). For labelling of cellular PIP_2 pools we expressed a GFP-fusion of the PH domain of phospholipase Cδ (GFP-PH-PLCδ; kindly provided by T. Meyer). HEK293 and endothelial cells were transfected by electroporation and Transfast^{TM} (Promega) reagent, respectively.

Fluorescence microscopy

Fluorescence imaging was performed on a Zeiss Axiovert 200 microscope, equipped with a MultiSpec Imager for beam splitting, Chroma filters and a CoolSNAP fx-HQ CCD-camera. Image deconvolution was performed using AutoDeblur 7.4.1. software.

Electrophysiology

All patch-clamp experiments were performed as described (Balzer et al 1999) using an extracellular solution containing (in mM) 137 NaCl, 5.4 KCl, 2 $CaCl_2$, 15 HEPES, pH adjusted to 7.4 with HCl. The pipette solution contained (in mM) 110 K-gluconate, 10 KCl, 5 $MgCl_2$, 10 HEPES, pH adjusted to 7.4 with N-methyl-D-glucamine and either 3 EGTA or 10 BAPTA as indicated.

Results and discussion

Overexpression of TRPC3 promotes a redox sensitive cation conductance in HEK293 cells

Measurement of membrane conductances in HEK293 cells revealed a remarkable resistance of wild-type and vector-transfected cells to strong oxidative stress. Up to 1 h exposure of cells to 400 μM of the lipophilic peroxide *tert*-butyl hydroperoxide (t-BHP; 400 μM) barely affected membrane conductance of HEK293 cells (Fig. 1). By contrast, cells overexpressing TRPC3 (T3–9; Fig. 1 or YT3–1; not shown), exhibited a t-BHP-induced increase in basal membrane conductance which was mostly carried by Na^+, as evident from a hyperpolarizing shift in reversal potential upon removal of extracellular Na^+ as shown for T3–9 cells in Fig. 1. The TRPC3-dependent conductance of HEK293 cells resembled the endogenous oxidant-activated conductance measured in pig aortic endothelial cells (ECAP, Fig. 1). These results are in line with our previous observation that the redox-sensitive cation conductance of endothelial cells is inhibited by expression of a dominant negative TRPC3 species (N-TRP3) (Balzer et al 1999). Taken together, these data indicate that TRPC3 contributes to the formation or regulation of a redox-sensitive cation channel. A similar redox-sensitive cation conductance was observed in HEK293 cells stably expressing TRPC4 ($n=6$), while cells expressing a C-terminal deletion mutant of TRPC3 (T3ΔC-4) were redox resistant ($n=9$, data not shown). We propose that TRPC3 and TRPC4 expression confers a specific cellular redox sensitivity which results in Na^+ loading and membrane depolarization during oxidative stress. These TRPC-mediated stress responses are expected to result in substantially modified Ca^{2+} signalling.

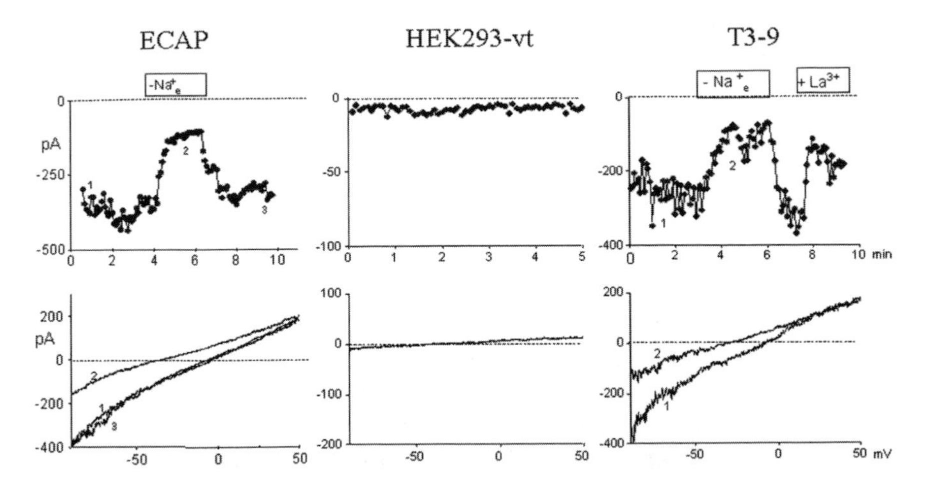

FIG. 1. Redox-activated Na^+ conductances in pig aortic endothelial cells (ECAP; left) and in TRPC3 overexpressing HEK293 cells (T3-9; right) as well as membrane currents recorded in vector-transfected HEK293 cells (HEK293-vt; middle). All cells were exposed to *tert*-buthylhydroperoxide (t-BHP; 400 μM; 1 h) before current recordings. Upper panel shows time courses of membrane currents measured at -70 mV. Removal of extracellular Na^+ ($-Na^+$, replaced by *N*-methyl-*D*-glucamine) and addition of the channel blocking ion La^{3+} is indicated. Lower panel shows individual current to voltage relations derived from voltage-ramp protocols at the times indicated in time courses.

Phospholipase C is involved in TRPC3 activation by peroxides

Since peroxide-induced stress is known to result in activation of PLCγ (Wang et al 2001, Qin et al 1995), we analysed the possible role of redox-activation of PLC enzymes by use of the inhibitor U73122. As shown in Fig. 2, t-BHP failed to activate significant cation currents in T3–9 cells in the presence of U73122 (40 μM). U73343 (40 μM), a structurally related compound, which barely affects phospholipase activity, exerted negligible inhibitory effects on redox activation of TRPC3. These data are in line with the concept that PLC is involved in the mechanism by which peroxide-induced stress activates TRPC3 cation channels. So far, the role of PLCγ in the redox response of cells is incompletely understood. Recent evidence suggests that PLCγ transduces signals that allow for cell survival during oxidative stress (Wang et al 2001). It remains to be elucidated if TRPC mediated cation fluxes play a role in this cell survival mechanism. In contrast to other Ca^{2+} entry pathways, TRPC3-mediated Ca^{2+} entry appears to lack tight functional coupling to mitochondria (Thyagarajan et al 2001) and may therefore serve specific Ca^{2+} signalling events without promoting mitochondrial dysfunction and apoptosis.

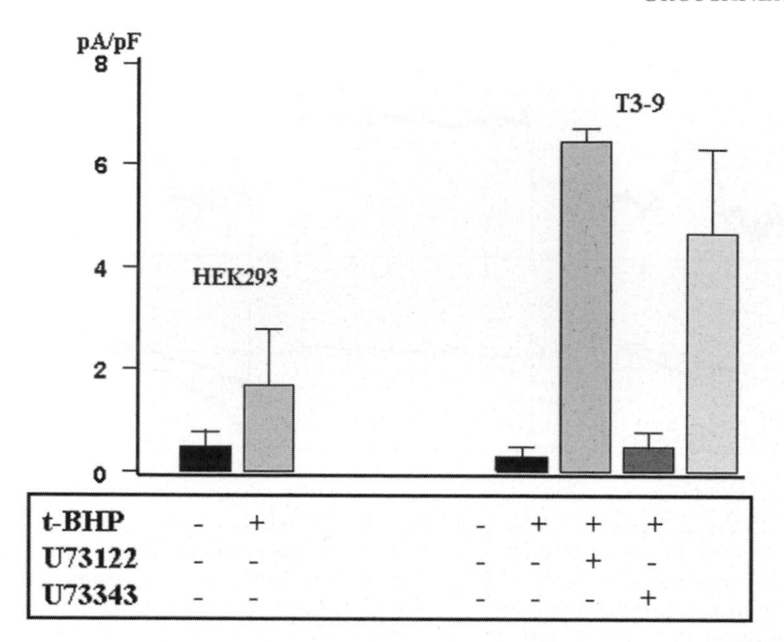

FIG. 2. U73122 inhibits oxidant-induced currents in TRPC3 overexpressing HEK293 cells (T3-9). Current density measured at −70 mV in wild-type HEK293 cells (HEK293; left) and in TRPC3 overexpressing HEK293 cells (T3-9; right) under control conditions (without exposure to *tert*-butylhydroperoxide; t-BHP) and after oxidative stress (t-BHP; 400 μM; 1 h) in the absence or presence of the phospholipase C inhibitor U73122 (40 μM) or the inactive derivative U73343 (40 μM) are shown. Mean values ± SEM from 8–12 experiments.

Oxidant-induced internalization of caveolin 1 and disruption of cholesterol-rich microdomains as a potential mechanism in redox activation of TRPC3

TRPC proteins have been proposed as a component of caveolin 1 (Cav1)-rich membrane microdomains (caveolae) (Lockwich et al 2001), which play an important role in PLC signalling. Since caveolae have been recognized as a redox sensitive structure (Smart et al 1994), we investigated the effects of oxidative stimuli on cellular localization of TRPC3 and Cav1 by monitoring the cellular localization of YFP-TRPC3 and YFP-Cav1 fusion proteins. As shown in Fig. 3, TRPC3 localization in the plasma membrane was not affected by t-BHP (400 μM; 1 h), while Cav1 was in large part translocated to intracellular compartments. A similar effect was obtained by selective oxidation of membrane cholesterol catalysed by extracellular cholesterol oxidase (0.5 U/ml, 1 h). Our results demonstrate that peroxide-induced stress as well as oxidation of membrane cholesterol severely disrupts the integrity of Cav1-rich lipid rafts, leaving TRPC3

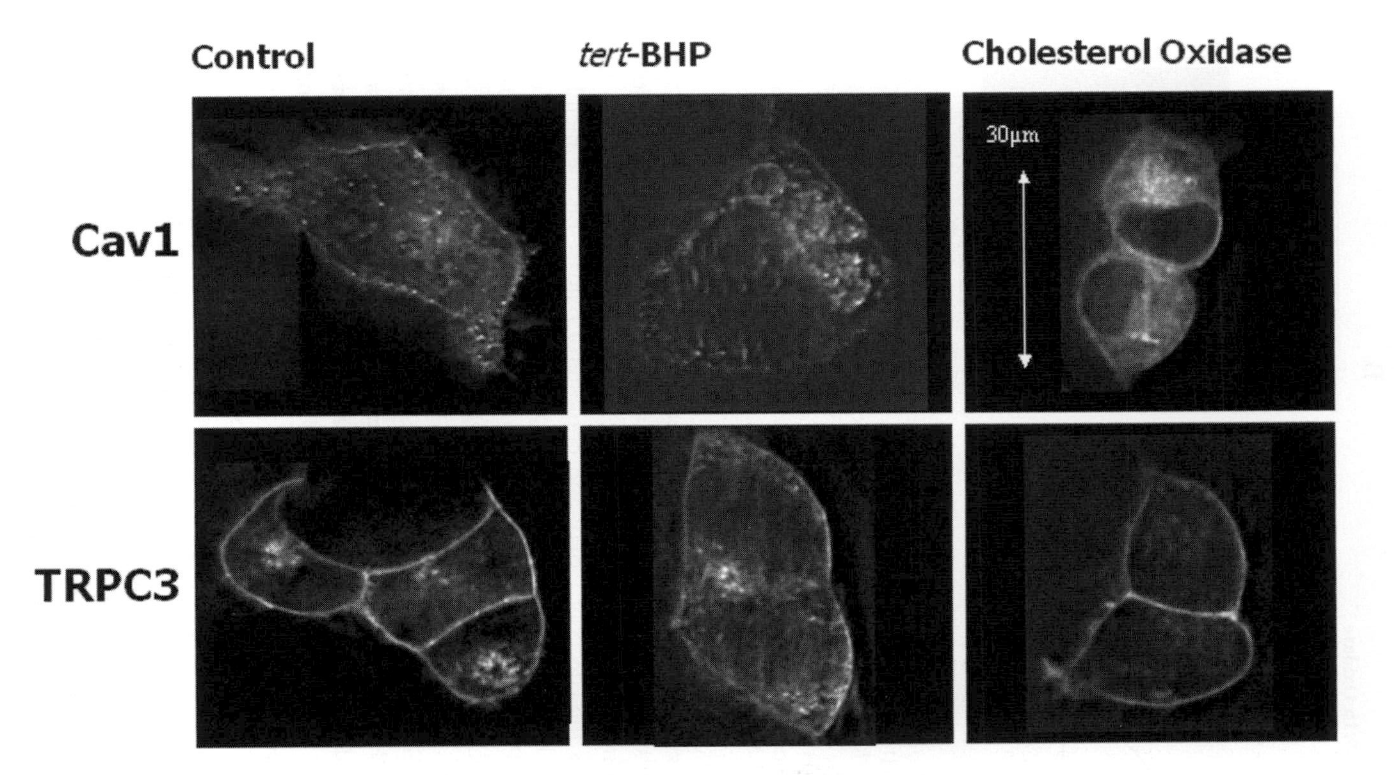

FIG. 3. Effects of oxidative stress on cellular localization of Cav1 and TRPC3. Deconvoluted micrographs from HEK293 cells expressing YFP fusion proteins of Cav1 and TRPC3, respectively. Both proteins display substantial membrane targeting under control conditions (no oxidative stress). The integrity of Cav1-rich lipid rafts is severely disrupted by peroxide-induced (t-BHP; 400 μM; 1 h) stress as well as by oxidation of membrane cholesterol (cholesterol oxidase; 0.5 U/ml; 1 h), resulting in internalization of Cav1. By contrast, TRPC3 localization is barely affected by oxidative stimuli.

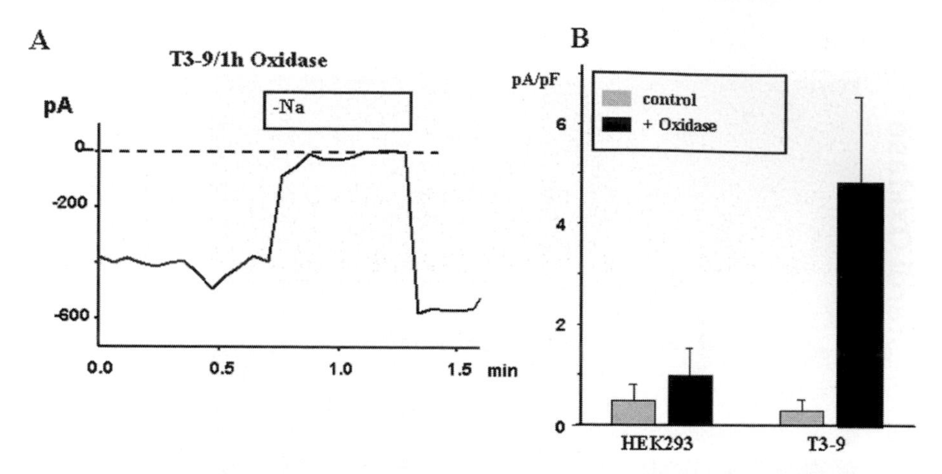

FIG. 4. Oxidation of membrane cholesterol activates a Na^+ conductance in TRPC3 overexpressing HEK293 cells. (A) Time course of membrane currents measured at -70 mV in a cell exposed to cholesterol oxidase (0.5 U/ml; 1 h). Removal of extracellular Na^+ (Na^+ replaced by *N*-methyl-*D*-glucamine; $-Na^+$) is indicated. (B) Current density measured at -70 mV in HEK293 cells (HEK293; left) and in TRPC3 overexpressing HEK293 cells (T3-9; right) under control conditions (control; open columns) and after exposure to cholesterol oxidase. Mean values $\pm$ SEM from 5–8 experiments are shown.

proteins in the plasma membrane. Importantly, both interventions activated TRPC3 cation channels in the HEK293 expression system. Figure 4 illustrates the cation conductance recorded in cholesterol oxidase-treated T3-9 cells. Figure 5 shows the effects of t-BHP and cholesterol oxidase on the membrane targeting of a phosphatidylinositol 4,5-bisphosphate (PIP_2)-binding GFP fusion protein (GFP-PH-PLCδ). Peroxide treatment resulted in a profound loss of membrane targeted GFP-PH-PLCδ indicating hydrolysis of PIP_2 during oxidative stress. Treatment of HEK293 cells with cholesterol oxidase, however, failed to affect GFP-PH-PLCδ targeting. This lack of effect of cholesterol oxidase on GFP-PH-PLCδ membrane targeting indicates that TRPC3 activation by cholesterol oxidase is associated with little or no PIP_2 hydrolysis and probably with relatively little disturbance of intracellular redox balance. It therefore is tempting to speculate that cholesterol-rich lipid rafts serve as an important primary redox sensor and that oxidation of membrane cholesterol by itself represents some kind of activating stimulus for TRPC3 channels. Recent reports suggest complex modulation of PLCγ signalling in response to disruption of lipid raft structures, resulting in uncoupling between tyrosine phosphorylation, membrane targeting and enzymatic activity (Jang et al 2001). Our results suggest that the oxidative disruption of the integrity of cholesterol rich microdomains is a key event in redox activation of TRPC species.

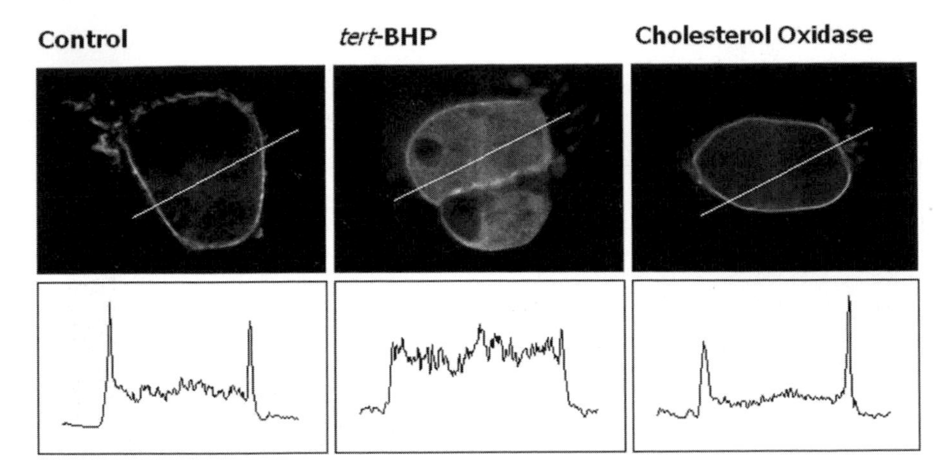

FIG. 5. Imaging of the cellular localization of PIP_2 in HEK293 cells. Cells were transfected to express a GFP-tagged PIP_2-binding protein (GFP-PH-PLCd). Membrane-targeting of the PIP_2-binding GFP fusion protein markedly disturbed by treatment with t-BHP (400 μM; 1 h; middle) indicating hydrolysis of PIP_2 during oxidative stress. Localization is not affected by extracellular cholesterol oxidase treatment (0.5 U/ml for 1 h; right). Lower panel displays the extent of membrane targeting by fluorescence intensity scans along the line in the corresponding image (upper panel). Lines correspond to 30 μm.

Summary and conclusions

Ion channels formed by TRPC species have been suggested as key players in cellular Ca^{2+} homeostasis. The main mechanism by which physiological signals control cation permeation through TRPC channels is receptor-dependent stimulation of PLC and the associated changes in the function of intracellular Ca^{2+} stores. Here we demonstrate that TRPC channels are able to sense redox stimuli which are considered of particular pathophysiological relevance. Our results provide evidence for redox sensitivity of TRPC3 channels based on a tight link between channel function and the integrity of cholesterol-rich membrane microdomains. We suggest the cross-talk between TRPC channels and raft associated signalling components as a mechanism which enables the transduction of redox signals by TRPC species. A role of TRPC-mediated cation fluxes in cellular stress responses appears therefore likely and makes TRPC proteins appear as a most attractive target for novel therapeutic strategies.

Acknowledgements

Supported by the Austrian Research Foundation (FWF), SFB-Biomembranes F715 and FWF-14950.

References

Balzer M, Lintschinger B, Groschner K 1999 Evidence for a role of Trp proteins in the oxidative stress-induced membrane conductances of porcine aortic endothelial cells. Cardiovasc Res 42:543–549

Berk BC 1999 Redox signals that regulate the vascular response to injury. Thromb Haemost 82:810–817

Chiamvimonvat N, O'Rourke B, Kamp TJ et al 1995 Functional consequences of sulfhydryl modification in the pore-forming subunits of cardiovascular Ca^{2+} and Na^{+} channels. Circ Res 76:325–334

Chiarugi P, Fiaschi T, Taddei ML et al 2001 Two vicinal cysteines confer a peculiar redox regulation to low molecular weight protein tyrosine phosphatase in response to platelet-derived growth factor receptor stimulation. J Biol Chem 276:33478–33487

Cooper CE, Patel RP, Brookes PS, Darley-Usmar VM 2002 Nanotransducers in cellular redox signaling: modification of thiols by reactive oxygen and nitrogen species. Trends Biochem Sci 27:489–492

Feng W, Liu G, Allen PD, Pessah IN 2000 Transmembrane redox sensor of ryanodine receptor complex. J Biol Chem 275:35902–35907

Finkel T, Holbrook NJ 2000 Oxidants, oxidative stress and the biology of ageing. Nature 408:239–247

Jang IH, Kim JH, Lee BD et al 2001 Localization of phospholipase C-gamma1 signaling in caveolae: importance in EGF-induced phosphoinositide hydrolysis but not in tyrosine phosphorylation. FEBS Lett 491:4–8

Kamata H, Hirata H 1999 Redox regulation of cellular signalling. Cell Signal 11: 1–14

Koliwad SK, Kunze DL, Elliott SJ 1996 Oxidant stress activates a non-selective cation channel responsible for membrane depolarization in calf vascular endothelial cells. J Physiol 491:1–12

Li A, Segui J, Heinemann SH, Hoshi T 1998 Oxidation regulates cloned neuronal voltage-dependent Ca^{2+} channels expressed in Xenopus oocytes. J Neurosci 18:6740–6747

Lockwich T, Singh BB, Liu X, Ambudkar IS 2001 Stabilization of cortical actin induces internalization of transient receptor potential 3 (Trp3)-associated caveolar Ca^{2+} signaling complex and loss of Ca^{2+} influx without disruption of Trp3-inositol trisphosphate receptor association. J Biol Chem 276:42401–42408

Martindale JL, Holbrook NJ 2002 Cellular response to oxidative stress: signaling for suicide and survival. J Cell Physiol 192:1–15

Qin S, Inazu T, Yamamura H 1995 Activation and tyrosine phosphorylation of p72syk as well as calcium mobilization after hydrogen peroxide stimulation in peripheral blood lymphocytes. Biochem J 308:347–352

Smart EJ, Ying YS, Conrad PA, Anderson RG 1994 Caveolin moves from caveolae to the Golgi apparatus in response to cholesterol oxidation. J Cell Biol 127:1185–1197

Thyagarajan B, Poteser M, Romanin C, Kahr H, Zhu MX, Groschner K 2001 Expression of Trp3 determines sensitivity of capacitative Ca^{2+} entry to nitric oxide and mitochondrial Ca^{2+} handling: evidence for a role of Trp3 as a subunit of capacitative Ca^{2+} entry channels. J Biol Chem 276:48149–48158

Vennekens R, Voets T, Bindels RJ, Droogmans G, Nilius B 2002 Current understanding of mammalian TRP homologues. Cell Calcium 31:253–264

Wang XT, McCullough KD, Wang XJ, Carpenter G, Holbrook NJ 2001 Oxidative stress-induced phospholipase C-gamma 1 activation enhances cell survival. J Biol Chem 276:28364–28371

DISCUSSION

Fleig: It is known that oxidative stress or peroxide can induce single channel-like phenomena in bilayers without any proteins (Mendez & Penner 1998). Have you had success in getting single channels comparing non-expressing and over-expressing inside-out patches?

Groschner: On the basis of our whole cell data, I would say that the membrane of native HEK cells does not respond to oxidants with single channel activity. Nonetheless, it is important that we look at this because we should be able to detect single channel events in the membranes of TRPC3 overexpressing cells.

Fleig: If you apply oxidative stress using peroxide, you necessarily introduce pH changes. Have you checked for pH change effects on this process?

Groschner: If we use cholesterol oxidase there is no change in pH, but there is some change with peroxide, as you point out. Quite recently we looked at the pH sensitivity of TRPC3 and TRPC6, and they are not terribly pH sensitive.

Fleig: Can you protect the channels from oxidative stress by using vitamins C or E?

Groschner: We looked at this in the endothelium very early on. There is some prevention of the oxidant effect by lipophilic antioxidants, but not complete.

Putney: Your conclusion is that the action of the oxidant requires a basal PLC activity, rather than activated PLC. I didn't quite see how you came to that conclusion. Have you actually looked by measuring inositol phosphate to see whether the oxidants increase this? Did you check to see whether oxidants activate PLC?

Groschner: Yes, but the problem is that although we see a clear effect when using the PH domain of PLCδ as a sensor, the effect is relatively small in conventional PI turnover measurements. I am not sure what we actually see with the PH-domain construct. It just shows us that there is something going on with the PIP_2 in the membrane. In conventional measurements of PI turnover, there is only some effect with strong oxidants. There is definitely nothing with cholesterol oxidase.

Putney: A related question is what the PH domain imaging experiments mean. It seemed that you were implying that PIP_2 was moving somewhere. As long as PIP_2 goes away, by whatever mechanism, PH will fall off. Or if IP_3 is formed, then the PH domain actually moves to the cytoplasm because it binds IP_3.

Groschner: That is something we were considering.

Putney: Do oxidants appear to increase the sensitivity of TRPC3 activation by added OAG?

Groschner: I can't answer that, because we haven't looked at it.

Westwick: If you add antioxidants in your TRPC3 expression system, does this modify the carbachol-induced effect that you see? In other words, do you need basal oxidase activity?

Groschner: We have never done this, so I can't answer that.

Nilius: Have you done cholesterol extraction experiments to get rid of the raft structure?

Groschner: Yes, but this does not perfectly mimic oxidant effects. The interesting thing is that there is normally quite a minor fraction of cholesterol — about 10% — that is converted to cholestenone in the presence of oxidants. This is relatively little reduction as compared to cholesterol extraction with cyclodextrins. Probably, what is happening in the lipid rafts is that if there is an increase by that amount of cholestenone it changes the state of the rafts. This has been shown in terms of impairment of platelet-derived growth factor signalling in rafts. The effect does not require complete conversion of cholesterol.

Nilius: Have you ever tried a similar sort of experiment in cells which do not have this raft structure, such as CaCo-2 cells? When you express TRPC3 in CaCo-2 cells there could be a dramatic change in the oxidant stress activation.

Groschner: I would indeed expect this if they don't have caveolae.

Ambudkar: You can use the binding of cholera toxin subunit to plasma membrane rafts, to see whether the rafts are still intact?

Groschner: What we see is that the localization of caveolin is changing.

Ambudkar: What about the non-caveolar lipid rafts? Cholesterol oxidation would disrupt all the rafts, not specifically the caveolar rafts.

Groschner: That is true.

Ambudkar: You tested TRPC3 function. Were these tests done in cells expressing full-length caveolin? In other words, was the oxidative stress-induced effect on channel activity seen in cells that were overexpressing caveolin?

Groschner: The redox sensitivity does not change with overexpression of caveolin.

Ambudkar: Does the basal activity of TRPC3 change? Have you examined carbachol-stimulated Ca^{2+} mobilization?

Groschner: We do indeed see some effect of caveolin-1 overexpression on the PLC-dependent stimulation of TRPC3.

Zhu: In terms of the channels that you saw in the endothelial cell, compared with the mouse cell, do you think they are more or less the same channels? They are both activated by oxidative stress.

Groschner: They are superficially similar, but there are definitely differences, such as in OAG sensitivity. I am not convinced they are the same.

Penner: I don't believe I have seen a record in which you have discontinued the application of t-BHQ to a patch clamped cell. What happens? Our observation with t-BHQ is that when we apply it we don't see large currents that activate. But as soon as we stop the application we get a huge current. We have always interpreted this as some inhibitory effect of t-BHQ on the non-saturable current.

Groschner: We tested for acute effects, but the effects we see on TRPC3 need exposure times of longer than 30 min. We don't have a quick response. It correlates quite well with the time course of changes in the lipid raft structure. We haven't looked at the effects of washing it out.

Nilius: The TRPM2 channel has this famous arachidonic acid recognition sequence, **IS**XX**TKE** arachidonate recognition sequence. This is considered to be important for sensing oxidative stress. Is there a similar sequence in TRPC3?

Groschner: I don't know.

Nilius: Have you thought about doing mutagenesis work to remove the sensitivity?

Groschner: No, not yet in terms of changing a specific motif.

Nilius: It is quite impressive in TRPM2 (see Hara et al 2002). If they delete this sequence it doesn't respond any more.

Scharenberg: But if you delete this motif those channels no longer go to the cell membrane. If we delete anything in the TRPM region of TRPM2, the resulting protein never goes to the surface.

Montell: The observation that oxidative stress leads to constitutive activation of certain TRP channels could be very interesting from the standpoint of integration of multiple stimuli. For example, it has been observed that pH potentiates the activation of TRPV1 by heat. It would be interesting to understand whether the effect of oxidative stress that you observe on TRPC3 and 4 is generalizable to lots of other channels within the TRP superfamily. This could be a prelude to addressing whether any TRP channel integrates the effects of oxidative stress with other stimuli.

Groschner: The main type of integration I can imagine is that there is a sensitization to PLC-derived signals. We should probably look at the sensitivity to this type of signal. Of course, there could be some change in response to store depletion as well. We probably need to go to a lower level of oxidative stress, perhaps to some kind of threshold level that introduces just some oxidation of cholesterol.

Montell: So as far as you know, every member of the TRP superfamily could have some response to oxidative stress and such an effect could potentiate the response to other stimuli.

Putney: It occurred to me that the TRPC4 might be significant here. In HEK cells, the role of TRPC4 depends a lot on who does it and which TRPC4 they use. Are you using shorter (β) TRPC4? In your hands, does that TRPC4 respond to PLC activation?

Groschner: We haven't done many experiments with TRPC4. It was the long form that generated a non-selective conductance. If you look at the current–voltage relationship there is a clear difference as compared to TRPC3.

Putney: It would be important if the oxidants can activate a channel which normally isn't easily activated, just by turning on PLC.

Groschner: We should use the oxidase instead. Lipophilic peroxides will definitely cause oxidation of critical thiol residues everywhere in the cell, so this might have an additional effect on the channel. The more gentle way to do this is to oxidise the lipid selectively.

Schilling: With regard to direct effects on the channel, have you tried to up- or down-regulate glutathione reductase, or change the levels of GSH in the cell, and then look at single-channel activity?

Groschner: No. I think the currents that we see in response to strong oxidants and cholesterol oxidase, at first glance look very similar, but if you look in more detail at the conductances, there are some differences. This may be because there are additional effects due to accumulation of oxidised glutathione, for example, so we have to separate these two types of effects. I agree that the best way would be to do that at the single channel level.

Nilius: Klaus, you have done a lot of work on the endothelium. You therefore know all about the famous Koliwad channel (Koliwad et al 1996). Have you compared this with your channel?

Groschner: We need to do this.

Putney: Your evidence that a TRPC is involved is based on the effects of your dominant negative. I remember when you first published on this you made the point that you can't be sure that it is specific for a particular TRP. Have you ever gone back and looked at other expressed TRPs? Does it suppress your TRP response in the HEK cells?

Groschner: We haven't done this, except for TRPC3.

Montell: It would be interesting to look specifically at subthreshold levels of oxidative stress stimuli in combination with other stimuli. Perhaps those TRP channels that are activated by heat or cold will exhibit some modulation of the temperature range in combination with oxidative stress.

Groschner: We haven't looked at temperature, but we have started to do this with pH.

Montell: In the cell there are lots of things that are going on at the same time. A given channel might not get activated by one type of stimulus. Some channels might be waiting for coincidence of two or three stimuli.

Benham: Klaus, what do you think about oxidative stress as a sort of ubiquitous mechanism for activating TRPCs? Do you think it really is a TRPC3-specific effect?

Groschner: It could apply to the other TRPC family members as well. We think it may indeed be a rather general mechanism.

Putney: As someone who has earned his living with dirty pharmacology, I know we all tend to trust the specificity of U73122. There is an isomer we use as a control, but it is not a stereoisomer: it is more like a reactivity isomer. U73122 completely knocks out I_{CRAC} under conditions where there is no other evidence that you need PLC for I_{CRAC}, so you come to the same conclusion, that there is some basal PLC activity that is needed. But I wonder about this.

Groschner: I would have loved to have seen both U73122 and the control compound work in the same way, to have something new in terms of channel modulation. But it turns out that it looks as if it is just again PLC.

Putney: If you think it is PLC, since we know PLC can regulate those channels, can it be that the effect of oxidants that is important for regulating the TRPs is just to lower the level of PIP_2? Is there anything in your data that would indicate that it can't be that simple?

Groschner: Cholesterol oxidase did not cause translocation of the PH domain at all, indicating that PIP_2 does not change much under these conditions.

Hardie: Do you know what happens to the ATP levels after prolonged oxidative stress? Is it enough to impair mitochondrial function and lead to a reduction in ATP?

Groschner: We didn't look at this, but I think there would be a change in mitochondrial function if you expose the cell for an hour to a strong oxidant. Nonetheless, I am convinced that the oxidase is clean, but we haven't checked for changes in mitochondrial function yet.

Hardie: A suppression of DAG kinase activity by ATP depletion or some other non-specific effect would also lead to an enhancement of any DAG-activated channels via higher basal PLC activity.

Scharenberg: In many tumour cell lines the mitochondria don't support ATP production. You can knock them out completely and have normal ATP levels for a long period, as long as glucose is present.

References

Hara Y, Wakamori M, Ishii M et al 2002 LTRPC2 Ca^{2+}-permeable channel activated by changes in redox status confers susceptibility to cell death. Mol Cell 9:163–173

Koliwad SK, Kunze DL, Elliott SJ 1996 Oxidant stress activates a non-selective cation channel responsible for membrane depolarization in calf vascular endothelial cells. J Physiol 491:1–12

Mendez F, Penner R 1998 Near-visible ultraviolet light induces a novel ubiquitous calcium-permeable cation current in mammalian cell lines. J Physiol 507:365–377

Distribution of TRPC channels in a visceral sensory pathway

Maria Buniel, Brian Wisnoskey*, Patricia A. Glazebrook, William P. Schilling* and Diana L. Kunze†[1]

*Rammelkamp Center for Education and Research, MetroHealth Campus of Case Western Reserve University and Departments of †Neurosciences and *Physiology and Biophysics, 2500 MetroHealth Drive, Cleveland, OH 44109-1998, USA*

Abstract. Until recently most of the published studies addressing the mechanisms of activation of TRPC channels have been carried out in heterologous expression systems. Lack of specific antagonists for the TRPC channels has hampered functional studies of endogenous channels. We approached the role of TRPC channels in native tissue with a study of the distribution of the channel proteins in the carotid chemosensory pathway in the rat. In a previous report we showed that TRPC3/4/5/6 and TRPC7 were present in neurons throughout the petrosal ganglion while TRPC1 was expressed in only a subpopulation of petrosal neurons, at least half of which projected to the carotid body. The TRPC proteins were differentially distributed to the branches of the axons that project centrally to the nucleus of the solitary tract and peripherally to the carotid body. The smallest unmyelinated sensory fibres projecting to the carotid body contained TRPC1/3/4/5 or TRPC6 but not TRPC7. TRPC1 and TRPC3 were concentrated in the larger diameter fibres. Interestingly, only TRPC1 and TRPC4 could be demonstrated in the final terminal endings within glomus cell clusters of the carotid body. In the central axon of the sensory neurons, both TRPC4 and TRPC5 were demonstrated in fibres exiting the solitary tract and projecting to the secondary relay neurons the nucleus of the solitary tract.

2004 Mammalian TRP channels as molecular targets. Wiley, Chichester (Novartis Foundation Symposium 258) p 236–247

Electrical activity of the visceral sensory afferent pathways is regulated by ligands that activate G protein-coupled membrane receptors, including those linked to activation of phospholipase C. In addition, receptor tyrosine kinases, such as TrkB, are present on many of the visceral sensory neurons (Barbacid 1994). Both of these receptor types have been linked to activation of transient receptor potential (TRP) channels, specifically the TRPC family (Harteneck et al 2000, Hofmann et al 2000, Montell 2001, Vennekens et al 2002).

[1]This paper was presented at the symposium by Diana L. Kunze to whom correspondence should be addressed.

The TRPC family members are important candidates for modulation of activity that occurs at the soma, the peripheral sensory receptor terminals and/or at the central pre-synaptic terminals of the sensory neurons. Using immunohistochemical techniques and confocal microscopy, we investigated the distribution of TRPC proteins in one visceral sensory pathway, the carotid body chemoreceptor pathway, as the first step towards identifying the role played by TRPC channels in sensory transduction and transmission.

The soma for the neurons of carotid chemoreceptor pathway lie in the petrosal ganglion. The cells are pseudo-unipolar with a single axon projecting from the soma forming a 'T', one branch of which projects peripherally to form a contact with the oxygen-sensing glomus cells in the carotid body and the other branch extends centrally to make a synaptic contact with the secondary relay neuron in the nucleus of the solitary tract. Approximately 10–15% of the chemoreceptor sensory neurons have myelinated fibres while the other 80–85% have unmyelinated fibres (Eyzaguirre & Uchizono 1961). The two populations have different discharge characteristics (Fidone & Sato 1969). We investigated the distribution of the TRPC proteins in the soma and its central and peripheral axon branches of both groups of fibres.

Methods

Antibodies

Rabbit polyclonal anti-TRPC antibodies used in this study were generated in collaboration with Su Li, Chris Poll and John Westwick of Novartis Respiratory Research Centre, Horsham, UK. These were characterized by Goel et al (2002). Further characterization in Sf9 cells expressing each of the TRPC channels was carried out as described in the present report.

Cell and tissue preparation

Spodoptera frugiperda (Sf9) cells were cultured and infected with TRPC constructs as described by Goel et al (2002). Cells were sparsely plated on poly-L-lysine coated chips and maintained for at least 24 h postinfection at 27 °C in humidified air. Cells were placed in fixative solution (3% paraformaldehyde and 0.1% Trition-X in 0.1 M phosphate buffer) for 10 minutes followed by brief rinses in phosphate buffered saline (PBS). Petrosal ganglia were obtained from Sprague-Dawley rats (100 g) in accordance with NIH guidelines and the animal protocols approved by Case Western Reserve University Animal Resource Center. Tissue preparation and immunohistochemical techniques for the sensory ganglia have been previously described (Buniel et al 2003). The TRPC1/3/4/5 rabbit polyclonal antibodies were used at a 1:10 000 dilution. The TRPC6 antibody was used at 1:5000 and

TRPC7 was used at 1:4000. The mouse monoclonal tyrosine hydroxylase antibody (Incstar) was used at 1:200. Both the donkey anti-rabbit-Rhodamine red and donkey anti-mouse-FITC secondary antibodies were obtained from Jackson Immunoresearch Laboratories and used at 1:800 and 1:500 respectively.

Preparation of the brain slices

Sprague-Dawley rats, 6 weeks old, are injected intraperitoneally (IP) with a Ketamine/Acepromazine/Xylazine cocktail mixture. After checking for lack of reflexes they were perfused with saline followed by 3% paraformaldehyde. The brains were removed and refrigerated in 3% paraformaldehyde with 5% sucrose overnight. The brains remained in 30% sucrose one day and then were quick frozen in isopentane on dry ice. 16 μm sections were cut and collected on Superfrost/Plus slides. The tissue was blocked with 5% normal donkey serum, 0.3% triton-x 100, 0.1% BSA in PBS for 30 min. The TRPC antibody was diluted in PBS containing 0.1% BSA and 0.3% TX-100. The primary antibody was placed on the slides overnight in humidified chamber in the refrigerator. Following rinses in PBS the secondary antibody was added for 90 min at room temperature. The donkey anti-rabbit Rodamine Red X (Jackson ImmunoResearch) was diluted in PBS containing 10% normal donkey serum and 0.3% Triton. Following rinses with PBS, coverslipping with Vectashield (Vector Laboratories), the sections were scanned on the Leica SP2 confocal microscope.

Results

The specificity of the individual TRPC antibodies used in the studies of the sensory pathway was confirmed in Sf9 cells transfected with each of the TRPC proteins as shown in Fig. 1. Each of the antibodies recognized its own protein but not that of the other family members.

Petrosal neurons express TRPC3/4/5/6 and 7 ubiquitously, although at different levels among the neurons within the ganglia. TRPC1, on the other hand, is found only in a subpopulation of neurons, at least half of which innervate the carotid body (Buniel et al 2003). This association with the chemoreceptor pathway is illustrated in Fig. 2 by the co-localization of tyrosine hydroxylase (TH) immunoreactivity with that of TRPC1 immunoreactivity. TH in the petrosal ganglion is confined to the neurons projecting to the carotid body (Finley et al 1992).

A critical question with important functional implications is whether all the TRPC proteins found in the sensory soma are distributed to both peripheral and central divisions of the axons of the sensory neurons. Examination of the axonal branch that projects peripherally to the carotid body revealed a differential

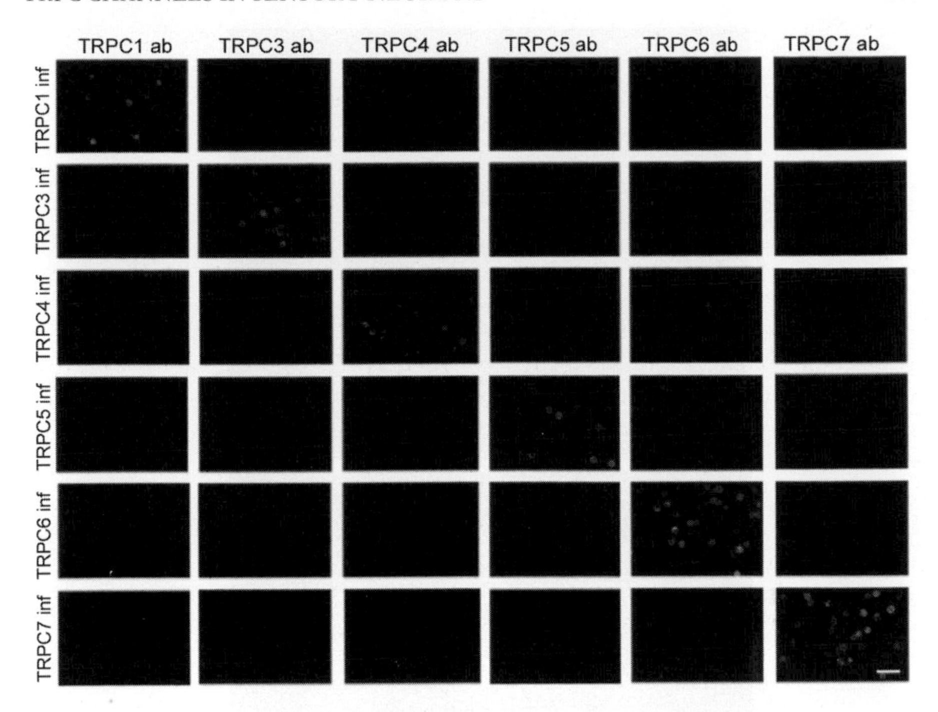

FIG. 1. TRPC antibody specificity in transfected Sf9 cells. Sf9 cells were transfected with TRPC 1/3/4/5/6/7 constructs and TRPC antibody specificity was tested. Rows represent cells infected with the different TRPC constructs and the TRPC antibodies applied are shown separately in each column. Immunoreactivity is only seen when the corresponding TRPC antibody is paired with cells transfected with the respective TRPC construct. The lack of immunoreactivity in cells stained with antibodies that did not correspond to the transfected TRPC construct illustrate the specificity of the TRPC antibodies and show that cross-reactivity of the antibodies among the different TRPC proteins does not occur. Scale bar represents 50 μm.

distribution of immunoreactivity between the smallest diameter unmyelinated fibres and the larger diameter unmyelinated and the myelinated fibres as determined by co-localization with a neurofilament (NF) cocktail that labels all but the finest unmyelinated fibres (Lasek et al 1983, Lawson et al 1993). TRPC1 and TRPC3 immunoreactivity was present in most fibres while TRPC4/5/6 immunoreactivity was found primarily in the smallest unmyelinated fibres that did not react with neurofilament antibodies. Interestingly, as the axons that penetrate the carotid body glomus cells clusters, only immunoreactivity for TRPC1 and TRPC4 was present in the fine terminal regions.

At the central terminal region we looked for the presence of the TRPC proteins in fibres of the solitary tract, the location of the entering visceral sensory fibres and in the nucleus of the solitary tract where the fibres form synaptic contacts with the

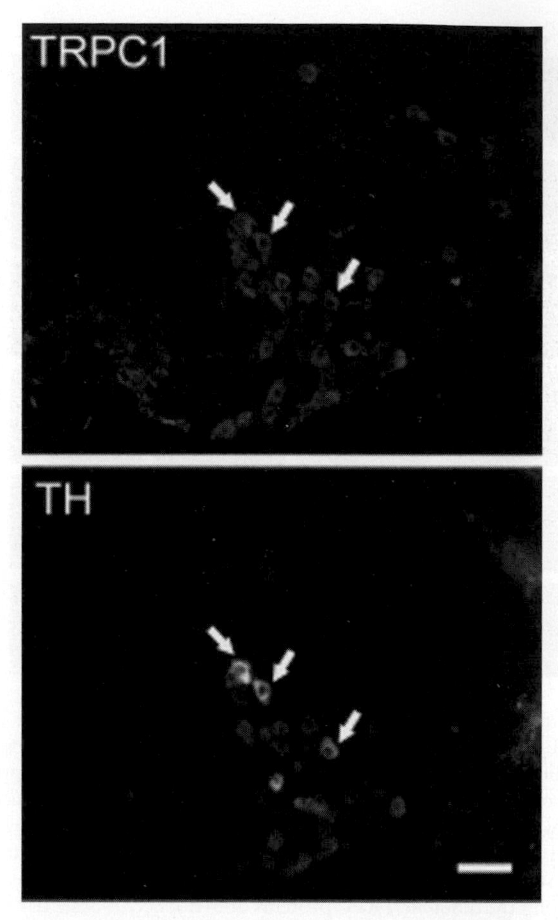

FIG. 2. Co-localization of TRPC1 and tyrosine hydroxylase (TH) in the petrosal ganglia. Sections of the petrosal ganglia were double-stained for TRPC1 and TH. The top confocal image shows TRPC1 immunoreactive neurons and TH immunoreactivity in the same section is shown in the middle confocal image. Co-localization of TRPC1 and TH is seen in the TH immunoreactive neurons, several of which are indicated by the arrows. Not visible in these images are many neurons unlabelled by either TRPC1 or TH antibodies. Scale bar represents 50 μm.

secondary relay neurons. TRPC4 and TRPC5 were detected within the tract. An example of the TRPC4 immunoreactivity is shown in Fig. 3. Both were also seen leaving the tract to encircle neurons in the immediate vicinity of the tract where the implication is that they play a role in neurotransmitter release. A precedent for this is found in the studies of Obukhov and Nowycky (2002) where TRPC4 has been implicated in calcium influx and subsequently the modulation of exocytosis in adrenal chromaffin cells and PC12 cells.

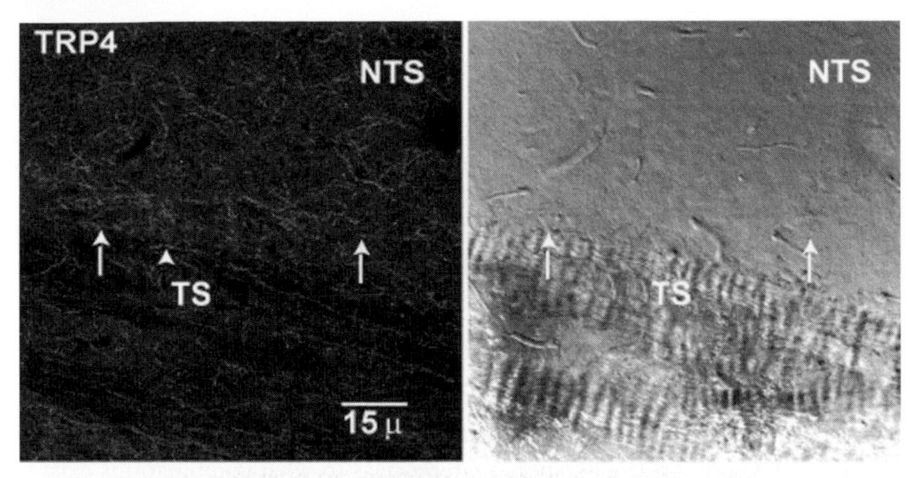

FIG. 3. Distribution of TRPC4 within the medial nucleus of the solitary tract (NTS). A longitudinal brain section that encompasses the solitary tract and its surrounding nuclei has been immunostained with the TRPC4 antibody and donkey anti-rabbit IgG Rodamine Red-X. The left confocal image shows fibres immunolabelled (arrowhead only) in the tract (TS) and in the NTS. The right picture is a differential interference contrast image of the same section. The arrows demonstrate examples of fibres in close contact with cell bodies in the NTS.

Discussion

A major finding of the studies of Buniel et al (2003) is that TRPC1 in the petrosal ganglion appears to be closely associated with the chemosensory pathway and is found in the peripheral axons and their final terminal regions innervating the Type I carotid glomus cells. The latter are the oxygen-sensing cells in the carotid body. In response to hypoxic stimuli the glomus cells release a variety of transmitters that are presumed to depolarize and activate the carotid afferent nerve terminals (Gauda 2002). The transmitters reported to be released from the glomus cells include several that activate G_q-coupled receptors. Many of these receptors are present on the sensory soma in the petrosal ganglion. However, the identification of appropriate receptors on the nerve endings that innervate the Type 1 glomus remains to be done. A key question of whether specific receptors are associated with particular TRPC channels may be answered with immunohistochemical co-localization.

A second major finding includes the ubiquitous expression of the TRPC proteins at the level of the sensory soma but a more selective distribution to the peripheral and central components of the axons of these neurons. The larger diameter axons have a restrictive TRPC distribution, expressing primarily TRPC1 and TRPC3. The fine unmyelinated sensory fibres that express all of the TRPCs found at the soma except TRPC7, were identified by the absence of the three neurofilament proteins, NF200, 160 and 68. Only TRPC1 and TRPC4 are present in the final terminals in our studies. These may be present as homomultimers or, alternatively, TRPC1 and TRPC4 may associate as heteromultimers (Goel et al 2002). Why are not all the TRPC proteins in the peripheral axons also seen in the final terminal regions where sensory transduction takes place? If the expression of the other TRPC channels is very weak our techniques may not resolve their presence. On the other hand, these fibres may not project into the glomus cell clusters and their functional role may be at some distance from the end of the fibre. For instance, a single axon branches, innervating several glomus cells (Kondo 1976). TRPC channels may regulate transmission at a region prior to the branching point. The results of these immunohistochemical studies now provide the basis for comparing the distribution of individual TRPC proteins with receptors for specific ligands that link through G_q or TrkB with special reference as to what combinations exist in myelinated versus unmyelinated fibres.

Finally, two of the TRPC channels that are thought to form hetermomultimers in native tissue, TRPC4 and TRPC5 (Goel et al 2002) are present in the central axons and terminals. Future experiments will ask whether, indeed, TRPC4 and TRPC5 form heteromultimers in the central fibres and whether any of the many ligands that provide modulation of transmitter release at these presynaptic terminals do so via the TRPC channels.

Acknowledgements

This work was supported by: NIH-HL25830 (DLK) and NIH-GM52019 (WPS).

References

Barbacid M 1994 The Trk family of neurotrophin receptors. J Neurobiol 25:1386–1403

Buniel M, Schilling WP, Kunze DL 2003 Distribution of transient receptor potential channels in the rat carotid chemosensory pathway. J Comp Neurol 464:404–413

Eyzaguirre C, Uchizono K 1961 Observations of the fiber content of nerves reaching the carotid body of the cat. J Physiol 159:268–281

Fidone SJ, Sato A 1969 A study of chemoreceptor and baroreceptor A and C-fibres in the cat carotid nerve. J Physiol 205:527–548

Finley JC, Polak J, Katz DM 1992 Transmitter diversity in carotid body afferent neurons: dopaminergic and peptidergic phenotypes. Neuroscience 51:973–987

Goel M, Sinking WG, Shilling WP 2002 Selective association of TRPC channel subunits in rat brain synaptosomes. J Biol Chem 277:48303–48310

Gauda EB 2002 Gene expression in peripheral arterial chemoreceptors. Micro Res Tech 59:153–167

Harteneck C, Plant TD, Schultz G 2000 From worm to man: three subfamilies for TRP channels. Trends Neurosci 23:159–166

Hofmann T, Schaefer M, Schultz G, Gudermann T 2000 Transient receptor potential channels as molecular substrates of receptor mediated cation entry. J Mol Med 78:14–25

Kondo H 1976 Innervation of the carotid body of the adult rat: a serial ultrathin section analysis. Cell Tissue Res 173:1–15

Lasek RJ, Oblinger MM, Drake PF 1983 Molecular biology of neuronal geometry: expression of neurofilament genes influence axonal diameter. Cold Spring Harb Symp Quant Biol 48:731–744

Lawson SN, Perry MJ, Prabhakar E, McCarthy PW 1993 Primary sensory neurones: neurofilament, neuropeptides, and conduction velocity. Brain Res Bull 30:239–243

Montell C 2001 Physiology, phylogeny, and functions of the TRP superfamily of cation channels. Sci STKE 90:RE1

Obukhov AG, Nowycky MC 2002 TRPC4 can be activated by G-protein-coupled receptors and provides sufficient Ca^{2+} to trigger exocytosis in neuroendocrine cells. J Biol Chem 277:16172–16178

Vennekens R, Voets T, Bindels RJ, Droogmans G, Nilius B 2002 Current understanding of mammalian TRP homologues. Cell Calcium 31:253–264

DISCUSSION

Putney: In the glomus, is it correct that you see channels distributed in a punctate manner through the cytoplasm?

Kunze: Yes, they could be in the endoplasmic reticulum (ER) and/or the Golgi apparatus. We can't really tell without doing co-localization studies.

Putney: You are presumably looking at the channels in their native distribution. We carry out experiments which might give us all kinds of aberrant distributions, such as overexpressing GFP fusions. We find that the TRP channels appear to have a much more expected and orthodox distribution primarily in the plasma membrane. Does that make you think about what the antibodies are actually seeing?

Kunze: No. Unsurprisingly, overexpression systems often give a very nice membrane-delineated expression. But, when we are dealing with the endogenous channels there are a lot fewer, and we often see a diffuse distribution that includes intracellular compartments as well as the membrane. They can be co-localized to regions or compartments such as the ER with specific markers for those compartments. We haven't done this yet for TRPC channels. To know the protein is actually inserted in the plasma membrane in the native tissue we would need an antibody that recognizes an extracellular epitope.

Putney: So is your interpretation that when we overexpress it, it is in the ER and places like that as well, but because we pile up so much on the membrane it actually looks more like what we expect?

Kunze: That's right.

Putney: This means that you are not sure that the punctate images you see in the cells represents the functional channel. The functional channel might be present at much lower concentration in the plasma membrane.

Kunze: I think that is right.

Muallem: We shouldn't worry too much about this. Those proteins are on their way to the plasma membrane. You would see them if the antibodies recognize the immature proteins.

Kunze: Some of them have to be there in the intracellular compartment.

Putney: Then the utility of the approach is to draw conclusions about which cells are expressing the channel rather than to study the subcellular functional distribution of the channel.

Kunze: No, I think you can do both. You can determine specifically what cells or cellular regions express the proteins. For instance, in neurons, you can gain insight to function if you know whether they were located in axons, at presynaptic regions or in dendrites. Or in epithelial cells you might want to know if they are distributed to apical or basolateral regions. But you can also learn something from their subcellular distribution.

Penner: It is not at all clear that these channels are going to the plasma membrane, as Shmuel Muallem asserts. We have to be careful about whether or not they are going to the plasma membrane. I have another question. Do the antibodies show any preference for monomers or tetramers?

Kunze: I don't know.

Schilling: We know that we can immunoprecipitate the appropriate TRPC protein with each antibody, and we know that they recognize the appropriate protein on Westerns. We now know the profile of immunoreactivity in neuronal tissue. It will be important to generate additional antibodies to different epitopes on each of the TRPC channel proteins and ask whether the same profile is obtained using all three of these approaches—immunoprecipitation, westerns and immunocytochemistry.

Putney: One of your validations with the antibodies is to show that they recognize exogenously expressed protein which is a tetramer. One of your tests of the anti-TRPC3 antibody is to express TRPC3 in HEK cells and verify whether the antibody sees it *in situ*.

Kunze: This has been done. We have expressed the TRPC channels in Sf9 cells and shown that each antibody recognizes only its own protein and not that of the other TRPC proteins.

Schilling: It is recognizing a protein. Whether it is a tetramer, dimer or monomer is another issue.

Kunze: One of the things we are now trying to do is to determine whether combinations of TRPC antibodies co-localize in native tissue.

Muallem: For ultimate validation you need to go to knockout mice. If you knockout *Trpc4* and still see staining, then you have a problem.

Scharenberg: If you are worried about epitope masking, there are ways to get around that.

Montell: The subcellular localization that we are all seeing is striking. The question is whether there is regulated movement to the plasma membrane, or are some TRPCs actually ER channels? If there is regulated movement, one possibility is that those channels, which are so often constitutively active *in vitro*, such as TRPC1, may in fact be constitutively active *in vivo* when they get to the plasma membrane. The regulation of such channels could be movement to the plasma membrane, rather than stimulation of PLC, for example. Has anyone here tried to look at whether any of these TRPC channels are Ca^{2+} release channels? Most of us don't entertain this as a serious thought.

Penner: Speak for yourself!

Montell: Certainly with polycystin 2 there has been debate about whether it is a release channel or a plasma membrane channel, and now more people believe it is a release channel.

Penner: We need to consider the release channel situation at the very least in the overexpression system. There is no way we can exclude that channels that are gated by lipids can be gated inside that compartment. If they are functional in there, they are Ca^{2+} permeable and they may release Ca^{2+} from whatever compartment they are in.

Gill: The experiments in the inositol trisphosphate (IP_3) knockouts show that overexpression of TRPCs causes no release at all. It is all entry. At the beginning we were worried about the possibility of release, but we haven't seen it.

Schilling: That doesn't mean there aren't specific mechanisms that could activate TRPC3 channels in the ER. There could be lipid modulation at the level of the ER.

Putney: Unless there is a very tiny specialized subcompartment, there are some predicted experimental outcomes if TRPC3 is functionally present in ER. We don't see these outcomes in experiments with TRPC3.

Montell: If we saw the intracellular localization of the TRPC channels presented here, without any preconceived notion as to whether TRPs are influx or release channels, we would first propose that the TRPCs are primarily intracellular channels.

Kunze: That wouldn't necessarily be correct. Assuming that the channels that appear to be at the membrane are actually inserted into the membrane, a punctate distribution of channels could very well be expected. I would expect that in a native cell you wouldn't have big currents anyway, so you are not going to expect many channels there. In fact, if you make the calculations when the complete surface is covered with functional channels, the currents would be huge if they all opened. Furthermore, there is a very good precedent for channels occurring in patches or clusters in the membrane surface in the nice studies of potassium channel distribution from J. Trimmer's group (for example, Antonucci et al 2001).

Montell: It would be interesting to prepare primary cells and look at the effects of receptor stimulation on TRPC translocation to the plasma membrane.

Barritt: The K^+ channels are fine, upstanding, well characterized channels. They are also found in the ER in a similar way to the TRP channels you are describing. This is an important point.

Kunze: Yes, some of the K^+ channels do have a distribution similar to this.

Ambudkar: When we first looked at endogenous TRPC1 with our antibody, we used HEK293 cells. We saw a lot of intracellular labelling but we also detected the protein in the plasma membrane. We purified plasma membranes and endogenous TRPC1 protein was enriched with plasma membrane markers. Thus, there is definitely a proportion of TRPC1 in the plasma membrane, although we could see quite a lot of it in the ER. In some cells it looks almost all intracellular, and then there are some cells where it is more in the plasma membrane. I don't know what this means. It could be a trafficking issue or due to the status of the cell, e.g. cell cycle.

Putney: It could be the antibody. Even though your antibody recognizes a specific TRPC1 band, it recognizes many things other than TRPC1. The other bands, which are all blocked by peptide, are much darker than the TRPC1 band. When you do histochemistry with an antibody like that you would expect that your image is primarily things other than TRPC1.

Ambudkar: I agree, that is possible. We also use tagged TRPC1 to confirm the localization. Anyhow, I have another slightly different question regarding the involvement of TRPCs in internal Ca^{2+} release. We have a rather strange observation. We have cells where OAG stimulates internal Ca^{2+} release at 50 μM. If we titrate the OAG concentration down to about 25 μM there is no internal Ca^{2+} release, but it stimulates influx.

Nilius: You say that this is a strange phenomenon, but it is well known arachidonic acid releases Ca^{2+} so I am not so surprised by your results (Oike et al 1994).

Authi: Most of the people here are looking at complete cells where protein synthesis occurs. There is therefore post-translational ER processing and then movement of the proteins to the plasma membrane. Even then, a lot of labelling is seen within intracellular compartments for TRPC proteins. I feel we need to consider a function related to Ca^{2+} of TRPC proteins within intracellular compartments. While I understand everyone here is working on cells where protein synthesis occurs, in platelets there is no protein synthesis and we see TRPC1 in intracellular membranes. This concept of this protein being made here, translated and then moved to the plasma membrane as a consequence of the life of that protein doesn't exist in platelets.

Montell: That's an important point.

Authi: We have to get away from this concept that the function of TRPC proteins is only in the plasma membrane. We have to consider also some role in intracellular compartments. With Mori's data on the DT40 cell (Mori et al 2002), can we discard the results showing that knockout of TRPC1 in these cells leads to a reduction in receptor or agonist-stimulated Ca^{2+} release? Don Gill, you are saying that without any IP_3 receptors there is no release at all. But perhaps the function of the TRPC protein in Ca^{2+} release also requires the presence of IP_3 receptors.

Gill: When we have overexpressed any of the TRPs in those cells without IP_3 receptors, we see no release. In terms of *Trpc1* knockout cells, we didn't see any significant difference in the entry or release in those cells.

Montell: One clear example *in vivo* of a regulated movement of a TRP-related channel was published by Paulsen, Huber and colleagues (Bähner et al 2002). They looked at TRPL distribution in photoreceptor cells maintained in the dark or exposed to light and found that there were light-dependent changes in the spatial distribution of TRPL. TRPL was concentrated in the rhabdomeres of light-treated flies. However, after maintaining the flies in the dark, TRPL translocated to the cell bodies. This dynamic movement of TRPL contributes to long-term adaptation. This is a clear *in vivo* example describing the regulated movements of a TRP channel, which is certainly an influx channel.

Nilius: Is there any cell in our bodies that doesn't have TRP channels?

Westwick: Good question.

Putney: The early guess was that hepatocytes didn't, but they are detected by PCR.

Montell: The bacteria in your gut don't have any TRPs!

References

Antonucci DE, Lim ST, Vassanelli S, Trimmer JS 2001 Dynamic localization and clustering of dendritic $K_v2.1$ voltage-dependent potassium channels in developing hippocampal neurons. Neuroscience 108:69–81

Bähner M, Frechter S, Da Silva N, Minke B, Paulsen R, Huber A 2002 Light-regulated subcellular translocation of Drosophila TRPL channels induces long-term adaptation and modifies the light-induced current. Neuron 34:83–93

Mori Y, Wakamori M, Miyakawa T et al 2002 Transient receptor potential 1 regulates capacitative Ca^{2+} entry and Ca^{2+} release from endoplasmic reticulum in B lymphocytes. J Exp Med 195:673–681

Oike M, Droogmans G, Nilius B 1994 Mechanosensitive Ca^{2+} transients in endothelial cells from human umbilical vein. Proc Natl Acad Sci USA 91:2940–2944

Emerging roles of TRPM channels

Andrea Fleig and Reinhold Penner[1]

Laboratory of Cell and Molecular Signaling, Center for Biomedical Research at The Queen's Medical Center and Department of Cell and Molecular Biology, John A. Burns School of Medicine at the University of Hawaii, Honolulu, HI 96813, USA

Abstract. The molecular characterization of the genes encoding the transient receptor potential (TRP) cation channels found in *Drosophila* photoreceptors gave rise to a systematic cloning strategy for mammalian isoforms. Using expressed sequence tag (EST) and genomic database searches, at least 20 new mammalian TRP-related genes have been cloned and the resulting channels extensively characterized. Here, we will focus on TRP channels from the TRPM subfamily. Although generally classified as non-selective cation channels, individual members of this family feature considerable functional diversity in terms of selectivity, specific expression pattern, as well as diverse gating and regulatory mechanisms. The functional characteristics of these channels have profound impact on the regulation of ion homeostasis that go beyond simple Ca^{2+} signalling. They activate and function in the context of a variety of physiological and pathological conditions, which make them exciting targets for drug discovery.

2004 Mammalian TRP channels as molecular targets. Wiley, Chichester (Novartis Foundation Symposium 258) p 248–262

TRP-related channels constitute a large and diverse superfamily of proteins that are expressed in a variety of organisms, tissues and cell types, including electrically excitable and nonexcitable cells (for reviews see Clapham et al 2001, Vennekens et al 2002). The TRP channels have been divided into three main subfamilies: TRPC for 'canonical', TRPM for 'melastatin-like' and TRPV for 'vanilloid-receptor-like' (Montell et al 2002). All TRP channel proteins are composed of six putative transmembrane domains, a slightly hydrophobic pore-forming region, while both N- and C-terminal domains are intracytoplasmic. Despite these similarities of structure, the functions of TRP channels are very different from one channel to another, even amongst the members of the same subfamily.

The human TRPM subfamily currently consists of eight members. The structural features of the TRPM subfamily of cation channels is defined by a large

[1] This paper was presented at the symposium by Reinhold Penner, to whom correspondence should be addressed.

conserved N-terminal region, an adjacent cation channel transmembrane spanning region as in all other TRP channels, and a short nearby region of coiled-coil character. The C-terminal domain distal to the coiled-coil region varies significantly between different TRPM family members and may contain structures that are important in controlling the ion channel activation mechanism. The activation mechanisms of several TRPMs have been elucidated and it seems that each channel has specific ion selectivity and a particular mechanism of activation.

TRPM1

The founding member of the TRPM subfamily, TRPM1 (melastatin), is discussed to be a tumour suppressor gene, due to the fact that it is down-regulated in highly metastatic melanoma cells (Duncan et al 1998). Down-regulation of melastatin mRNA in primary cutaneous tumours is a prognostic marker for metastasis in patients with localized malignant melanoma (Duncan et al 2001, Fang & Setaluri 2000). Although melastatin has been reported to mediate Ca^{2+} entry when expressed in HEK293 cells (Xu et al 2001), TRPM1 has not yet been characterized electrophysiologically and it is therefore not clear whether the protein forms a functional ion channel in either heterologous expression systems or in a native context.

TRPM2

TRPM2 is an ion channel, whose C-terminus is characterized by a Nudix domain that contains nucleotide pyrophosphatase activity (Perraud et al 2001, Sano et al 2001). Whole-cell and single-channel analysis of HEK293 cells expressing TRPM2 demonstrate that the protein functions as a 60 pS Ca^{2+}-permeable cation channel that is highly specifically gated by free ADP-ribose (Fig. 1) and it appears that TRPM2's enzymatic domain is responsible for ADP-ribose-mediated gating of the channel (Perraud et al 2001). Intracellular Ca^{2+} appears to be an important modulator and co-factor of TRPM2, as elevated $[Ca^{2+}]_i$ can significantly increase the sensitivity of TRPM2 towards ADP ribose, enabling it to gate the channel at lower concentrations (McHugh et al 2003, Perraud et al 2001). TRPM2 is dominantly expressed in the brain, but is also detected in many other tissues, including bone marrow, spleen, heart, leukocytes, liver and lung. Native TRPM2 currents have been recorded from the U937 monocyte cell line (Perraud et al 2001), neutrophils (Heiner et al 2003), and CRI-G1 insulinoma cells (Inamura et al 2003), where ADPR induces large cation currents (designated I_{ADPR}) that closely match those mediated by recombinant TRPM2. The channel can also be gated by H_2O_2 (Hara et al 2002) and NAD (Hara et al 2002, Heiner et al 2003,

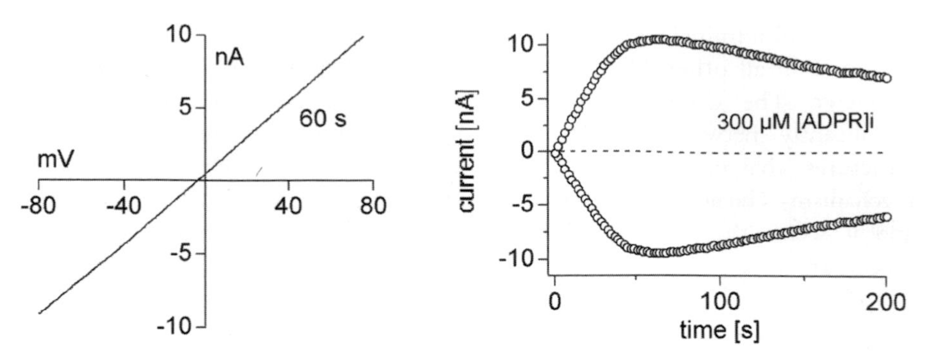

FIG. 1. Current–voltage relationship and whole-cell currents of TRPM2. The right panel illustrates the typical time course of TRPM2 whole-cell currents in HEK293 cells expressing TRPM2 recorded at +80 and −80 mV, respectively. Currents were induced by 300 μM ADP-ribose. The left panel shows the typical linear current-voltage (I/V) relationship of TRPM2 evoked by a voltage ramp of 50 ms duration at the peak of the current (60 s into the experiment).

Sano et al 2001), suggesting that it might be involved in sensing the cell's redox state. Based on these properties, TRPM2 provides a positive feedback on Ca^{2+} influx and it is therefore hypothesized to play a role in apoptosis.

TRPM3

Structurally, TRPM3 is closest to melastatin (TRPM1), but presently, there is no information about the functional properties of this protein.

TRPM4

TRPM4, which does not contain any obvious enzymatic domain, has the distinct properties of a Ca^{2+}-activated non-selective (CAN) cation channel (Launay et al 2002). It is a 25 pS ion channel that is specific for monovalent cations and does not carry any significant Ca^{2+} (Fig. 2). Its activation by $[Ca^{2+}]_i$ is characterized by a short delay and it seems that Ca^{2+} influx is considerably more effective in activating the channel than release of Ca^{2+} from intracellular stores. Nevertheless, TRPM4 has significant impact on $[Ca^{2+}]_i$, as it provides a mechanism that allows cells to depolarize in a Ca^{2+}-dependent manner. CAN channels have been observed in numerous electrically excitable and non-excitable cells (Partridge et al 1994, Petersen 2002). In non-excitable cells that lack voltage-dependent Ca^{2+} channels, this depolarization would decrease the driving force for Ca^{2+} influx through store-operated Ca^{2+} channels, whereas in excitable cells this channel could be important to shape action potential duration and spiking frequency and thereby supporting Ca^{2+} influx through voltage-dependent Ca^{2+} channels. Thus, TRPM4 activation

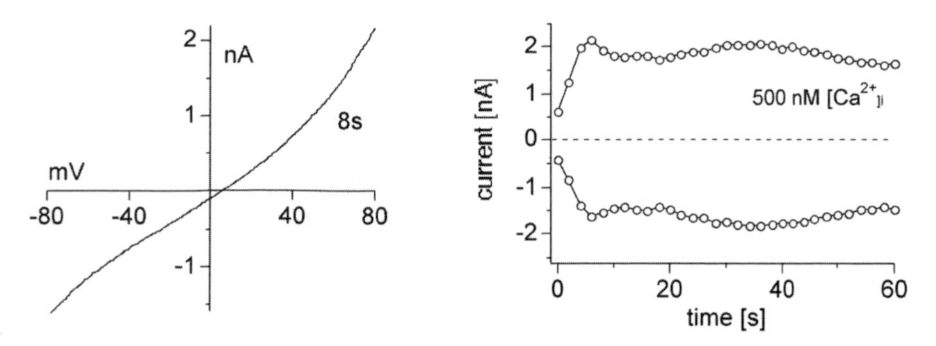

FIG. 2. Current–voltage relationship and whole-cell currents of TRPM4. The right panel illustrates the typical time course of TRPM4 whole-cell currents in HEK293 cells overexpressing TRPM4 recorded at +80 and −80 mV, respectively. Currents were induced by perfusing cells with intracellular solutions in which Ca^{2+} was buffered to 500 nM. The left panel shows the typical current-voltage (I/V) relationship of TRPM4 evoked by a voltage ramp of 50 ms duration at the peak of the current (8 s into the experiment). The I/V shows slight rectification at extreme negative and positive voltages.

can have significant impact on $[Ca^{2+}]_i$ without itself transporting Ca^{2+} ions in that it may either suppress or promote Ca^{2+} influx depending on the cell type in which it is expressed.

TRPM5

The TRPM5 gene was identified during functional analysis of the chromosomal region (11p15.5) associated with loss of heterozygosity in a variety of childhood and adult tumours and the Beckwith-Wiedemann-Syndrome (Prawitt et al 2000). It is expressed as a 4.5 kb transcript in a variety of fetal and adult human and murine tissues and is structurally related to TRPM4. TRPM5 has been reported to be a Ca^{2+}-permeable ion channel that is activated following store depletion and has been proposed to function as a sensor for bitter taste in sensory neurons (Perez et al 2002). Another study has proposed a receptor-mediated mechanism that depends on PLC activation and proceeds in a Ca^{2+}-independent manner (Zhang et al 2003).

In our own studies, we find no evidence for a store-operated activation mechanism of TRPM5 nor do we see Ca^{2+}-independent activation of TRPM5. Instead, we find that the protein is directly activated by elevated $[Ca^{2+}]_i$ in both whole-cell recordings and in excised membrane patches (Fig. 3). TRPM5 is a monovalent-specific ion channel of 25 pS conductance that is directly activated by a fast increase in $[Ca^{2+}]_i$ in response to inositol 1,4,5-trisphosophate (IP_3)-producing receptor agonists (Prawitt et al 2003). It therefore shares the activation

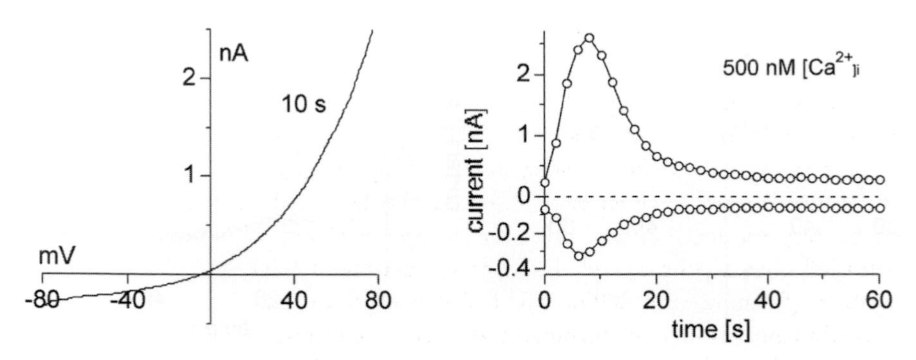

FIG. 3. Current–voltage relationship and whole-cell currents of TRPM5. The right panel illustrates the typical time course of TRPM5 whole-cell currents in HEK293 cells expressing TRPM5 recorded at +80 and −80 mV, respectively (note the different scale of the y-axis for inward and outward currents). Currents were induced by perfusing cells with intracellular solutions in which Ca^{2+} was buffered to 500 nM. The left panel shows the typical current-voltage (I/V) relationship of TRPM5 evoked by a voltage ramp of 50 ms duration at the peak of the current (10 s into the experiment). The I/V shows significant voltage-dependent outward rectification at positive voltages.

mechanism as well as selectivity with the Ca^{2+}-activated cation channel TRPM4, but unlike TRPM4 (which does not respond to Ca^{2+} release and requires Ca^{2+} influx for maximal activation), TRPM5 is strongly activated by receptor-mediated Ca^{2+} release. Moreover, TRPM5 does not simply mirror changes in Ca^{2+}, but requires *fast* changes in Ca^{2+} to activate. This unique property of TRPM5, combined with its transient activation kinetics, provides a compelling mechanism that allows taste cells to translate a receptor-mediated elevation in $[Ca^{2+}]_i$ into an electrical response that ultimately results in transmitter release. We also find that TRPM5 expression is not limited to taste receptor cells. The presence of TRPM5 in a variety of tissues, including pancreas, and the measurement of TRPM5 currents in a β cell line argue for a more generalized role of the channel as a tool that couples agonist-induced intracellular Ca^{2+} release to electrical activity and subsequent cellular responses such as neurotransmitter or insulin release.

TRPM6

TRPM6 is closely related to TRPM7, as its primary structure suggests that it also contains a kinase domain. TRPM6 appears to be responsible for hypomagnesemia with secondary hypocalcaemia when mutated (Schlingmann et al 2002, Walder et al 2002). Although TRPM6 has not yet been characterized as a functional ion channel, its similarity to TRPM7, which is able to transport a range of divalent cations, including Ca^{2+} and Mg^{2+}, suggests that it may function as a Mg^{2+}-permeable channel.

TRPM7

TRPM7 is a widely expressed protein found in virtually all mammalian cells (Nadler et al 2001, Runnels et al 2001). It is the only ion channel presently known to be essential for cellular viability, as knocking the protein out in DT40 cells results in cell death (Nadler et al 2001). However, these TRPM7-deficient cells can be rescued and remain viable by supplementation of extracellular Mg^{2+}, indicating that a primary cell biological function of TRPM7 relates to Mg^{2+} transport (Schmitz et al 2003). TRPM7 is notable in that it contains a protein kinase domain within its C-terminal sequence. In contrast to TRPM2, where a C-terminal nudix hydrolase domain has been clearly implicated in channel activation, the relevance of TRPM7's kinase domain to channel function remains controversial: TRPM7 channel activation has been suggested to be dependent on the phosphotransferase activity of the intrinsic kinase domain (Runnels et al 2001), while suppression of activation/channel deactivation has been shown to occur in response to Mg^{2+}-nucleotide complexes or Mg^{2+} alone (Nadler et al 2001), G-protein activation (Hermosura et al 2002, Takezawa et al 2003), and phosphatidylinositol 4,5-bisphosphate (PIP_2) hydrolysis (Runnels et al 2002). In addition, conflicting data have been presented regarding TRPM7 channel permeation characteristics, with data suggesting both non-selective conduction of Na^+ and Ca^{2+} and complex permeation with selectivity towards divalent cations (Nadler et al 2001, Runnels et al 2001).

Our own work suggests that TRPM7 is highly selective for divalent cations and is regulated by both intracellular Mg·ATP and cytoplasmic levels of free $[Mg^{2+}]_i$, which is why we have named endogenous currents with TRPM7 properties MagNuM for Magnesium-Nucleotide-regulated Metal ion currents (Nadler et al 2001). In resting cells, physiological levels of these molecules strongly suppress the activity of TRPM7 channels and only a small constitutive activity remains, sufficient to maintain basal divalent cation fluxes. In whole-cell patch-clamp experiments, intracellular solutions that lack added Mg·ATP or are reduced in free Mg^{2+} lead to activation of TRPM7-mediated currents that exhibit a characteristic highly non-linear current-voltage relationship with pronounced outward rectification (Fig. 4). The large outward currents at positive potentials are carried by monovalent ions (e.g. Cs^+ or K^+), whereas the small inward currents at more physiological, negative potentials are carried by divalent ions such as Ca^{2+} and Mg^{2+}. The channel also carries a range of other divalent ions such as Zn^{2+}, Ni^{2+}, Co^{2+}, Ba^{2+}, Sr^{2+} and Cd^{2+} (Monteilh-Zoller et al 2003).

TRPM7-mediated, endogenous MagNuM currents share some features with the store-operated current I_{CRAC}, most notably its ability to conduct large monovalent currents in the absence of divalent charge carriers such as Ca^{2+} and Mg^{2+}. Furthermore, MagNuM is activated under experimental conditions that have

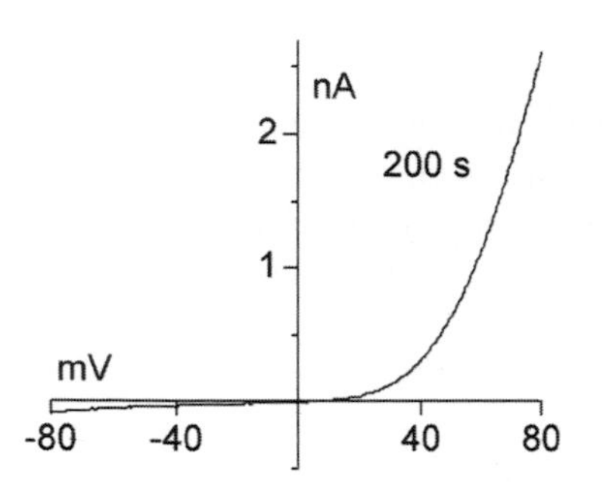

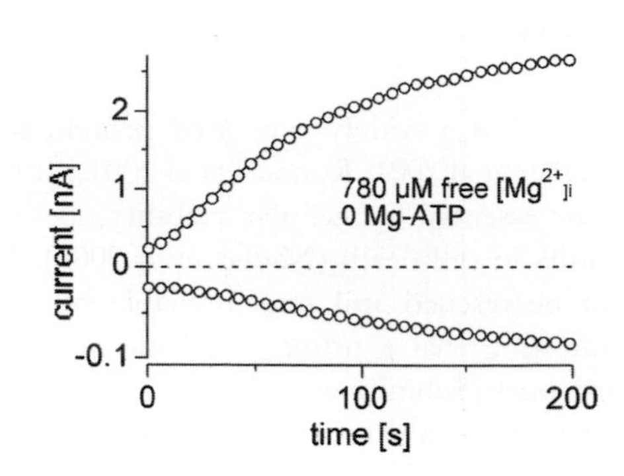

FIG. 4. Current–voltage relationship and whole-cell currents of TRPM7. The right panel illustrates the typical time course of TRPM7 whole-cell currents in HEK293 cells over-expressing TRPM7 recorded at +80 and −80 mV, respectively (note the different scale of the y-axis for inward and outward currents). Currents were induced by perfusing cells with intracellular solutions that lacked ATP and in which $[Mg^{2+}]_i$ was buffered to 780 μM. The left panel shows the typical current-voltage (I/V) relationship of TRPM7 evoked by a voltage ramp of 50 ms duration at the peak of the current (200 s into the experiment). The I/V shows significant outward rectification at positive voltages due to permeation block of divalent ions at negative membrane potentials.

traditionally been used to study I_{CRAC} at the single channel level and it is now generally accepted that the 40 pS channels observed under these conditions do not represent CRAC channels but are in fact attributable to TRPM7 channels (Hermosura et al 2002, Kozak et al 2002, Prakriya & Lewis 2002).

While variations in cellular Mg·ATP levels may provide an important 'passive' regulatory mechanism of TRPM7, we recently found evidence that TRPM7 activity can be 'actively' modulated by intracellular levels of cAMP induced by G_i- and G_s-coupled receptors, respectively (Takezawa et al 2003). This modulation requires both functional protein kinase A as well as a functional TRPM7 kinase domain. In addition, analyses of mutant human TRPM7 proteins reveal that while the protein's C-terminal kinase domain is not essential for channel activation, it regulates not only the active receptor-mediated regulation of channel activity but also the passive constitutive activity in that it determines the sensitivity of the channel to intracellular levels of Mg^{2+} and Mg·ATP (Schmitz et al 2003). By virtue of its sensitivity to physiological Mg·ATP levels, TRPM7 may be involved in a fundamental process that adjusts plasma membrane divalent cation fluxes according to the metabolic state of the cell and may play an important role in pathophysiological circumstances such as ischaemia. In addition, the receptor-mediated regulation of TRPM7 may support Ca^{2+} and Mg^{2+} fluxes following agonist stimulation.

TABLE 1 Properties of TRPM ion channels

Channel	*Gating*	*γ*	*Selectivity*	*Expression*	*Function*	*Reference*
TRPM1	n.d.	n.d.	n.d.	Down-regulated in malignant melanoma	n.d.	Duncan et al 1998
TRPM2	ADPR, NAD, H_2O_2	60 pS	Non-selective (Na^+, K^+, Cs^+, Ca^{2+})	Brain, neutrophils, pancreatic cells	Apoptosis, membrane depolarization	Perraud et al 2001, Sano et al 2001
TRPM3	n.d.	n.d.	n.d.	n.d.	n.d.	—
TRPM4	$[Ca^{2+}]_i$	25 pS	Monovalents (Na^+, K^+, Cs^+)	Heart, liver	Membrane depolarization	Launay et al 2002
TRPM5	$[Ca^{2+}]_i$ SOC?, PLCβ?	25 pS	Monovalents (Na^+, K^+, Cs^+)	Taste cells, pancreatic cells, immune cells	Depolarizing Ca^{2+} release sensor, taste transduction, insulin release	Perez et al 2002, Zhang et al 2003, Prawitt et al 2003
TRPM6	n.d.	n.d.	n.d.	Kidney	Involved in hypomagnesemia	Schlingmann et al 2002, Walder et al 2002
TRPM7	MgATP, Mg^{2+}	40 pS	Divalents (e.g. Ca^{2+}, Mg^{2+}, Zn^{2+}, Ni^{2+})	Ubiquitous	Mg^{2+} transporter, essential for cell proliferation	Nadler et al 2001, Runnels et al 2001,
TRPM8	Cold, menthol	83 pS	Non-selective (Na^+, K^+, Cs^+, Ca^{2+})	Sensory neurons, prostate	Thermosensation	Peier et al 2002, McKemy et al 2002

Missing information is indicated by n.d. (not determined). Single-channel conductance is indicated by γ. See text for details.

TRPM8

TRPM8 is expressed in a subset of pain- and temperature-sensing neurons (McKemy et al 2002, Peier et al 2002) and is also found in prostate tissue (Tsavaler et al 2001). Cells overexpressing TRPM8 channel can be activated by cold temperatures and by a cooling agent, menthol (McKemy et al 2002, Peier et al 2002). It thus appears to be at least functionally related to several members of the TRPV subfamily, which respond to various temperature ranges (Benham et al 2002).

Conclusions

As discussed above, the TRPM subfamily represents a fairly heterogeneous group of ion channels. This heterogeneity is evident in practically all biophysical parameters that define ion channel function (Table 1). Although all TRPM channels characterized so far can be classified as second messenger-operated channels, their individual activation mechanisms as well as the kinetics of activation and inactivation are very different. Similarly, the selectivities of individual channels vary widely with some members being strictly monovalent with no Ca^{2+} permeation (TRPM4 and TRPM5), others being non-selective and Ca^{2+}-permeable (TRPM2 and TRPM8) and one member being exquisitely divalent-specific (TRPM7). Nevertheless, all of these channels have significant impact on Ca^{2+} signalling, either directly by permeating the ion or indirectly by controlling the membrane potential and thereby setting the driving force for Ca^{2+} influx and/or regulating electrical activity. Moreover, one of the TRPM subfamily members, TRPM7 (and possibly its closest relative TRPM6), appear to control the flux of Mg^{2+} and presumably other divalent ions. Since most of the TRPM family members play an important role in ion homeostasis and Ca^{2+} signalling in a variety of cell types and have also been implicated in various pathophysiological contexts, these ion channels offer great potential for drug discovery efforts.

Acknowledgements

This work was supported by grants R01-GM065360 to A.F. and R01-NS040927 to R.P.

References

Benham CD, Davis JB, Randall AD 2002 Vanilloid and TRP channels: a family of lipid-gated cation channels. Neuropharmacology 42:873–888

Clapham DE, Runnels LW, Strubing C 2001 The TRP ion channel family. Nat Rev Neurosci 2:387–396

Duncan LM, Deeds J, Hunter J et al 1998 Down-regulation of the novel gene melastatin correlates with potential for melanoma metastasis. Cancer Res 58:1515–1520

Duncan LM, Deeds J, Cronin FE et al 2001 Melastatin expression and prognosis in cutaneous malignant melanoma. J Clin Oncol 19:568–576

Fang D, Setaluri V 2000 Expression and Up-regulation of alternatively spliced transcripts of melastatin, a melanoma metastasis-related gene, in human melanoma cells. Biochem Biophys Res Commun 279:53–61

Hara Y, Wakamori M, Ishii M et al 2002 LTRPC2 Ca^{2+}-permeable channel activated by changes in redox status confers susceptibility to cell death. Mol Cell 9:163–173

Heiner I, Eisfeld J, Halaszovich CR et al 2003 Expression profile of the transient receptor potential (TRP) family in neutrophil granulocytes: evidence for currents through long TRP channel 2 induced by ADP-ribose and NAD. Biochem J 371:1045–1053

Hermosura MC, Monteilh-Zoller MK, Scharenberg AM, Penner R, Fleig A 2002 Dissociation of the store-operated calcium current I_{CRAC} and the Mg-nucleotide-regulated metal ion current MagNuM. J Physiol 539:445–458

Inamura K, Sano Y, Mochizuki S et al 2003 Response to ADP-ribose by activation of TRPM2 in the CRI-G1 insulinoma cell line. J Membr Biol 191:201–207

Kozak JA, Kerschbaum HH, Cahalan MD 2002 Distinct properties of CRAC and MIC channels in RBL cells. J Gen Physiol 120:221–235

Launay P, Fleig A, Perraud AL, Scharenberg AM, Penner R, Kinet JP 2002 TRPM4 is a Ca^{2+}-activated nonselective cation channel mediating cell membrane depolarization. Cell 109: 397–407

McHugh D, Flemming R, Xu SZ, Perraud AL, Beech DJ 2003 Critical intracellular Ca^{2+}-dependence of transient receptor potential melastatin 2 (TRPM2) cation channel activation. J Biol Chem 278:11002–11006

McKemy DD, Neuhausser WM, Julius D 2002 Identification of a cold receptor reveals a general role for TRP channels in thermosensation. Nature 416:52–58

Monteilh-Zoller MK, Hermosura MC, Nadler MJ, Scharenberg AM, Penner R, Fleig A 2003 TRPM7 provides an ion channel mechanism for cellular entry of trace metal ions. J Gen Physiol 121:49–60

Montell C, Birnbaumer L, Flockerzi V et al 2002 A unified nomenclature for the superfamily of TRP cation channels. Mol Cell 9:229–231

Nadler MJ, Hermosura MC, Inabe K et al 2001 LTRPC7 is a MgATP-regulated divalent cation channel required for cell viability. Nature 411:590–595

Partridge LD, Muller TH, Swandulla D 1994 Calcium-activated non-selective channels in the nervous system. Brain Res Brain Res Rev 19:319–325

Peier AM, Moqrich A, Hergarden AC et al 2002 A TRP channel that senses cold stimuli and menthol. Cell 108:705–715

Perez CA, Huang L, Rong M et al 2002 A transient receptor potential channel expressed in taste receptor cells. Nat Neurosci 5:1169–1176

Perraud AL, Fleig A, Dunn CA et al 2001 ADP-ribose gating of the calcium-permeable LTRPC2 channel revealed by Nudix motif homology. Nature 411:595–599

Petersen OH 2002 Cation channels: homing in on the elusive CAN channels. Curr Biol 12:R520–R522

Prakriya M, Lewis RS 2002 Separation and characterization of currents through store-operated CRAC channels and Mg^{2+}-inhibited cation (MIC) channels. J Gen Physiol 119: 487–507

Prawitt D, Enklaar T, Klemm G et al 2000 Identification and characterization of MTR1, a novel gene with homology to melastatin (MLSN1) and the trp gene family located in the BWS-WT2 critical region on chromosome 11p15.5 and showing allele-specific expression. Hum Mol Genet 9:203–216

Prawitt D, Monteilh-Zoller MK, Brixel L et al 2003 TRPM5 is a transient calcium-activated cation channel responding to rapid changes in $[Ca^{2+}]_i$. Proc Natl Acad Sci USA 100: 15166–15171

Runnels LW, Yue L, Clapham DE 2001 TRP-PLIK, a bifunctional protein with kinase and ion channel activities. Science 291:1043–1047

Runnels LW, Yue L, Clapham DE 2002 The TRPM7 channel is inactivated by PIP_2 hydrolysis. Nat Cell Biol 4:329–336

Sano Y, Inamura K, Miyake A et al 2001 Immunocyte Ca^{2+} influx system mediated by LTRPC2. Science 293:1327–1330

Schlingmann KP, Weber S, Peters M et al 2002 Hypomagnesemia with secondary hypocalcemia is caused by mutations in TRPM6, a new member of the TRPM gene family. Nat Genet 31:166–170

Schmitz C, Perraud AL, Johnson CO et al 2003 Regulation of vertebrate cellular Mg^{2+} homeostasis by TRPM7. Cell 114:191–200

Takezawa R, Schmitz C, Scharenberg AM, Penner R, Fleig A 2003 Receptor-mediated regulation of TRPM7 through its endogenous protein kinase domain. Submitted

Tsavaler L, Shapero MH, Morkowski S, Laus R 2001 Trp-p8, a novel prostate-specific gene, is up-regulated in prostate cancer and other malignancies and shares high homology with transient receptor potential calcium channel proteins. Cancer Res 61:3760–3769

Vennekens R, Voets T, Bindels RJ, Droogmans G, Nilius B 2002 Current understanding of mammalian TRP homologues. Cell Calcium 31:253–264

Walder RY, Landau D, Meyer P et al 2002 Mutation of TRPM6 causes familial hypomagnesemia with secondary hypocalcemia. Nat Genet 31:171–174

Xu XZ, Moebius F, Gill DL, Montell C 2001 Regulation of melastatin, a TRP-related protein, through interaction with a cytoplasmic isoform. Proc Natl Acad Sci USA 98:10692–10697

Zhang Y, Hoon MA, Chandrashekar J et al 2003 Coding of sweet, bitter, and umami tastes: different receptor cells sharing similar signaling pathways. Cell 112:293–301

DISCUSSION

Hardie: TRPM5 is supposed to be Ca^{2+} independent, according to Zhang et al (2003). However, the channels actually behave very similarly to *Drosophila* TRP channels when they are in a constitutively active state. The TRP channel in *Drosophila* is not directly activated by Ca^{2+}, but as soon as it is activated by DAG, for example, it then becomes sensitive to Ca^{2+} modulation in very much the same way: i.e. there is then sequential facilitation and inhibition and if you change the Ca^{2+} rapidly you will get rapid excitation followed by rapid inhibition. But it is not actually activated by Ca^{2+} in the first place. I wonder if this apparent differentiating property and being directly activated by Ca^{2+} might only be a property of the TRPM5 channel when it is in a constitutively activated state, whereas if it was in the closed state it would give the appearance of being a Ca^{2+} independent activation gating mechanism.

Penner: If our channels were constitutively active in the absence of an elevated Ca^{2+} signal, I would say yes. But we have never been able to see any constitutive activity of this ion channel unless we elevate Ca^{2+} above resting. If you use plain, simple cells in unbuffered physiological conditions, like our experiment with thrombin, when we add thrombin we stimulate the release of Ca^{2+}, you see the response. It is really not the receptor that activates this ion channel because we can excise the channel and simply gate it just by puffing on Ca^{2+}, without any agonist.

Scharenberg: Can you exclude receptor operation in that case?

Penner: No, but Ca^{2+} alone will gate the channel in an excised patch.

Putney: What if you do it with high buffer in the pipette?

Penner: We don't see anything. If we put it in BAPTA we don't see it. In fact, confirming the discrepancy between Zuker's experiments (in which they actually elevated Ca^{2+}) and ours, it is important to note that their Ca^{2+} elevations occurred quite slowly. If you raise the Ca^{2+} slowly you won't see the channel. The only real discrepancy between Zuker's data and ours is that they claim that BAPTA doesn't do anything, and they can still activate the channel by receptor in the presence of BAPTA. We don't see that.

Nilius: Are you sure that your TRPM5 is not TRPM4?

Penner: Yes.

Nilius: In our hands TRPM4 behaves in exactly the same way as your TRPM5. We see inactivation, probably Ca-dependent. Our EC_{50} for activation by Ca^{2+} is much higher.

Putney: Are you also using HEK cells?

Nilius: Yes, but our HEK cells are completely insensitive to ATP.

Fleig: One of the main differences between TRPM5 and TRPM4 activation is the source of Ca^{2+}. Ca^{2+} influx, for example, seems to be crucial in activating TRPM4. However, these channels seem to be less responsive to Ca^{2+} release than TRPM5.

Penner: The response times of TRPM4 and TRPM5 to Ca^{2+} are orders of magnitude apart. TRPM5 responds instantly to Ca^{2+} changes. TRPM4 has a delay. You have seen this in your endothelial cells.

Nilius: I was sure that the 25 pS conductance in endothelial cells is TRPM4. I am not any more. Why? Because this is typical voltage dependence. If this is endogenous TRPM4 in endothelium it loses for any reason this dramatic voltage dependence: currents are almost linear and do not show inactivation at negative and slow activation at positive potentials. This is a mystery to me.

Penner: The inactivation that we observe with TRPM5 is maintained if you excise the patch.

Nilius: With TRPM4 in high Ca^{2+} concentrations it goes down again. We always see inactivation for TRPM4 which is faster the higher $[Ca^{2+}]_i$.

Penner: We don't see that, at least up to 1.8 μM Ca^{2+}. This significantly depresses TRPM5 but does nothing to TRPM4.

Scharenberg: Are you using a stable HEK cell line? This could be a contributing factor.

Nilius: Yes, we got the cells from Jean-Pierre Kinet after some painful administration events.

Penner: Bernd has our TRPM4-expressing cells now.

Nilius: More or less. They had to be sent twice. It proved challenging to send cells from Boston to Brussels, which is now a difficult task.

Muallem: Aren't you a little quick to conclude that TRPM5 has differentiating properties? You conclude that the channel functions as a differentiating channel simply based on its response to the very fast Ca^{2+} rise. You need more rigorous evidence to conclude that it is actually differentiating. With thrombin you are going to get a much higher local Ca^{2+} if it is going to happen close to the plasma membrane, whereas with CPA the $[Ca^{2+}]_i$ rise at this location will be lower. You don't need a fancy mechanism to explain the difference between the results with thrombin and CPA.

Nilius: I concur.

Penner: You have seen the CPA data. CPA actually elevates cytosolic Ca^{2+} a lot higher than receptor stimulation, but it is a slow process that does not produce a significant TRPM5 current.

Muallem: You are talking about TRPM5 as differentiating. That is, it is reading the rate of Ca^{2+} change. I am not saying this doesn't happen, just that your data aren't sufficient for you to make this kind of strong conclusion.

Hardie: If it was really a differentiating channel, responding to the rate of rise, then if you took Ca^{2+} up to 600 nM and then changed it rapidly, you should still see a difference: is this the case? I think it may just be a case of sequential excitation and inhibition. In a certain Ca^{2+} range this will give it a transient response.

Penner: I agree with the differentiation in terms of the whole cell recordings. In fact, the inactivation is probably what accounts for this. It is like the sodium channel: if you change the membrane potential of a neuron or muscle cell very slowly, you will never trigger an action potential. The sodium channels will activate but they will also inactivate so you never build up enough current to actually trigger an action potential. It is very similar to that situation.

Putney: The explanation is the inactivation. This results in the phenomenology that looks like it is differentiating. Shmuel Muallem may be correct: you need a very high Ca^{2+}, but you can't test it by setting the Ca^{2+} very high because you get the inactivation.

Muallem: There are ways to test this. You could load the cells with a caged compound and try to get different rates of $[Ca^{2+}]_i$ increase.

Putney: When you uncage Ca^{2+} you are providing a rapid rate of Ca^{2+} change, so this is just as consistent with his interpretation as another.

Muallem: I don't think so. Very high local Ca^{2+} increase at different cellular locations can be established by uncaging.

Authi: I have a question about the TRPM7 activation. You routinely have high BAPTA in the pipette in order to see the current. Have you done this in physiological levels of buffer? Is there a Ca^{2+}-dependent gating or inactivation to TRPM7?

Penner: Yes to both questions. We have done it in the most physiological way that we can imagine, including no buffer at all. We see the same type of activation of the current. Ca^{2+} does appear to inhibit the activity of these channels from the intracellular side, so if you don't buffer you don't get quite as large a current, but it can be seen clearly.

Authi: Is there a difference between the change of rate of release and store-operated entry in the Ca^{2+} inactivation of TRPM7? I am trying to get at the idea of spatial gradients.

Penner: We haven't really looked at the Ca^{2+}-dependent inactivation under unbuffered conditions. It is conceivable that the Ca^{2+} that enters during the store-operated Ca^{2+} entry phase of the stimulation may inhibit TRPM7. This is possible.

Nilius: You mentioned that there is a voltage-dependent inactivation of TRPM2. Can you comment on this in more detail?

Penner: If we do a normal pulse protocol in which we step to various membrane potentials in normal K^+ or Cs^+ glutamate solutions, we see pretty much rectangular current pulses. However, if you do the same experiment in a cell in which you perfuse sodium glutamate into the cell, the outward currents are completely maintained and flat but the inward currents progressively inactivate. So when we step to negative membrane potentials in the presence of high sodium inside the cell, we see a massive inactivation of TRPM2.

Nilius: When does a cell see 100 μM ADP ribose levels?

Penner: That is a good question. We can dramatically sensitize the cells to ADP ribose if we have a coincident increase in intracellular Ca^{2+}. It goes down to 10 μM.

Nilius: So your channel only plays a role when Ca^{2+} is high.

Penner: No, because no one really knows what the concentration of ADP ribose is, just as we don't know the exact Ca^{2+} concentration underneath the plasma membrane.

Scharenberg: The only clue we have concerning the major cellular sources of ADP ribose is the enzyme NUDT9. It is very specifically localized to mitochondria and has a K_m of about 100 μM. Presumably, then, ADP ribose levels of around 100 μM can be built up within mitochondria. We don't know under which metabolic circumstances this might occur. We have tried to knock this enzyme out, but it is cellular lethal. Poly(ADP-ribose) glycohydrolase is another potential source of ADP ribose. Also, it has recently been shown that a class of protein deacetylases forms an acetylated form of ADP ribose (reviewed in Denu 2003). In this regard, we have a crystal structure of NUDT9 and some models we have made of the C-terminus of

TRPM2 which indicate that there probably is additional room around the ribose binding pocket. Both adenosine and ribose groups bind in very deep pockets in the enzyme. In the model we have made of the channel domain there is additional room in the pocket which would probably accommodate an acetyl group or something of that size. Nothing on the order of a nicotinamide group would be likely to fit in.

Penner: Unfortunately, there is little information about ADP ribose in the literature.

References

Denu JM 2003 Linking chromatin function with metabolic networks: Sir2 family of NAD(+)-dependent deacetylases. Trends Biochem Sci 28:41–48

ZhangY, Hoon MA, Chandrashekar J et al 2003 Coding of sweet, bitter, and umami tastes: different receptor cells sharing similar signaling pathways. Cell 112:293–301

Final general discussion

Putney: In this final discussion I'd like us to focus on some general issues. Let me begin by putting out a statement and seeing whether I can elicit some comments. Let's put on the table that there is considerable indirect evidence that TRPs are involved in store-operated Ca^{2+} entry.

Muallem: Over the last days I think it has become clear that there is enough evidence that TRP channels are doing something to Ca^{2+} entry that is important enough that we should try to uncover the details. Is it I_{CRAC} or not? There are several modes of Ca^{2+} entry that can occur in cells and the TRP channels may be important in this respect in mediating Ca^{2+} entry modes that are not classical store-operated channels (SOCs). From my perspective the most important thing is emphasizing how important the TRPs are going to be in the future in terms of their physiological role.

Putney: We brought together people studying TRP channels in this meeting. A lot of us started doing this with the hope that it had something to do with store-operated entry. We put our latest findings on the table here, and this subject hasn't come up very much. It hasn't made itself that apparent in everyone's data. Indu Ambudkar has some of the best expression data on store-operated channels.

Ambudkar: I think we can all, at the very least, agree that some of these TRP channels are activated (or can be activated) in response to agonist stimulation of cells, which is a primary stimulus for Ca^{2+} entry. We have seen enough data to convince us that TRPCs are part of this signalling cascade. It is true that we still do not fully understand the mechanism by which the channels are activated, more so in the case of the SOCs. Important questions that we need to sort out in the future are whether agonist stimulation leads to activation of more than one type of channel and whether such 'agonist-activated' channels are customized to facilitate a specific function in each cell type.

Westwick: What is the evidence that the removal of any one of these TRPs affects Ca^{2+} influx?

Putney: For TRPC1 there are a number of independent studies with knockdown giving a partial reduction of store-operated entry. In addition, there is work by Indu Ambudkar using antisense in HSG cells (Liu et al 2000).

Muallem: There are also RNAi data.

Putney: It is interesting that there is best support for TRPC1 from knockdown data. But then, with the exception of the HSGs, it has been one of the most difficult TRPs to study by expression.

Zhu: From the practical point of view, in addition to thinking about the different kinds of TRP channels and whether they form SOCs or not, or how they form as a heteromer or a homotetramer, from the discussion here about TRPV and TRPM I have come to realize that depending on distribution of the channels, there is a profound effect on Ca^{2+} influx. There will be many other channels affecting the Ca^{2+} influx. If one of these is changed this may have some effect on store-operated entry (for example, a change in TRPM4 activity will affect membrane potential thereby altering the driving force of Ca^{2+} entry). Whether they are directly store-operated or not is a different issue.

Montell: Even though many people started working on TRP channels due to interest in store-operated Ca^{2+} entry, as time goes on it has become increasingly apparent that TRP channels are very important for fluxing a variety of cations other than Ca^{2+}. This is best illustrated by the work on the TRPMs. Now we are hearing that one of the TRPMs permeates zinc the best. There have been so many surprises with the TRPs, not just from the standpoint of activation, but also with respect to the permeation properties of these channels. In the end, Ca^{2+} influx might prove to be just a small part of the TRP story.

Putney: The philosophy that we have adopted in our lab is that we have two projects: trying to understand store-operated entry, and trying to understand TRP channels. They might or might not come together at some point in the future. These are physiologically important questions.

Groschner: Some of the difficulties in understanding the role of TRP channels in Ca^{2+} signalling come from the fact that we are dealing in most cases with quite non-selective channels. The TRPM Ca^{2+} signalling data show nicely that the permeation of other ions will of course affect Ca^{2+} entry pathways. This introduces substantial difficulties in interpretation.

Fleig: We have to be careful to work out where the TRP channels are located and functional. Whether there is a signalling system that can approach or gate those channels is a separate issue. But locating and gating of TRP channels in a particular cell compartment is an important subject. With regard to the non-selective behaviour of some heterologously expressed TRP channels, let's not forget that in a cell-free system inositol 1,4,5-trisphosphate (IP_3) receptors, too, behave like non-selective cation channels. However, physiologically they conduct Ca^{2+} due to the Ca^{2+} concentration differences between the endoplasmic reticulum (ER) and cytosol. A similar scenario could be imagined for some TRP channels, as well.

Schilling: Where these channels are localized is going to turn out to be very important. The structural components and additional accessory proteins bound to, or in close proximity to the channels could play an important role in understanding how they are regulated by receptor stimulation or any other mechanism. There seems to be growing evidence that where they are located or tethered in the cell somehow determines how they are actually activated. In the end we need to work hard on the structural components and the targeting.

Muallem: I want to come back to the importance of store-operated channels. Over the last 10–15 years there has been an enormous body of evidence that Ca^{2+} influx (not just release) is vital to many physiological processes. Some of these have been examined in the TRP4 knockout mouse. I think we have already established that there is a correlation between Ca^{2+} influx and TRP channels and physiological functions that can be attributed to Ca^{2+} influx. Having this scenario convinces me that it is quite important to continue studying them not only for physiological reasons, but also for their direct involvement in both Ca^{2+} influx and the specific physiological functions of Ca^{2+} influx.

Kunze: As a person who studies neurons I am really uncomfortable with so many cation channels so ubiquitously expressed with such high conductance! I think tight regulation has to be extremely important.

Putney: We came here to discuss a family of ion channels, but the actual relatedness of their sequences isn't that high, and the relatedness of their functions isn't all that high either. There are a few take home messages that I would categorize as good news, puzzling news and bad news. The good news is that over the past few years there have been a lot of discrepant findings in the literature, and we are starting now to deal with those rationally. We realise that these discrepant findings have a scientific and not a personal basis. Understanding the scientific basis of these discrepancies is actually teaching us more about how TRP channels work. Another piece of good news is that we are starting to move from the easy, convenient expression systems towards work that tells us what the TRP channels are doing in their native environments. The puzzling news is this issue of TRPs acting in places in the cell other than the plasma membrane. There are two perspectives on this. One is the nature of the experiments that we do. We might create artefacts through TRPs acting in places other than the plasma membrane. The more interesting idea is that the TRPs might act physiologically in membrane compartments other than the plasma membrane. The bad news is that despite the considerable optimism years ago when TRPs were first identified that this would solve the SOC problem, it hasn't. But one of the encouraging things is that there are not only a lot of TRPs, there is more than one kind of SOC. At some point we may get a match. Have we made any progress in solving some of the problems I outlined in my introduction? One was whether any TRPs are really store operated. The answer to that is 'could be'. Another problem was the signal for TRPC3, 6 and 7. Is it IP_3 or diacylglycerol (DAG)? The answer to that is 'yes'. A third was the signal for TRPC4 and 5.

Here we have some options. Is it (a) DAG, (b) IP_3, or (c) G protein? The answer is (d) — none of the above! The stunning conclusion of the meeting is that TRPs are complicated, and when overexpressed they can't be trusted. My final thought is an analogy. TRPs are like bad children: they may associate with the wrong kind, they'll turn on you in a heartbeat, and eventually they may break your heart. But like your children, in the end, you've got to love them!

Reference

Liu X, Wang W, Singh BB et al 2000 Trp1, a candidate protein for the store-operated Ca^{2+} influx mechanism in salivary gland cells. J Biol Chem 275:3403–3411

Index of contributors

Non-participating co-authors are indicated by asterisks. Entries in bold indicate papers; other entries refer to discussion contributions.

Subject index

T

U

V